Lecture Notes in Computer Science 4891

Commenced Publication in 1973
Founding and Former Series Editors:
Gerhard Goos, Juris Hartmanis, and Jan van

Tharam S. Dillon   Elizabeth Chang
Robert Meersman   Katia Sycara (Eds.)

Advances in Web Semantics I

Ontologies, Web Services and Applied Semantic Web

 Springer

Volume Editors

Tharam S. Dillon
DEBI Institute Curtin Business School
Curtin University of Technology
Perth, WA, Australia
E-mail: t.dillon@curtin.edu.au

Elizabeth Chang
DEBI Institute Curtin Business School
Curtin University of Technology
Perth, WA, Australia
E-mail: Elizabeth.Chang@cbs.curtin.edu.au

Robert Meersman
Vrije Universiteit Brussel (VUB), STARLab
Brussels, Belgium
E-mail: robert.meersman@vub.ac.be

Katia Sycara
School of Computer Science, Carnegie Mellon University
Pittsburgh, PA, USA
E-mail: katia@cs.cmu.edu

Library of Congress Control Number: Applied for

CR Subject Classification (1998): D.1.0, D.2, D.2.1, D.2.6, D.2.13, D.3.1, D.4.5, F.3.1, F.4.1, F.4.3

LNCS Sublibrary: SL 3 – Information Systems and Application, incl. Internet/Web and HCI

ISSN   0302-9743
ISBN-10   3-540-89783-6 Springer Berlin Heidelberg New York
ISBN-13   978-3-540-89783-5 Springer Berlin Heidelberg New York

Springer is a part of Springer Science+Business Media

springer.com

© Springer-Verlag Berlin Heidelberg 2008

Typesetting: Camera-ready by author, data conversion by Scientific Publishing Services, Chennai, India
Printed on acid-free paper   SPIN: 12574834   06/3180   5 4 3 2 1 0

Preface

The all pervasive web is influencing all aspects of human endeavour. In order to strengthen the description of web resources, so that they are more meaningful to both humans and machines, web semantics have been proposed. These allow better annotation, understanding, search, interpretation and composition of these resources. The growing importance of these has brought about a great increase in research into these issues.

We propose a series of books that will address key issues in web semantics on an annual basis. This book series can be considered as an extended journal published annually. The series will combine theoretical results, standards, and their realizations in applications and implementations. The series is titled "Advances in Web Semantics" and will be published periodically by Springer to promote emerging Semantic Web technologies. It will contain the cream of the collective contribution of the International Federation for Information Processing (IFIP) Web Semantics Working Group; WG 2.12 & WG 12.4. This book, addressing the current state of the art, is the first in the series. In subsequent years, books will address a particular theme, topic or issue where the greatest advances are being made. Examples of such topics include: (i) process semantics, (ii) web services, (iii) ontologies, (iv) workflows, (v) trust and reputation, (vi) web applications, etc. Periodically, perhaps every five years, there will be a scene-setting state of the art volume. This will provide a collection of work by the top experts and minds in the field, in an area of the utmost importance to business, IT and industry.

The IFIP Web Semantics Working Group 2.12 & 12.4 (http://www.ifipsemanticweb.curtin.edu.au/) constitutes a timely active international community of scientists, engineers, and practitioners dedicated to advancing the state of the art of research and practice in the emerging field of the Semantic Web, and meanwhile providing input and guidance on the direction, scope, and importance of different aspects of artificial intelligence, data modeling, and software theory and practice in web semantics. It is unique in that it seeks to synthesize concepts from these diverse fields in a comprehensive fashion in the context of the semantics of the Web.

The primary goal of this book series is to investigate, present and promote core concepts, ideas and exemplary technologies for the next generation of semantic web research, from a range of state-of-the-art research institutions (both academic and industrial).

Each book will be carefully written to provide maximum exposure to collective knowledge, well thought out, providing insights into ground breaking research results and present emerging trends for the present and the future of the Semantic Web. The series will promote and present emerging semantic web technologies to governments, researchers and industry and commercial practitioners. This book series, in contrast to other available literature, will focus not only on the existing concepts and ideas but also on the future of the Semantic Web. Moreover, concepts covered in the book will serve as a good reference point for academic and industrial researchers who want to familiarize themselves with emerging trends in semantic web research, technologies and applications.

October 2008

Tharam Dillon
Elizabeth Chang
Robert Meersman
Katia Sycara

Table of Contents

Introduction.. 1
   Tharam S. Dillon, Elizabeth Chang, Robert Meersman, and
   Katia Sycara

Part 1: Ontologies and Knowledge Sharing

Ontology Engineering – The DOGMA Approach 7
   Mustafa Jarrar and Robert Meersman

Process Mining towards Semantics 35
   A.K. Alves de Medeiros and W.M.P. van der Aalst

Models of Interaction as a Grounding for Peer to Peer Knowledge
Sharing.. 81
   David Robertson, Adam Barker, Paolo Besana, Alan Bundy,
   Yun Heh Chen-Burger, David Dupplaw, Fausto Giunchiglia,
   Frank van Harmelen, Fadzil Hassan, Spyros Kotoulas,
   David Lambert, Guo Li, Jarred McGinnis, Fiona McNeill,
   Nardine Osman, Adrian Perreau de Pinninck, Ronny Siebes,
   Carles Sierra, and Chris Walton

Extraction Process Specification for Materialized Ontology Views 130
   Carlo Wouters, Tharam S. Dillon, Wenny Rahayu,
   Robert Meersman, and Elizabeth Chang

Advances in Ontology Matching 176
   Avigdor Gal and Pavel Shvaiko

Multi-site Software Engineering Ontology Instantiations Management
Using Reputation Based Decision Making 199
   Pornpit Wongthongtham, Farookh Khadeer Hussain,
   Elizabeth Chang, and Tharam S. Dillon

Part 2: Applied Semantic Web

Representing and Validating Digital Business Processes 219
   Lianne Bodenstaff, Paolo Ceravolo, Ernesto Damiani,
   Cristiano Fugazza, Karl Reed, and Andreas Wombacher

Towards Automated Privacy Compliance in the Information Life
Cycle... 247
   Rema Ananthanarayanan, Ajay Gupta, and Mukesh Mohania

Web Semantics for Intelligent and Dynamic Information Retrieval
Illustrated within the Mental Health Domain 260
  Maja Hadzic and Elizabeth Chang

Ontologies for Production Automation 276
  Jose L. Martinez Lastra and Ivan M. Delamer

Determining the Failure Level for Risk Analysis in an e-Commerce
Interaction 290
  Omar Khadeer Hussain, Elizabeth Chang,
  Farookh Khadeer Hussain, and Tharam S. Dillon

Part 3: Web Services

Process Mediation of OWL-S Web Services 324
  Katia Sycara and Roman Vaculín

Latent Semantic Analysis – The Dynamics of Semantics Web Services
Discovery 346
  Chen Wu, Vidyasagar Potdar, and Elizabeth Chang

Semantic Web Services for Satisfying SOA Requirements 374
  Sami Bhiri, Walid Gaaloul, Mohsen Rouached, and
  Manfred Hauswirth

Author Index 397

Introduction

Tharam S. Dillon[1], Elizabeth Chang[1], Robert Meersman[2], and Katia Sycara[3]

[1] Digital Ecosystems and Business Intelligence Institute
Curtin University of Technology
G.P.O. Box U1987, Perth, WA 6845, Australia
{Tharam.Dillon,Elizabeth.Chang}@cbs.curtin.edu.au
[2] STARLab, Vrije Universiteit Brussel, Belgium
meersman@vub.ac.be
[3] The Robotics Institute, Carnegie Mellon University
katia@cs.cmu.edu

Abstract. The Web has now been in existence for quite some time and it is pervasive in its influence on all aspects of society and commerce. It has also produced a major shift in our thinking on the nature and scope of information processing. However in its technological nature and its supporting theoretical foundations, it was relatively rudimentary, being largely suitable for information dissemination. It is rapidly moving away from this, to application deployment and knowledge deployment that require complex interactions and properly structured underlying semantics. This has been a sudden upsurge of research activity in the problems associated with adding semantics to the Web. This work on semantics will involve data, knowledge, and process semantics.

1 Introduction

The emergence of Semantic Web (SW) and the related technologies promise to make the web a meaning full experience. Tim Berners-Lee[1] envisioned that the Semantic Web paradigm promises a meaningful web for both humans and the machines (agents). It provides representation, integration and access to information using semi-structured data models where data and meta-data semantics are represented in a way that is easily interpreted by autonomous agents. However, high level modeling, design and querying techniques prove to be challenging tasks for organizations that are hoping to utilize the SW paradigm for their commercial and industrial applications.

2 Themes

In order to help understand and interpret the research issues that are arising in this challenging area and the proposals for their solution, we have proposed a series of books, which is an extended journal format. The series will have periodic state of the art volumes and intermediate volumes directed towards hot button issues. The series will address a number of themes and these include but are not limited to:

---

[1] Berners-Lee, T. *Weaving the Web*. San Francisco: Harper, 1999.

T.S. Dillon et al. (Eds.): Advances in Web Semantics I, LNCS 4891, pp. 1–6, 2008.

Theme 1: Semantics and Concepts for Semantic Web
- Conceptual semantics
- Formal models
- Theoretical models for SW
- Semantics of agent and Web interaction
- Security and trust for the semantic Web
- Database technologies for the semantic Web
- Impact of semantic Web computing on organizations and society
- Semantics for ubiquitous computing
- Conceptual correctness in SW modeling

Theme 2: Emerging New Trends
- Query models and languages
- Ontology dynamics and evolution
- Ontology-Driven IS
- Human centered aspects for SW
- Trust & Security

Theme 3: Ontological Engineering
- Metadata and knowledge markup
- Ontology-Driven IS and Applications
- Ontological correctness in SW modeling
- New and emerging Standards
- Query models for knowledge extraction
- Formal models and language specification

Theme 4: Language for Semantic Web
- Extensions to languages
- Languages for knowledge extraction and integration
- Content-based information and knowledge retrieval
- Database technologies for the semantic Web
- Tools & Prototypes

Theme 5: Interoperability, Integration and Composition
- Concepts, Frameworks and Platforms
- Interoperability of data and Web services
- Ontology-Driven Interoperability and Integration for IS

Theme 6: Applications of Semantic Web in Domains
- Formal and Practical knowledge representation and inference for the semantic Web
- Human centered aspects specifically for the semantic Web
- Ontology-Driven IS Applications
- Health Information systems
- Bioinformatics

Theme 7: SOA and Web Semantics
- Web Services and SW
- Ontology bases and SOA

Theme 8: Industrial Applications & Experiences
- Prototypes
- Experience
- Security and trust for the semantic Web
- Multimodality and visualization technologies for the semantic Web
- Applications on mobile devices
- Application in Bioinformatics and Health domains

This is the first book in this series and is titled "Advances in Web Semantics: A State-of-the Art". In addition, there will be one (or more) periodic theme driven book/(s) issues, one specific theme (such as semantics and concepts for semantic web, process semantics for the web, etc), that is of utmost importance to the advancement of the web semantics and its future.

3 Author Contributions

The contributions in this book are divided into three broad categories namely: Ontologies, Web Services and Applied Semantic Web. There are 6 chapters in Ontologies and Knowledge Sharing, 3 chapters in Web Services and 5 chapters in Applied Semantic Web respectively. Each chapter of the current book is carefully written and presented by selected authors and by the IFIP workgroup members, who are leaders in their field of expertise from wide ranging domains. The author selection process is by invitation only and the selection process is highly selective and should be approved by the editors and the IFIP workgroups. Each Chapter was then peer reviewed by at least two members of the working group and revised in accordance with the input from these reviews.

Chapter 2 presents a methodological framework for ontology engineering (called DOGMA), which helps in building ontologies that are highly reusable and usable, easier to manage and to maintain. The authors discuss the major challenges in ontology engineering. They also discuss ontology reusability verses ontology usability. Authors present the DOGMA approach, which prescribes that ontology, be built as separate domain axiomatization and application axiomatizations. While a domain axiomatization focuses on the characterization of the intended meaning of a vocabulary at the domain level, application axiomatizations focus on the usability of this vocabulary according to certain perspectives. In this chapter, the authors also show how ontology specification languages can be effectively (re)used in ontology engineering.

Process mining assists automatic discovery of information about process models in organizations. In chapter 3, the authors show how the analysis provided by current techniques can be improved by including semantic data in event logs. The chapter is divided into two main parts. The first part illustrates current process mining techniques and shows usage of the process mining tool ProM to answer concrete questions.

The second part utilizes usage scenarios to depict benefits of semantic annotated event logs and they define a concrete semantic log format for ProM.

Current large scale knowledge sharing methods have relied on pre-engineering of content. In chapter 4 the authors describe how to focus on semantics related to interactions. Their method is based on interaction models that are mobile in the sense that they may be transferred to other components, this being a mechanism for service composition. By shifting the emphasis to interaction knowledge sharing of sufficient quality for sustainable communities can be obtained without the barrier of complex meta-data provision.

Extracting high quality ontology from a given base ontology in needed in versioning, distribution and maintenance of ontologies. Chapter 5 presents a formalism one to extract the sub-ontology from a base ontology. In addition to extracting the sub-ontology it carries optimization of its structure using different optimization schemes. Authors discuss the overview of the formalism demonstrating several optimization schemes. Examples of how the formalism is deployed to reach a high-quality result, called a materialized ontology view, are covered in the chapter. The methodology provides a foundation for further developments, and shows the possibility of obtaining usable ontologies by automation.

Ontologies are one of the basic operations of semantic heterogeneity reconciliation. The aim of Chapter 6 is to motivate the need for ontology matching, introduce the basics of ontology matching, and then discuss several promising themes in the area. In particular, authors focus on such themes as uncertainty in ontology matching, matching ensembles, and matcher self-tuning. Finally, authors outline some important directions for future research.

In Chapter 7 the authors explore the development of systems for software engineering ontology instantiations management in the methodology for multisite distributed software development. In a multisite distributed development environment, team members in the software engineering projects must naturally interact with each other and share lots of project data/agreements amongst themselves. Since they do not always reside at the same place and face-to-face meetings are expensive and very infrequent, there is a need for a methodology and tools that facilitate effective communication for efficient collaboration. Whilst multi-site distributed teams collaborate, there are a lot of shared project data updated or created that leads to a large volume of project information and issues. Systematic management of Software engineering knowledge is represented in the software engineering ontology.

The Web infrastructure and the availability of Semantic Web-style metadata have enabled new types of inter-organizational business processes, which in turn need new business process models. Current approaches focus on producing taxonomies or categorizations, or on stating what aspects of business process models should be included in (or excluded from) modeling. In Chapter 8, authors discussed the different facets such a representation must possess, starting from standard ones like workflow representation and arriving to the capability of defining the structure of a company's value chain and describing the position of the process actors within the value network. While authors believe a visual approach to modeling is preferable from the modeler's point of view, they focused on the logics-based models underlying each facet of the process representation. They build on the idea that the different facets of a business process model require different formalizations, involving different inference techniques, some of them Semantic Web-style, others closer to classical Prolog reasoning. Also using a simple business process model, they discussed how well our multi-faceted representation can be integrated by a hybrid approach to reasoning, transferring knowledge and metrics obtained reasoning on each part of the model to the other parts.

In Chapter 9, the authors look at some of the issues associated with the problem of privacy protection, and present some directions of work that would help automate the solution to the problem. They have also presented architecture for privacy protection at the back end that comprises of simple rules that specify which user items can access which data items or categories.

Current search engines that are based on syntactic keywords are fundamental for retrieving information often leaving the search results meaningless. Increasing semantics of web content would solve this problem and enable the web to reach its full potential. This would allow machines to access web content, understand it, retrieve and process the data automatically rather than only display it. In chapter 10 authors illustrate how the ontology and multi-agent system technologies, which underpin this vision, can be effectively implemented within the mental health domain.

The manufacturing sector is currently under pressure to swiftly accommodate new products by quickly setting up new factories or retrofitting existing ones. In order to achieve this goal, engineering tasks currently performed manually need to be automated by using ontologies. Chapter 11 reviews current approaches to use ontologies in the manufacturing domain, which include use for vocabulary definition in multi-agent systems and use for describing processes using Semantic Web Services. In addition, current and emerging research trends are identified.

Before initiating a financial e-commerce interaction over the World Wide Web, the initiating agent would like to analyze the possible Risk in interacting with an agent, to ascertain the level to which it will not achieve its desired outcomes in the interaction. To determine the possible risk in an interaction, the initiating agent has to determine the probability of failure and the possible consequences of failure to its resources involved in the interaction. In Chapter 12 as a step towards risk analysis, authors propose a methodology by which the initiating agent can determine beforehand the probability of failure in interacting with an agent, to achieve its desired outcomes.

The ability to deal with incompatibilities of service requesters and providers is a critical factor for achieving smooth interoperability in dynamic environments. Achieving interoperability of existing web services is a complicated process including a lot of development and integration effort, which is far from being automated. Semantic Web Services frameworks strive to facilitate flexible dynamic web services discovery, invocation and composition and to support automation of these processes. In chapter 13, authors focus on mediation and brokering mechanisms of OWL-S Web Services, which overcomes various types of problems and incompatibilities.

The recent proliferation of SOA applications exacerbates this issue by allowing loosely-coupled services to dynamically collaborate with each other, each of which might maintain a different set of ontologies. Chapter 14 presents the fundamental mechanism of Latent Semantic Analysis (LSA), an extended vector space model for Information Retrieval (IR), and its application in semantic web services discovery, selection, and aggregation for digital ecosystems. Firstly, the authors explore the nature of current semantic web services within the principle of ubiquity and simplicity. This is followed by a succinct literature overview of current approaches for semantic services/software component discovery and the motivation for introducing LSA into the user-driven scenarios for service discovery and aggregation. They then direct the readers to the mathematical foundation of LSA – SVD of data matrices for calculating statistics distribution and thus capturing the 'hidden' semantics of web services concepts. Some existing applications of LSA in various research fields are briefly presented, which gives rise to the analysis of the uniqueness (i.e. strength, limitations, parameter settings) of LSA application in semantic web services. They also provide a conceptual level solution with a proof-of-concept prototype to address such uniqueness. Finally they propose an LSA-enabled semantic web services architecture fostering service discovery, selection, and aggregation in a digital ecosystem.

Web services have been emerging as the lead implementation of SOA upon the Web, adding a new level of functionality for services description, publication, discovery, composition and coordination. We still need to combine web services in a useful manner, which limits scalability. A common agreement is the need to semantically enrich Web Services description. Semantic web services have been emerging with promising potential to satisfy SOA requirements. Chapter 15 is divided into three parts. Firstly, authors depict the concepts and principles of SOA. They highlight the dynamicity, flexibility and the challenges it poses. In the second part, the authors present Web services as the key technology implementing SOA principles. They state in particular the main standards and technologies and investigate how they fit in SOA and how far they satisfy its requirements. Finally, authors exhibit SWS conception and foundation. They consider the two SWS languages WSMO and OWL-S show how these promise to overcome SOA challenges.

Ontology Engineering – The DOGMA Approach

Mustafa Jarrar and Robert Meersman

STARLab, Vrije Universiteit Brussel, Belgium
{mjarrar,meersman}@vub.ac.be

Abstract. This chapter presents a methodological framework for ontology engineering (called DOGMA), which is aimed to guide ontology builders towards building ontologies that are both highly reusable and usable, easier to build and to maintain. We survey the main foundational challenges in ontology engineering and analyse to what extent one can build an ontology independently of application requirements at hand. We discuss ontology reusability verses ontology usability and present the DOGMA approach, its philosophy and formalization, which prescribe that an ontology be built as separate domain axiomatization and application axiomatizations. While a domain axiomatization focuses on the characterization of the intended meaning (i.e. intended models) of a vocabulary at the domain level, application axiomatizations focus on the usability of this vocabulary according to certain application/usability perspectives and specify the legal models (a subset of the intended models) of the application(s)' interest. We show how specification languages (such as ORM, UML, EER, and OWL) can be effectively (re)used in ontology engineering.

1 Introduction and Motivation

The Internet and other open connectivity environments create a strong demand for sharing the semantics of data. *Ontologies* are becoming increasingly essential for nearly all computer science applications. Organizations are looking towards them as vital machine-processable semantic resources for many application areas. An ontology is an agreed understanding (i.e. semantics) of a certain domain, axiomatized and represented formally as logical theory in the form of a computer-based resource. By sharing an ontology, autonomous and distributed applications can *meaningfully* communicate to exchange data and thus make transactions *interoperate* independently of their internal technologies.

Research on ontologies has turned into an interdisciplinary subject. It combines elements of Philosophy, Linguistics, Logics, and Computer Science. Within computer science, the research on ontologies emerged "mainly" within two subcommunities: artificial intelligence (among scientists largely committed to building shared knowledge bases) and database (among scientists and members of industry who are largely committed to building conceptual data schemes, also called semantic data models [V82]). This legacy in computer science brings indeed successful techniques and methods to enrich the art of ontology engineering. However, some confusion on how to reuse these techniques is witnessed. For example, many researchers have confused ontologies with data schemes, knowledge bases, or even logic programs.

T.S. Dillon et al. (Eds.): Advances in Web Semantics I, LNCS 4891, pp. 7–34, 2008.

Unlike a conceptual data schema or a "classical" knowledge base that captures semantics for a given enterprise application, the main and fundamental advantage of an ontology is that it captures domain knowledge highly independently of any particular application or task [M99b] [JDM03]. A consensus on ontological content is the main requirement in ontology engineering, and this is what mainly distinguishes it from conceptual data modeling. Neither an ontology nor its development process is a single person enterprise [KN03].

The main goal of this chapter is to present a methodological framework for ontology engineering (called DOGMA[1] [M01] [J05]) to guide ontology builders towards building ontologies that are both highly reusable and usable, easier to build, and smoother to maintain. Some parts of this research has been published in earlier articles such as [M96] [M99a] [J05a] [M99b] [JM02a] [JDM03] [J06] [JH07] [J07c]. This chapter provides an up-to-date specification of DOGMA based on [J05]. A comprehensive view on DOGMA, its formalisms, applications, and supportive tools can be found in [J05]. Others have also enriched our approach with special techniques for ontology evolution [DDM07] [DM05], community-driven ontologies [DDM06], ontology integration [DSM04], and visualization [P05] [TV06].

Before presenting the DOGMA methodological framework, we investigate and illustrate (in section 2) the main *foundational* challenges in ontology engineering. We argue that different usability perspectives (i.e. different purposes of what an ontology is made for and how it will be used) hamper reusability and lead to different or even to conflicting ontologies, although these ontologies might intuitively be in agreement at the domain level. The more an ontology is independent of application perspectives, the less usable it will be. In contrast, the closer an ontology is to application perspectives, the less reusable it will be. From a methodological viewpoint, notice that if a methodology emphasizes usability perspectives, or evaluates ontologies based only on how they fulfill specific application requirements, the resultant ontology will be similar to a *conceptual data schema* (or a classical knowledge base) containing specific – and thus less reusable– knowledge. Likewise, if a methodology emphasizes only on the independence of the knowledge and ignores application perspectives, the resultant ontology will be less usable.

To tackle such a foundational challenge, we propose a methodological framework, called DOGMA, in section 3. The idea of DOGMA, in nutshell, is that: an ontology is doubly articulated into a *domain axiomatization* and *application axiomatization*[2]. While a domain axiomatization is mainly concerned with characterizing the "intended meanings" of domain vocabulary (typically shared and public), an application axiomatization (typically local) is mainly concerned with the usability of these vocabularies. The double articulation implies that all concepts and relationships introduced in an application axiomatization are predefined in its domain axiomatization. Multiple application

---

[1] Developing Ontology-Grounded Methods and Applications.

[2] We use the term axiomatization in this chapter to mean an articulation or specification of knowledge, as a set of axioms about a certain subject-matter. We interchange this term with the term ontology in order to avoid the any misunderstanding of what an ontology is. For example, a conceptual data schema or an application's knowledge base is being named as ontologies, which in our opinion are only application axiomatizations. We preserve the term ontology to mean a domain axiomatization, which captures knowledge at the domain level.

axiomatizations (e.g. that reflect different usability perspectives, and that are more usable) share and reuse the same intended meanings in a domain axiomatization.

To illustrate this framework, one can imagine WordNet as a domain axiomatization, and an application axiomatization built in OWL (or RDF, ORM, UML, etc). The DOGMA methodology then suggests that all vocabulary in the OWL axiomatization should be linked with word-senses (i.e. concepts) in WordNet. In this way, we gain more consensus about application axiomatizations; we improve the usability of application axiomatizations and the reusability of domain axiomatizations; application ontologies that are built in the same way (i.e. commit to the same domain axiomatizations) will be easier to integrate, etc.

The DOGMA approach goes farther and suggests the notion of "ontology base" for representing domain axiomatizations in a manner easy to evolve and use (see section 3). The basic building block of ontology base is called *lexon*, a content-specific lexical rendering of a binary conceptual relation. As we shall explain later application axiomatizations then *commit* to an ontology base containing those lexons.

2   Fundamental Challenges in Ontology Engineering

In this section, we investigate and specify several challenges in ontology engineering. We examine to what extent one can build an ontology independently of application requirements. We discuss ontology reusability verses ontology usability, then we present the work done by other researchers in relation to these challenges. To end, we draw some important requirements for ontology engineering.

Ontologies are supposed to capture knowledge at the domain level independently of application requirements [G97] [CJB99] [M99a] [SMJ02]. We may argue this is in fact the main and most fundamental asset of an ontology. The greater the extent to which an ontology is independent of application requirements, the greater its reusability[3], and hence, the ease at which a consensus can be reached about it. Guarino indeed in [G97] posited that*"Reusability across multiple tasks or methods should be systematically pursued even when modeling knowledge related to a single task or method: the more this reusability is pursued, the closer we get to the intrinsic, task-independent aspects of a given piece of reality (at least, in the commonsense perception of a human agent)."*

Ontology application-independence is not limited to the independence of *implementation* requirements - it should also be considered at the *conceptual level*. For example, notice that application-independence is the main disparity between an ontology and a classical *data schema* (e.g. EER, ORM, UML, etc.) although each captures knowledge at the conceptual level [JDM03]. Unlike ontologies, when building a data schema the modeling decisions depend on the specific needs and tasks that are planned to be performed within a certain enterprise, i.e. for "in-house" usage.

The problem is that when building an ontology, there always will be intended or expected usability requirements -"at hand"- which influence the independency level

---

[3] *Notice that ontology usability is subtly different from ontology reusability. Increasing the reusability of knowledge implies the maximization of its usage among several kinds of tasks. Increasing ontology usability could just mean maximizing the number of different applications using an ontology for the same kind of task.*

of ontology axioms. In the problem-solving research community, this is called the *interaction problem*. Bylander and Chandrasekaran[BC88] argue that *"Representing knowledge for the purpose of solving some problem is strongly affected by the nature of the problem and the inference strategy to be applied to the problem."*

The main challenge of usability influence is that different usability perspectives (i.e. differing purposes of *what an ontology is made for* and *how it will be used*) lead to different –and sometimes conflicting– application axiomatizations although these might agree at the domain level.

Example

The following example illustrates the influence of some usability perspectives when modeling ontologies in the Bibliography domain. Ontology A in Figure 1 and ontology B in Figure 2 are assumed to have been built autonomously; ontology A is built and used within a community of bookstores, and ontology B is built and used within a community of libraries[4]. Although both ontologies "intuitively" agree at the domain level, they differ formally because of the differences in their communities' usability perspectives. Obviously, building ontologies under strong influence of usability perspectives leads to more application-dependent, and thus less reusable ontologies.

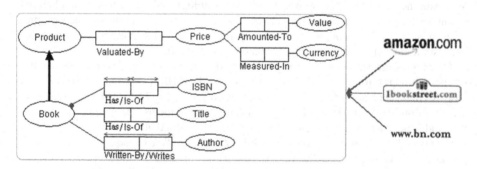

Fig. 1. Ontology A

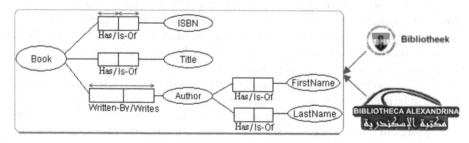

Fig. 2. Ontology B

---

[4] Notice that the goal of this example is neither to discuss the Bibliography domain itself, nor to present adequate an ontology - we use it only for illustration purposes.

In the following, we examine the influence of usability perspectives on the modeling decisions of both conceptual relations and ontology rules, respectively.

Modeling conceptual relations. The concept 'Author' in ontology B is attributed with the 'First Name' and the 'Last Name' concepts. Such details (i.e. *granularity*) are not *relevant* to bookstore applications; they are not specified in ontology A. Similarly, unlike ontology A, the pricing relations {Valuated-By(Book, Price), Amounted-To(Price, Value), Measured-In(Price, Currency)} are not *relevant* for library applications, so they are not specified in ontology B.

From such differences, one can see that deciding the *granularity level* and the *scope boundaries* depend on the relevance to the intended (or expected) usability. Although such differences do not necessarily constitute a disagreement between both axiomatizations, they hamper the reusability of both ontologies. In order to reuse such ontologies, the reusing applications need to make some adaptations, *viz. introducing* the incomplete knowledge and *dismissing* the "useless" knowledge that normally distracts and scales down the reasoning/computational processes.

Modeling ontology rules. Notice that both ontologies in the example above do not agree on the notion of what is a "Book". Although both ontologies agree that the ISBN is a unique property for the concept book (see the uniqueness rules[5]), they disagree whether this property is mandatory for each instance of a book. Unlike ontology B, ontology A axiomatizes that each instance of a book *must* have an ISBN value (see the mandatory rule[6]). This rule implies for example that "PhD Theses" or "Manuals", etc. would not be considered instances of books in ontology A because they do not have an ISBN, while they would be under ontology B.

One can see from this example that modeling the ISBN as mandatory property for all instances of the concept book is naturally affected by bookstores' business perspective. Obviously, bookstores communicate only the books "that can be sold" and thus "commercially" should have ISBN, rather than perusing the notion of book at the domain level. Nevertheless, at the domain level, both bookstore and library applications intuitively share the same concept of what is really a book. For example, suppose that one assigns an ISBN for an instance of a "PhD Thesis". This instance can then be considered as a book for bookstores. If however, the ISBN is removed for an instance of a book, then this instance will no longer be a book, even though it still refers to the same real life object and is still being referred to and used as a book.

Accordingly, as ontology rules are supposed to formally specify/constrain the permitted models[7] that can necessarily hold for a given domain, determining such rules, in practice is dominated by "what is permitted and what is not" for the *intended* or *expected* usability.

Furthermore, besides the modeling decisions of ontology rules, the determination of the number and the type of these rules (the reasoning scenario) are also influenced by usability perspectives. For example, a light-weight axiomatization (e.g. with a minimum number of rules or formalities) might be sufficient if the ontology is to be

---

[5] The uniqueness rule in ORM is equivalent to 0:1 cardinality restriction. (notation: '⊸'), it can be verbalized as "each book must have at most one ISBN".

[6] The mandatory rule in ORM is equivalent to 1-m cardinality restriction. (notation: '●'), it can be verbalized as "each book must have at least one ISBN".

[7] Also called "ontology models".

accessed and used by people (i.e. not computers). Depending on the application scenario other *types* of ontology rules (i.e. modeling primitives/constructs) might be preferred over the ORM set of rules (which are easier to reason for database and XML based applications).

At this point, we conclude that even application-types might intuitively agree on the same semantics at the domain level, but the usability influence on axiomatizing this semantics may lead to different (or even conflicting) axiomatizations. An axiomatization might be more relevant for some applications than others, due to the difference of their usability perspectives. This issue presents an important challenge to the nature and the foundation of ontology engineering.

Related Work

A general overview on ontology engineering methodologies is provided by [GFC04: pp.113-153], including short descriptions of the methods. A very recent list is given by [PT06], who do not provide details of the various methods but rather shortly situate them. It is nevertheless a very good starting point for further bibliographic research.

Guarino et al. have argued (in e.g. [G98a]) that in order to capture knowledge at the domain level, the notion of what is an ontology should be more precisely defined. Gruber's commonly used but somewhat schematic definition [G95] of an ontology is "an explicit specification of a conceptualization", and refers to an *extensional* ("Tarski-like") notion of a conceptualization as found e.g. in [GN87]. Guarino et al. point out that this definition *per se* does not adequately fit the purposes of an ontology. They argue, correctly in our opinion, that a conceptualization *should not be extensional* because a conceptualization benefits from invariance under changes that occur at the instance level and from transitions between different "states of affairs"[8] in a domain. They propose instead a conceptualization as *an intensional* semantic structure i.e. abstracting from the instance level, which encodes *implicit rules* constraining the structure of a piece of reality. Therefore in their words, "an ontology only indirectly accounts for a conceptualization". Put differently, an ontology becomes a logical theory which possesses a conceptualization as an explicit, partial model. The resulting OntoClean methodology for evaluating ontological decisions [GW02] consists of a set of formal notions that are drawn from Analytical Philosophy and called *meta-properties*. Such meta-properties include *rigidity, essence, identity, unity,* and *dependence*. The idea of these notions is to focus on the *intrinsic properties* of concepts, assumed to be *application-independent.*

Clearly in the previous example the two axiomatizations should therefore not be seen as different ontologies since they only differ on their description of extensions i.e. states of affairs. Both axiomatizations implicitly share the same intensional semantic structure or conceptualization. Furthermore, the ISBN is an *extrinsic* property since it is not rigid[9] for *all* instances of the concept "book". Therefore, it cannot be used to specify the intended meaning of a book at the domain level.

---

[8] A state of affairs refers to a particular instance of reality, or also called *a* possible world.
[9] "A property is rigid if it is essential to all its possible instances; an instance of a rigid property cannot stop being an instance of that property in a different world" [WG01].

An important problem of the OntoClean methodology, in our opinion, is its applicability. First of all it relies on deep philosophical notions that in practice are not easy or intuitive to teach and utilize –at least for "nonintellectual" domain experts; and secondly it only focuses on the intrinsic properties of concepts and such properties are often difficult to articulate. For example, how to formally and explicitly articulate the identity criteria of a book (or person, brain, table, conference, love, etc.)? Guarino and Welty state in [WG01]: *"We may claim as part of our analysis that people are uniquely identified by their brain, but this information would not appear in the final system we are designing"*. In short, it would seem that OntoClean can be applied mainly by highly trained intellectuals for domain analysis and ontological checks[10].

Ontology usability is also important. This is another factor that should not be ignored, especially with regards to the philosophically inspired research on ontologies (or the so-called "philosophical ontology" as in [S03a]). In keeping with current views in the field of information technology, ontologies are to be shared and used collaboratively in software applications. This gives even more weight to the importance of *ontology usability*.

Conclusions

The closer an axiomatization is to certain application perspectives, the more usable it will be. In contrast, the more an axiomatization is independent of application perspectives, the more reusable it will be. In other words, *there is a tradeoff between ontology usability and ontology reusability.*

From a *methodological viewpoint*, if a methodology emphasizes usability perspectives or evaluates ontologies based on how they fulfill specific application requirements, the resulting ontology will be similar to a conceptual data schema (or a classical knowledge base) containing application specific and thus, less reusable knowledge. Likewise, if a methodology emphasizes the independency of the knowledge, the resulting ontology in general will be less *usable*, since it has no intended use by ignoring application perspectives.

Based on the above, we conclude the following ontology engineering requirement:

The influence of usability perspectives on ontology axioms should be well articulated, pursuing both reusability and usability.

3 The DOGMA Approach

In this section, we present the DOGMA approach for ontology engineering, which aims to tackle the engineering challenges stated in the previous section. In section 3.1 we schematically illustrate the DOGMA approach. The DOGMA philosophy is presented in section 3.2. Section 3.3 presents the formalization of the notion of Ontology Base for representing domain axiomatization, and section 3.4 presents how applications axiomatizations can built and used.

---

[10] See [GGO02] for a successful application of OntoClean on cleaning up WordNet.

3.1 Overview

As we mentioned before, in DOGMA, we introduce the notion of *ontology base* for capturing domain axiomatizations; and we introduce the notion of application axiomatization, by which particular applications commit to an ontology base, i.e. a *domain axiomatization and its application axiomatizations*. see figure 3.

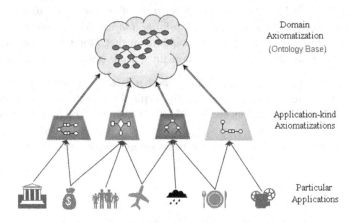

Fig. 3. DOGMA framework

The ontology base is intended to capture "plausible" domain axiomatizations. It basically consists of a set of binary conceptual relations. The lexical rendering of a binary conceptual relation is called *lexon*. A lexon is described as a tuple of the form <γ: Term1, Role, InvRole, Term2>, where Term1 and Term2 are linguistic terms. γ is a context identifier, used to *bound* the interpretation of a linguistic term: notably, for each context γ and term T, the pair (γ, T) is assumed to refer to a uniquely identifiable *concept*. Role and InvRole are lexicalizations of the paired roles in any binary relationship, for example WorkingFor/Employing, or HasType/IsTypeOf.

Particular applications now may *commit* to the ontology base through an application axiomatization. Such a commitment is called application's *ontological commitment*[11]. Each application axiomatization consists of (1) a selected set of lexons from an ontology base; (2) a specified set of rules to constrain the usability of these lexons.

Example

Resuming the Bibliography example above, the table at the left of Figure 4 shows a Bibliography ontology base. The (ORM) graphs at the right show two application axiomatizations (Bookstore and Library axiomatizations) by which particular applications

---

[11] We sometimes use the notion of application's "ontological commitment" and the notion "application axiomatization" interchangbly in this chapter. It is also worth to note that the notion of "ontological commitment" as found in [GG95] generally refers to a "conceptualization", literally, it is defined as "a partial semantic account of the intended conceptualization of a logical theory."

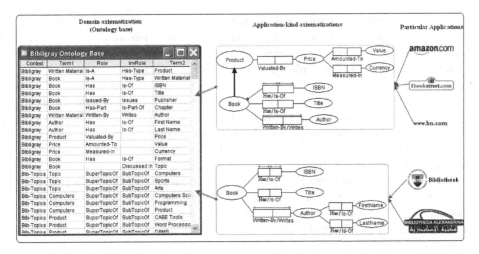

Fig. 4. Particular applications commit to a DOGMA ontology base through their respective application axiomatizations

might make a commitment to and share this same Bibliography ontology base. *Notice that all conceptual relations in both application axiomatizations correspond to (or are derived from) lexons in the Bibliography ontology base. In this way, different application axiomatizations share and reuse the same intended meaning of domain concepts.* As we shall show later, application axiomatizations can be represented not only in ORM (as shown in figure 4), but in any specification language such as OWL, EML, EEU, etc.

3.2 The DOGMA Philosophy (Application vs. Domain Axiomatization)

This section presents the fundamental idea of DOGMA. We introduce the notion of domain and application axiomatizations, and the formal relationship between them (called *double-articulation*). We discuss and formalize which type of knowledge should be captured in a domain verses application axiomatization. The translation of this philosophy into a software implementation is presented in section 3.3 and 3.4.

As we have discussed in section 2, decreasing the influence of usability perspectives is a principal engineering requirement when axiomatizing *domain concepts*. To capture knowledge at the domain level, one should focus on characterizing the *intended meaning* of domain vocabularies (i.e. domain concepts), rather than on how and why these concepts will be used. A domain axiomatization becomes an axiomatic theory that captures the axioms that account for (i.e. characterizes) the intended meaning of the domain vocabularies.

This motivates us to understand the relationship between a domain vocabulary and the specification of its intended meaning in a logical theory.

In general, it is not possible to build a logical theory to specify *the complete and exact intended meaning* of a domain vocabulary[12]. Usually, the level of detail that is

---

[12] This is because of the large number of axioms and details that need to be intensively captured and investigated, such detailed axiomatizations are difficult -for both humans and machines- to compute and reason on, and might holds "trivial" assumptions.

appropriate to explicitly capture and represent it is subject to what is reasonable and plausible for domain applications. Other details will have to remain implicit assumptions. These assumptions are usually denoted in linguistic terms that we use to lexicalize concepts, and this implicit character follows from our interpretation of these linguistic terms.

Incidentally, the study of the relationship between concepts and their linguistic terms is an ancient one; for example Avicenna (980-1037) [Q91] already argued that *"There is a strong relationship/dependence between concepts and their linguistic terms, change on linguistic aspects may affect the intended meaning... Therefore logicians should consider linguistic aspects 'as they are'. ..."[13].*

Indeed, the linguistic terms that we usually use to name symbols in a logical theory convey some important assumptions, which are part of the conceptualization that underlie the logical theory. We believe that these assumptions should not be excluded or ignored (at least by definition) as indeed *they implicitly are part of our conceptualization.*

Hence, we share Guarino and Giaretta's viewpoint [GG95], that an ontology (as explicit domain axiomatization) only *approximates* its underlying conceptualization; and that a domain axiomatization should be *interpreted intensionally*, referring to the intensional notion of a conceptualization. They point out that Gruber's [G95] earlier mentioned definition does not adequately fit the purposes of an ontology since a mere re-arrangement of domain objects (i.e. different state of affairs) would correspond to *different* conceptualizations. Guarino and Giaretta argue that a conceptualization benefits from invariance under changes that occur at the instance level by transitions between merely different "states of affairs" in a domain, and thus *should not be extensional*. Instead, they propose a conceptualization as *an intensional semantic structure* (i.e. abstracting from the instance level), *which encodes implicit rules* constraining the structure of a piece of reality. Indeed, this definition allows for the focus on the meaning of domain vocabularies (by capturing their intuitions) independently of a state of affairs. See [G98a] for the details and formalisms.

3.2.1 Definition (Double Articulation, Intended Models, Legal Models)

Given a concept **C** as *a set of rules (i.e. axioms) in our mind about a certain thing in reality*, the set **I** of "all possible" instances that comply with these rules are called the *intended models* of the concept **C**. Such concepts are captured at the domain axiomatization level. An application A_i that is interested in a subset I_{Ai} of the set **I** (according to its usability perspectives), is supposed to provide some rules to specialize **I**. In other words, every instance in I_{Ai} must also be an instance in **I**:

$$I_{Ai} \subseteq I$$

We call the subset I_{Ai}: the *legal models* (or extensions) of the application's concept C_{Ai}. Such application rules are captured at the application axiomatization level. Both domain and application axiomatizations can be seen (or expressed) for example as sentences in first order logic.

---

[13] This is an approximated translation from Arabic to English.

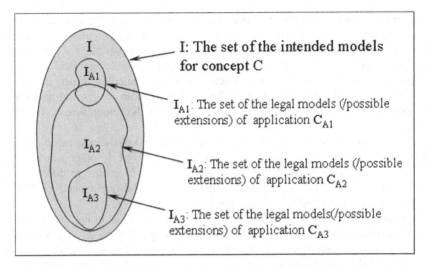

Fig. 5. An example of three different applications specializing a domain concept

We call the relationship (defined above) between a domain axiomatization and an application axiomatization: *double articulation*[14].

As we have illustrated in the previous section, bookstore applications that are interested *only* in the instances of the concept 'book' (that can be sold) need to declare the Mandatory rule that each instance of book must have an ISBN value.

In Figure 5 we show three kinds of applications specializing a domain concept.

The differences between the legal models of these application-types illustrate their different usability perspectives:

- The intersection between the legal models of C_{A2} and the legal models C_{A3} shows that I_{A3} is a subset of I_{A2}. An example of this case could be the difference between notions of 'book' in the axiomatization of bookstores and libraries: all legal instances of the bookstores' notion are legal instances for the libraries, but not vice versa. For libraries, the instances of e.g. 'Manual' or 'Master Thesis' can be instances of a 'book'; however, they cannot be instances of 'book' for bookstores, unless they are published with an 'ISBN'.
- The difference between I_{A1} and I_{A3} shows an extreme case: two types of applications sharing the same concept C while their legal models are completely disjoint according to their usability perspectives. An example of this case could be the difference between notions of 'book' in the axiomatization of bookstores' and museums': Museums are interested in exhibiting and exchanging instances of old 'books', while bookstores are not interested in such 'books', unless for example, they are re-edited and published in a modern style.

---

[14] The term "double articulation" in this chapter simply means *expressing knowledge in a twofold axiomatization*. The term "articulation" in WordNet means: "Expressing in coherent verbal form", "The shape or manner in which things come together and a connection is made", etc. In the semiotics and linguistics literature, the term "double articulation" has been introduced by [N90][M55] (which has a different meaning and usage than ours) to refer to the distinction between lexical and functional unites of language or between content and expression.

One may wonder how *domain concepts* can be agreed upon because of the difficulty in gaining an objective insight into the nuances of another person's thoughts. Many researchers admit that a conceptualization reflects a particular viewpoint and that it is entirely possible that every person has his/her "own" (linguistic manner of referring to) concepts. For example, Bench-Capon and Malcolm argued in [BM99] that conceptualizations are likely to be influenced by personal tastes and may reflect fundamental disagreements. In our opinion, herein lies the importance of linguistic terms.

Linguistic resources (such as lexicons, dictionaries, and glossaries) can be used as consensus references to *root* ontology concepts [J06] [M99a]. In other words, ontology concepts and axioms can be investigated using such linguistic resources and it can be determined whether a concept is influenced by usability perspectives. We explain this idea further in the following paragraphs.

The importance of using linguistic resources in this way lies in the fact that a linguistic resource renders/contains the intended meaning of a linguistic term as it is commonly "agreed" among the community of its language. The set of concepts that a language lexicalizes through its set of word-forms is generally an agreed conceptualization[15] [T00]. For example, when we use the English word 'book', we actually refer to the set of implicit rules that are common to English-speaking people for distinguishing 'books' from other objects. Such implicit rules (i.e. concepts) are learned and agreed from the repeated use of word-forms and their referents. Usually, lexicographers and lexicon developers investigate the repeated use of a word-form (e.g. based on a comprehensive corpus) to determine its underlying concept(s). See [J06] for more details about the incorporation of linguistic resources in DOGMA. Other researchers have also used linguistic resources in ontology engineering in other ways [BSZS06] [PS06].

3.2.2 On Representing Domain Axiomatizations
In this section, we discuss some choices that are relevant for *representing* domain axiomatizations.

A domain axiomatization merely cannot be a list of linguistic terms, and their intended meanings cannot be completely implicit. The intended meaning of linguistic terms should be axiomatized and represented by means of a formal language.

From a methodological viewpoint, such a formal language should be content-oriented rather than syntax-oriented. This language should serve as a theoretical tool which guides ontology builders through its primitives, and restrict them to focus *only* on and represent the "kinds" of axioms that account for the intended meaning of domain vocabularies.

By analogy, the conceptual "data" modeling languages ORM and EER provide database designers a set of primitives with which they can be guided to build a normalized database schema. Indeed, ORM and EER can be seen as content-oriented languages, because they *restrict* the focus of database designers to the *integrity of data models*.

An example of the difference between conceptual data modeling primitives and the kind of primitives that account for the intended meaning of a vocabulary[16] is the

---

[15] Thus, we may view a lexicon of a language as an informal ontology for its community.
[16] I.e. conceptual data modeling vs. conceptual domain modeling.

difference between the "Rigid" and "Mandatory". Something can be mandatory but not rigid, as in the case of 'ISBN' which is not a rigid property for every instance of a 'book' but could be mandatory for some applications. In other words, to model something as a rigid property, it should be rigid in *all possible* applications, while what can be mandatory for an application might not be mandatory for another. See [JDM03][GHW02] for more discussions on such issues.

Current research trends on ontology languages within the Semantic Web and the description logic communities are mainly concerned with improving *logical consistency* and inference services. Such services in our opinion are more suitable for building knowledge base applications or expert systems rather than axiomatizing "domain concepts". Significant results within the description logic community have indeed been achieved in the development of expressive and decidable logics, such as DLR [CGDL01], SHOIN [HST99], etc., yet less attention has been given to the quality of ontological *content*.

"...I was annoyed by the fact that knowledge representation research was more and more focusing on reasoning issues, while the core problems of getting the right representations were not receiving that much attention...". (Nicola Guarino[17]).

An example of a modeling primitive in the SHOQ description logic which in our opinion, should not be allowed in axiomatizing domain concepts since it does not account for meaning, is *datatypes* [P04]. Such a primitive belongs mainly to the symbolic level. In short, description logics (and their derivative languages such as DAML+OIL, or OWL) seem to play a useful role in specifying application (rather than domain) axiomatizations.

We shall return, in section 3.4 to the use of both conceptual data modeling languages and description logic based languages, for modeling and representing application axiomatizations.

We observe two possible ways to capture formal domain axiomatizations: (1) as an arbitrary set of axioms, e.g. using description logic, or (2) through a knowledge representation model (e.g. a database). The first case is common within the Semantic Web and Artificial Intelligence communities; in this case ontology builders are responsible (i.e. unguided) to decide whether an axiom accounts for the intended meaning of a vocabulary. This way offers ontology builders more freedom and expressiveness, but the risk of encoding usability perspectives is still high. In the second case, ontology builders are restricted only to capturing and storing the kind of axioms that account for factual meaning; assuming that the representation model is well studied and designed to pursue such axioms. This way is less expressive than the first one, but it reduces the risk of mixing domain and application axioms. The second way offers scalability in accessing and retrieving axioms, which is usually a problematic issue in the first way. The second way is mostly used within the lexical semantics community, e.g. WordNet [MBFGM90], Termintography [KTT03]. Notice that both ways are (or should be) well formalized and map-able to first order logic, and thus can be seen as logical theories.

We have chosen the second way for our approach. As we will show in section 3.3, we have developed a data model for capturing domain axiomatizations called an *ontology base*.

---

[17] An interview with Nicola Guarino and Christopher Welty (9 June 2004):
http://esi-topics.com/erf/2004/june04-ChristopherWelty.html

3.2.3 Summary: Properties of Domain Axiomatization

In this section, we summarize the basic properties of a domain axiomatization: it is (1) an axiomatized theory (2) that accounts for the intended meaning of domain vocabularies; (3) it is intended to be shared and used as a vocabulary space for application axiomatizations. It is supposed to be (4) interpreted intensionally, (5) and investigated and rooted at a human language conceptualization.

3.3 The Notion of an (Ontology Base), for Capturing Domain Axiomatizations

An ontology base is a knowledge representation model for capturing domain axiomatizations. This notion is used as a core component in the DOGMA approach.

Basically, an ontology base consists of a set of lexons. A lexon is a plausible binary relationship between context-specific linguistic terms, or in other words, a lexical rendering of a –plausible– binary conceptual relation.

3.3.1 Definition (Lexon)

A lexon is a 5-tuple of the form:

$$< \gamma: T_1, r, r', T_2 >$$

Where:

γ is a context identifier (the notion of context will be defined shortly).

T_1 and T_2 are linguistic terms from a language L.

r and r' are lexicalizations of the paired roles in a binary conceptual relationship R; the role r' is the inverse of the role r. One can verbalize a lexon as $(T_1 r T_2)$, and $(T_2 r' T_1)$. For example, the role pair of a *subsumption* relationship could be: "Is_type_of" and "Has_type"; the role pair of a *parthood* relationship could be: "Is_part_of" and "Has_part", and so forth.

The following is a set of lexons, as a simple example of part of an ontology base:

```
<Commerce: Person, Issues, Issued by, Order>
<Commerce: Order, Settled Via, Settles, Payment Method>
<Commerce: Money Order, Is a type of, Has type, Payment Method>
<Commerce: Check, Is a type of, Has type, Payment Method>
<Commerce: Payment Card, Is a type of, Has type, Payment Method>
<Commerce: Credit Card, Is a type of, Has type, Payment Card>
<Commerce: Credit Card, Has, Is of, Expiration Date>
```

3.3.2 Definition (Concept)

A term T within a context γ is assumed to refer to uniquely identified *concept* C:

$$(\gamma, T) \rightarrow C$$

Notice, for example, that within the context 'Commerce', the linguistic term 'Order' refers to "A commercial document used to request someone to supply something in return for payment". It may refer to other concepts within other contexts, e.g. within the context 'Military', the term 'Order' refers to "A command given by a superior that must be obeyed"[18]. Further detail about the notion of context will be discussed in the next section.

As we have discussed earlier, *a concept is circumscribed by a set of rules in our mind about a certain thing in reality.* The notion of *intended meaning* (or word meaning/sense) can be used alternatively with the notion of concept to denote something. The set of all possible instances (i.e. in all possible stats of affairs) that comply with these rules are called *intended models*.

As part of the context for each concept in DOGMA, there must be a *gloss*. A gloss is an auxiliary informal account for the commonsense perception of humans of the intended meaning of a linguistic term. The purpose of a gloss is *not* to provide or catalogue general information and comments about a concept, as conventional dictionaries and encyclopedias do [MBFGM90]. A gloss, for formal ontology engineering purposes, is supposed to render factual knowledge that is critical to understanding a concept, but that is unreasonable or very difficult to formalize and/or articulate explicitly. Although a gloss is not intended to be processed by machines, but its content is controlled by a set of well-defined guidelines (see [J06] for more details), such as: It should start with the *principal/super type* of the concept being defined; It should be written in the form of propositions; it should focus on distinguishing characteristics and intrinsic properties that differentiate the concept from other concepts; It should be consistent with the lexons and formal definitions, etc.

3.3.3 Definition (Role)

A *role* is an axiomatic entity lexically expressing how a concept (referred by a term within a context) relates to another concept. A lexon being a binary relationship always involves two roles.

A role within a context is not intended to refer to a concept; thus, $(\gamma, r) \rightarrow C$ is improper. In other words, our notion of role does not refer to a "stand alone" unary (or binary) concept. Rather, roles only lexicalize the participation of a "unary concept" in an n-ary conceptual relationship. As the notion of a lexon is a lexical rendering of a binary conceptual relationship, we formalize a lexon as two context-specific terms playing mutual roles, that both refers to a *binary concept (typically called binary conceptual relation)*:

$$< (\gamma, T, r), (\gamma, T, r) > \quad \rightarrow \quad C^2$$

The notation of a context-specific term playing a role (γ, T, r) is called *concept-role*.

For practical purposes, we shall not require for both roles to be explicitly lexicalized within a lexon. We assume that at least one role is to be lexicalized, represented as <Bibliography, Book, is-a, , Written Material>.

A DOGMA ontology base contains only binary relationships. This does not deny the existence of ternary (or *n*-ary) relationships. Ternary resp. *n*-ary relationships may always be converted into an equivalent set of 3, resp. *n*, binary relationships, possibly with the introduction of new terms/concepts. In practice relationships are however mainly binary.

---

[18] These two definitions of the term "Order" are taken from WordNet, (May 2004) http://www. cogsci.princeton.edu/cgi-bin/webwn.

3.3.4 Definition (Mapping Lexons into First Order Logic)

With each lexon $< \gamma: T_1, r, r', T_2 >$ in the ontology base there correspond three statements in first order logic, as follows:

$$\forall x\, T_1(x) \rightarrow (\forall y\, r(x, y) \rightarrow T_2(y))$$
$$\forall y\, T_2(y) \rightarrow (\forall x\, r'(y, x) \rightarrow T_1(x))$$
$$\forall x, y\, r(x, y) \leftrightarrow r'(y, x)$$

For example, the mapping of the lexon <Commerce: Person, Issues, IssuedBy, Order> into first order logic produces:

$$\forall x\, Person(x) \rightarrow (\forall y\, Issues(x, y) \rightarrow Order(y))$$
$$\forall y\, Order(y) \rightarrow (\forall x\, IssuedBy(y, x) \rightarrow Person(x))$$
$$\forall x, y\, Issues(x, y) \leftrightarrow IssuedBy(y, x)$$

Notice that Context is not part of our formal mapping of lexons. As we shall discuss in the next section, a context for our purposes here is a mostly *informal* notion used to link unambiguously (i.e. bound) the interpretation of a linguistic term to a concept. Linguistic terms, e.g. 'Person', 'Order', etc. can be seen as unambiguous terms (i.e. concepts) within the lexon formal mapping. A lexon (or its formal mapping) is assumed to be plausible (i.e. to be a *weak axiom*) within its context, see section 3.3.5. In section 3.3.6 we shall discuss how to introduce further formal axiomatizations at the ontology base level, for targeting systematic ontological quality.

Finally, our formal lexon mapping assumes unique role names. Each role label (or InvRole) should be unique within the formal mapping of lexons. As this is might not be the case in practice, one can provide an "internal" naming convention, for example, by renaming 'Issues' as 'Issues_Order' and 'IssuedBy' as 'IssuedBy_Person'.

At this point, we have established how lexons are the basic building blocks of an ontology base and that they express basic domain facts. *The principal role of an ontology base is to be a shared vocabulary space for application axiomatizations.* As sharing lexons means sharing the same concepts and their intended models, semantic interoperability between classes of autonomous applications can be achieved, basically, by sharing a certain set of lexons[19] and agreeing on their interpretation (*commitments*, see below in 3.4).

3.3.5 The Notion of Context

The notion of context has been, and still is, the subject of occasionally intense study, notably in the field of Artificial Intelligence. It has received different interpretations. Commonly, the notion of context has been realized as a set of formal axioms (i.e. a theory) about concepts. It has been used among other things: to localize or encode a particular party's view of a domain, cf. C-OWL [BHGSS03]; as a background, micro-theory, or higher-order theory for the interpretation of certain states of affairs [M93]; and to facilitate the translation of facts from one context to another, as in KIF [PFP+92].

---

[19] As we shall show in section 3.4, a class of interoperating applications may need to agree on and share some rules that constrain the use of a concept, i.e. share the same legal models.

In our approach, we shall use the notion of context to play a *"scoping"* role at the ontology base level. We say a term *within* a context refers to a concept, or in other words, that *context is an abstract identifier that refers to implicit (or maybe tacit[20]) assumptions, in which the interpretation of a term is bounded to a concept.*

Notice that a context in our approach is not *explicit* formal knowledge. In practice, we define context by referring to a source (e.g. a set of documents, laws and regulations, informal description of "best practice", etc.), which, by *human understanding*, is assumed to "contain" those assumptions. Lexons are assumed (by that same human understanding) to be "plausible within their context's source". Hence, a lexon is seen as a (weak) domain *axiom*.

Note. For ease of readability, we will in lexons continue to use a (unique) mnemonic label such as Commerce or Bibliography to denote a context rather than an abstract context identifier (pointer to a resource) in the representations of lexons below. To wit, Bibliography will point to a document in which a human interpreter, possibly assisted by programs, finds sufficient "context" to at least disambiguate both terms used in the lexon.

3.3.6 Further Formal Axiomatizations (Incorporating Upper Level Ontologies)

In order to achieve *a systematic ontological quality and precision*[21] on the specification of the intended meanings of linguistic terms, these specifications might need to receive more formal restrictions, than just mapping lexons into logical statements.

For example, without introducing further formal restrictions to the following lexons:

<Bibliography: Man, Is-a, Person>
<Bibliography: Author, Is-a, Person>
<Bibliography: John, Is-a, Person>

The ontological difference (or rather, the misuse of 'is-a') cannot be systematically detected[22]. In this section, we discuss how a formal axiomatic system can be introduced into an ontology base.

As we have chosen to represent formal domain axiomatization in a data model (i.e. ontology base), arbitrary and expressive formal definitions are restricted (see our discussion on this issue in section 3.2.3). Therefore, we extend the ontology base model to incorporate primitives of upper level ontologies. Our incorporation of upper level ontologies in this chapter is fairly simplistic; the deeper philosophical arguments that are necessary for such incorporation are presented schematically for the sake of completeness

---

[20] The difference between implicit and tacit assumptions, is that the implicit assumptions can, in principle, be articulated but still they have not, while tacit assumptions are the knowledge that cannot be articulated. it consists partially of technical skills -the kind of informal, hard-to-pin-down skills captured in terms like "know-how", and "we know more than we can tell or put in words". However, even though tacit assumptions cannot be articulated, they can be transferred through other means over than verbal or formal descriptions [Inn+03] [N94].

[21] The notion of "ontological precision" is defined by Aldo Gangemi in [G04] as *"the ability to catch all and only the intended meaning"*.

[22] By assuming that the 'is-a' refers to a subsumption relationship (i.e. Sub-Type of), only the first lexon is correct. The 'is-a' in the second lexon should interpreted as "is role of", because 'Author' is a role of 'Person' and not a type of a 'Person'; and obviously, the last lexon refers to 'is instance of'. See [GW02] for more details on this issue.

only. It is important to note that upper ontologies are still very much exponents of work in progress. Upper level ontologies are formal axiomatic systems that describe the most general categories of reality. Such ontologies are not only application and task independent but represent also domain (and possibly language) independent axiomatizations [DHHS01] [G98b].

Based on the literature of upper level ontologies as found for example in [DHHS01] [G98b], we introduce in our approach the notion of *upper-form*. Each term within a context should have an upper-form, likewise, each lexon should have an upper-form.

Term upper-forms

Term upper-forms are superior types of concepts, such as substantial, feature, abstract, region, event, process, type, role, property, particular, etc. The notation of a term upper-form declaration is:

$$\gamma(T) : < UpperFormName >$$

Examples:
Bibliography(Person):Substantial,
Bibliography(Author):Substantial,
Bibliography(First-Name):Property

A term can have several upper-forms, denoted $\gamma(T) : \{UpperForm, ...\}$. Examples:

Bibliography(Person):{Substantial, Type},
Bibliography(Author):{Substantial, Role},
Bibliography(John):{Substantial, Instance}.

Lexon upper-forms

Lexon upper-forms are relationship kinds, also called "basic primitive relations", such as parthood, dependence, property-of, attribution, subsumption, etc. Such relationship kinds are carefully and formally axiomatized in upper level ontologies, and they are general enough to be applied in multiple domains. Our notation of a lexon upper-form declaration is

$$< \gamma : T_1, r, r', T_2 >: < UpperFormName >$$

For instance, the lexon "<Bibliography: Book, Is-a, HasType, Written Material>: Subsumption" is declared as a subsumption relationship where the concept 'Book' formally subsumes the concept 'Written Material'. Similarly the declaration "<Bibliography: Book, Has-Part, Is-Part-Of, Chapter>: Parthood" now states that the lexon lexically expresses a particular *parthood* (meronymy) relationship, where within the context of Bibliography, an instance of the concept 'chapter' is a part of an instance of the concept 'Book'. And "<Bibliography: Author, Has, Is-Of, Name>: Property" declares the lexon as a *property-of* relationship, where the concept 'Name' is a property of the concept 'Author', and so forth.

Upper-forms may carry with them a formal axiomatization and reasoning mechanism, as defined in an associated *upper level ontology*,.They may thus be used to add a *formal account* to lexons. For example, a formal account of the lexon "<Bibliography, John, instance-of, Author>: Instantiation" may include or be induced from a (partial) formal axiomatization of the instantiation relationship such as found in [GGMO01]:

Instantiation(x,y) $\rightarrow$ $\neg$Instantiation(y,x)

(Instantiation(x,y) $\wedge$ Instantiation(x,z)) $\rightarrow$ ($\neg$Instantiation(y,z) $\wedge$ $\neg$Instantiation(y,z))

Particular(x) $=_{def}$ $\neg\exists y$ (Instantiation(y,x))

Universal(x) $=_{def}$ $\neg$Particular(x)

Fig. 6. A formal axiomatization of Instantiation, from [GGMO01]

Similarly Parthood, Subsumption etc. may become interpreted from their formal axiomatizations in [GGMO01]. For instance, under the formal axiomatization of the 'Subsumption' relationship in [GGMO01], the following lexon would be inadmissible because its declared upper-form conflicts with the adopted axiomatization of the term upper-forms, where Role may no longer subsume Type:

> Bibliography(Person):{Substantial, Type}
> Bibliography(Author):{Substantial, Role}
> <Bibliography: Author, Is-a, , Person>: Subsumption

Notice that formal axiomatizations of such upper forms are not necessarily to be used at runtime by applications that use or share lexons. Their main purpose could be as theoretical tools to help achieve verifiable quality during development and maintenance time of an ontology.

Note that our methodological principles and their implementation prototypes are not dependent on a particular choice of upper level ontology. This is left to ontology builders. However, libraries of upper-ontology plug-ins may be envisaged complete with predefined reasoning, for ontology builders to apply on selected sets of lexons as part of so-called *commitments*. This constitutes the foundational principle of *application axiomatization* in DOGMA as explained in the following section.

3.4 Application Axiomatization

The notion of an ontology base was introduced in order to capture domain axiomatizations independently of usability perspectives. In this section, we introduce the second part of the double articulation: *application axiomatizations*. First, we discuss the general properties of these axiomatizations; then, we introduce the key notion of an *application's ontological commitment*.

While the axiomatization of domain knowledge is mainly concerned with the characterization of the "intended models" of concepts, the axiomatization of application knowledge is mainly concerned with the characterization of the "legal models" of these concepts (see Fig. 5). Typically, as domain axiomatizations are intended to be shared, public, and highly reusable at the domain level, application axiomatizations are intended to be local and highly usable at the task/application-kind level.

As we have discussed earlier, applications that are interested only in a subset of the intended models of a concept (according to their usability perspective) are supposed to provide some rules to specialize these intended models. Such a specialization is called an *application axiomatization*. The vocabulary of concepts used in application axiomatization is restricted to the vocabulary defined in its domain axiomatization. (Note that this "specialization" should therefore not be confused with the subsumption relationship discussed earlier.) As clarified below, an application axiomatization comes defined as a set of rules that constrain the use of the domain vocabulary. More specifically, these rules declare what must necessarily hold in any possible interpretation for a given class of applications.

A particular application is said to *commit ontologically* to an intended meaning of a domain vocabulary (stored as lexons in an ontology base) through its application axiomatization., i.e. the latter may be considered a formal specification of such a commitment.

An application axiomatization typically consists of: (1) an *ontological view* that specifies which domain concepts in an ontology base are relevant to include in this axiomatization. These concepts can be explicit lexons or derived from lexons, (2) a set of rules that defines the legal models of the ontological view in the classical model-theoretic semantics sense, i.e. it formally specifies what must or must not hold in any possible world (interpretation) for the applications sharing this axiomatization.

We say that a particular extension of an application (i.e. a set of instances) commits to an ontology base through an application axiomatization if it conforms to or is consistent with the ontological view and the rules declared in this axiomatization (cf. model-theoretic semantics).

Speaking in operational terms, the eventual execution of an ontological commitment by an interpreter (or reasoner) provides an enforcer for the constraints on that application's intended behavior with its concepts, as laid down in the application's axiomatization. It is important to realize at this point that applications can run meaningfully *without* such an executed formal commitment to an ontology case –in fact most of today's legacy applications do– but clearly this happens "at their own risk" of producing –literally– meaningless results now and then.

3.4.1 Example

This example is based on that presented in Section 3.1. In this application scenario software agents wish to interoperate through a semantic mediator in order to exchange data messages and share business transactions (see Figure 7). The interoperation is enabled by the sharing of the same Bookstore axiomatization, i.e. as a global and legal data model[23]. The data source (or its "export schema" [ZD04]) of each agent is mapped into the shared axiomatization. All exchanged data messages (e.g. those formed in XML, RDF, etc.) may be validated whether they conform to the rules and the ontological view declared in the Bookstore axiomatization, for example under classical first-order model-theoretic semantics [R88].

The ontological view of the above bookstore axiomatization specifies which concepts are relevant for the task(s) of this application scenario. These concepts correspond to explicit lexons in the ontology base, or they might be derived from these lexons. One can see in the ontology base that a 'Book' is not explicitly a 'subtype of' a 'Product' as specified in the Bookstore axiomatization. This subsumption is derived

---

[23] This way of sharing and using axiomatizations (as global schema) seems more applicable to data integration and mediation systems [ZD04]. They can also be used to describe web services [NM02]. For example, an axiomatization could be specified for each web service (to describe the "static" information provided to/by a web service), so that all agents accessing a web service share the same axiomatization.

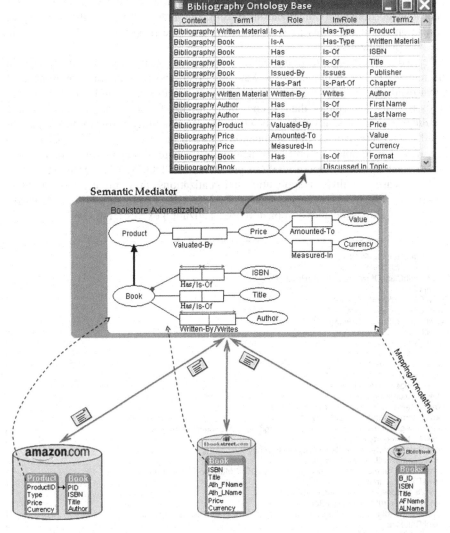

Fig. 7. Meaningful semantic interoperation between Bookstore applications

from the two lexons {<Bibliography: Book, Is-A, Has-Type, Written Material>, <Bibliography: Written Material, Is-A, Has-Type, Product>}. Based on these subsumptions, some inheritance also might be derived; for example, 'Book' inherits the relationship <Bibliography: Book, Written-By, Writes, Author> from its Written Material supertype. The choice of which concepts and relations should be committed to is an application-specific issue or may be subject to a usability perspective. See our discussion on this in Section 2.

In the Bookstore axiomatization, four rules are declared and may be conveniently verbalized: 1) each Book Has at least one ISBN; 2) each Book Has at most one ISBN; 3) each ISBN Is-Of at most one Book; 4) a Book may be Written-by several Author/s)and 5) an Author may Write/s several Book/s.

Notice that this approach enables usability perspectives to be encountered and encoded outside domain axiomatization. In turn, this indeed increases the usability of application axiomatizations as well as it increases the reusability of the underlying domain axiomatization.

Depending on the application scenario, application axiomatizations may be used in different ways. For example, in the Semantic Web and information search/retrieval scenarios, declaring rules might be not important because the main idea of these scenarios is to expand (rather than to constrain) queries. Filtering the unwanted results (i.e. illegal models) usually then falls within the responsibility of the application itself. In [J05] we presented such an application scenario of an ontology-based user interface, in which complaint web forms meaningfully share, through explicit ontological commitments, a vocabulary of relationships stored as a DOGMA Server lexon base.

To increase usability of application axiomatizations, they might be specified in multiple specification languages, such as DAML+OIL, OWL, RuleML, EER, UML, etc. Figure 8 shows the above Bookstore axiomatization expressed in OWL.

```
....
<owl:Class rdf:ID="Product" />
<owl:Class rdf:ID="Book">
 <rdfs:subClassOf rdf:resource="#Product" />
</owl:Class>
<owl:Class rdf:ID="Price" />
<owl:Class rdf:ID="Value" />
<owl:Class rdf:ID="Currency" />
<owl:Class rdf:ID="Title" />
<owl:Class rdf:ID="ISBN" />
<owl:Class rdf:ID="Author" />
<owl:ObjectProperty rdf:ID="Valuated-By">
 <rdfs:domain rdf:resource="#Product" />
 <rdfs:range  rdf:resource="#Price" />
</owl:ObjectProperty>
<owl:DataProperty rdf:ID=" Amounted-To .Value">
 <rdfs:domain rdf:resource="#Price" />
 <rdfs:range rdf:resource="http://www.w3.org/2001/XMLSchema#string"/>
</owl:DataProperty>
<owl:DataProperty rdf:ID="Measured-In.Currency">
 <rdfs:domain rdf:resource="#Price" />
 <rdfs:range rdf:resource="http://www.w3.org/2001/XMLSchema#string"/>
</owl:DataProperty>
<owl:DataProperty rdf:ID="Has.ISBN">
 <rdfs:domain rdf:resource="#Book" />
 <rdfs:range rdf:resource="http://www.w3.org/2001/XMLSchema#integer "/>
</owl:DataProperty>
<owl:DataProperty rdf:ID="Has.Title">
 <rdfs:domain rdf:resource="#Title" />
 <rdfs:range rdf:resource="http://www.w3.org/2001/XMLSchema#string"/>
</owl:DataProperty>
<owl:ObjectProperty rdf:ID="Written-By">
 <owl:inverseOf rdf:resource="#Writes "/>
 <rdfs:domain rdf:resource="#Book" />
 <rdfs:range  rdf:resource="#Author" />
</owl:ObjectProperty>
<owl:Restriction>
 <owl:onProperty rdf:resource="# Has.ISBN " />
 <owl:cardinality rdf:datatype="&xsd;nonNegativeInteger">1</owl:cardinality>
</owl:Restriction>
....
```

Fig. 8. An OWL representation of the Bookstore ontological commitment

Although both representations share the same intended meaning of concepts at the domain (/ontology base) level, notice the disparities between ORM and OWL in representing the Bookstore axiomatization. For example, ORM does not distinguish between DataProperties and ObjectProperties as does OWL. This is an example of an epistemological difference[24]. The ORM uniqueness constraint that spans over "Written-By/Writes" cannot be expressed in OWL, as it is implied by definition[25]. The other uniqueness and mandatory constraints are all expressed as a cardinality restriction in OWL.

The logical and epistemological disparities described above (which are induced by the difference between the formalizations and the constructs of both languages) illustrate different ways of characterizing the legal models of application axiomatizations. The choice of which language is more suitable for specifying application axiomatizations depends on the application scenario and perspectives. For example, ORM and EER are mainly suitable for database and XML (-based) application scenarios since they are comprehensive in their treatments of the integrity of data sets. For inference and reasoning application scenarios, description logic based languages (such as OWL, DAML, etc.) seem to be more applicable than other languages, as they focus on the expressiveness and the decidability of axioms. See [J07] [J07b] for the complete formalization of ORM in description logics. In this work we also identify which ORM constructs cannot be mapped into OWL. We have also implemented this formalization to reason about ORM diagrams using Racer (See [JD06]).

Allowing different languages, optimized techniques, or methodologies to be deployed at the application axiomatization level will indeed increase the usability of these axiomatizations. A recent application axiomatization language called Ω-RIDL [VDM04] has been developed within the DOGMA framework. It is claimed to better suited to the database applications' commitment to an ontology base.

4   Discussion and Conclusions

In this chapter, we have presented a comprehensive view of the DOGMA approach for ontology engineering. We have shown how application verses domain axiomatizations can be well articulated. We have introduced the notion of an ontology base for capturing domain axiomatizations, and the notion of application axiomatizations by which particular applications commit to the intended meaning of domain vocabulary.

In the following we summarize the main advantages of our approach:

Increase reusability of domain axiomatization, as well as usability of application axiomatizations. As we have shown in this chapter, the application-independence of an ontology is increased by separating domain and application axiomatizations. Usability perspectives have a negligible influence on the independence of a domain

---

[24] Epistemology level: The level that deals with the knowledge structuring primitives (e.g. concept types, structuring relations, etc.). [G94].

[25] The formalization of ObjectProperties in OWL does not allow the same tuple to appear twice in the same set, such as Written-By = {<author1, book1>, < author1, book1>,...}.

axiomatization, because ontology builders are prevented from encoding their application-specific axioms. In other words, domain axiomatizations are mainly concerned with the characterization of the "intended models" of concepts, while application axiomatizations are mainly concerned with the characterization of the "legal models" of these concepts.

Allows different communities to create and maintain domain axiomatization (typically public) and application axiomatizations (typically local). Indeed, domain experts, lexicographers, knowledge engineers, and even philosophers, may contribute to the development, maintenance, and review phases of domain axiomatizations. It is needless for them to know why and how these axiomatizations will be used. Application-oriented experts can also contribute to and focus on the development phases of application axiomatizations, without needing to know about the correctness of domain axioms. Hence, we offer an ontology representation model that is capable of distributed and collaborative development.

Allows the deployment of differently optimized technologies and methodologies to each articulation. For example, relational database management systems can be used (with high scalability and performance) to store and retrieve large-scale ontology bases. Natural language parsing and understanding techniques can be employed for extracting lexons from texts. Different specification languages can be used to specify application axiomatizations and these increase the usability of these axiomatizations.

Furthermore, *the importance of linguistic terms in ontology engineering is observed and incorporated* in our approach [SD04]. Not coincidentally, our approach allows for the adoption and reuse of many available lexical resources to support (or to serve as) domain axiomatizations. Lexical recourses (such as lexicons, glossaries, thesauruses and dictionaries) are indeed important recourses of domain concepts. Some resources focus mainly on the morphological issues of terms, rather than categorizing and clearly describing their intended meanings. Depending on its description of term meaning(s), its accuracy, and maybe its formality[26], a lexical resource can play an important role in ontology engineering.

An important lexical resource that is organized by word meanings (i.e. concepts, or called synsets) is WordNet [MBFGM90]. WordNet offers a machine-readable and comprehensive conceptual system for English words. Currently, a number of initiatives and efforts in the lexical semantic community have been started to extend WordNet to cover multiple languages. As we have discussed in section 3.2.1, the consensus about domain concepts can be gained and realized by investigating these concepts at the level of a human language conceptualization. This can be practically accomplished e.g. by adopting the informal description of term meanings that can be found in lexical resources such as WordNet, *as glosses.*

Acknowledgement. We are in debt to all present and former colleagues in STARLab for their comments, discussion, and suggestions on the earlier version of this work.

---

[26] i.e., the discrimination of term meanings in a machine-referable manner.

References

[BC88] Bylander, T., Chandrasekaran, B.: Generic tasks in knowledge-based reasoning: The right level of abstraction for knowledge acquisition. In: Knowledge Acquisition for Knowledge Based Systems, vol. 1, pp. 65–77. Academic Press, London (1988)

[BHGSS03] Bouquet, P., van Harmelen, F., Giunchiglia, F., Serafini, L., Stuckenschmidt, H.: C-OWL: Contextualizing ontologies. In: Fensel, D., Sycara, K.P., Mylopoulos, J. (eds.) ISWC 2003. LNCS, vol. 2870, pp. 164–179. Springer, Heidelberg (2003)

[BM99] Bench-Capon, T.J.M., Malcolm, G.: Formalising Ontologies and Their Relations. In: Bench-Capon, T.J.M., Soda, G., Tjoa, A.M. (eds.) DEXA 1999. LNCS, vol. 1677, pp. 250–259. Springer, Heidelberg (1999)

[BSZS06] Bouquet, P., Serafini, L., Zanobini, S., Sceffer, S.: Bootstrapping semantics on the web: meaning elicitation from schemas. In: Proceedings of the World Wide Web Conf (WWW 2006), pp. 505–512. ACM Press, New York (2006)

[CGDL01] Calvanese, D., De Giacomo, G., Lenzerini, M.: Identification constraints and functional dependencies in description logics. In: Proceedings of the 17th Int. Joint Conf. on Artificial Intelligence (IJCAI 2001), pp. 155–160 (2001)

[CJB99] Chandrasekaran, B., Johnson, R., Benjamins, R.: Ontologies: what are they? why do we need, them? IEEE Intelligent Systems and Their Applications 14(1), 20–26 (1999); Special Issue on Ontologies

[DDM06] De Moor, A., De Leenheer, P., Meersman, M.: DOGMA-MESS: A Meaning Evolution Support System for Interorganizational Ontology Engineering. In: Schärfe, H., Hitzler, P., Øhrstrøm, P. (eds.) ICCS 2006. LNCS, vol. 4068, pp. 189–202. Springer, Heidelberg (2006)

[DDM07] De Leenheer, P., de Moor, A., Meersman, R.: Context Dependency Management in Ontology Engineering: a Formal Approach. In: Spaccapietra, S., Atzeni, P., Fages, F., Hacid, M.-S., Kifer, M., Mylopoulos, J., Pernici, B., Shvaiko, P., Trujillo, J., Zaihrayeu, I. (eds.) Journal on Data Semantics VIII. LNCS, vol. 4380, pp. 26–56. Springer, Heidelberg (2007)

[DHHS01] Degen, W., Heller, B., Herre, H., Smith, B.: GOL: Towards an Axiomatized Upper-Level Ontology. In: Formal Ontology in Information Systems. Proceedings of the FOIS 2001, pp. 34–46. ACM Press, New York (2001)

[DM05] De Leenheer, P., Meersman, R.: Towards a formal foundation of DOGMA ontology Part I: Lexon base and concept definition server, TR STAR-2005-06, Brussel (2005)

[DSM04] De Bo, J., Spyns, P., Meersman, R.: Assisting Ontology Integration with Existing Thesauri. In: Meersman, R., Tari, Z. (eds.) OTM 2004. LNCS, vol. 3290, pp. 801–818. Springer, Heidelberg (2004)

[G04] Gangemi, A.: Some design patterns for domain ontology building and analysis. An online presentation (April 2004), http://www.loa-cnr.it/Tutorials/OntologyDesignPatterns.zip

[G94] Guarino, N.: The Ontological Level. In: Casati, R., Smith, B., White, G. (eds.) Philosophy and the Cognitive Science, pp. 443–456. Hölder-Pichler-Tempsky, Vienna (1994)

[G95] Gruber, T.: Toward principles for the design of ontologies used for knowledge sharing. International Journal of Human-Computer Studies 43(5/6) (1995)

[G98a] Guarino, N.: Formal Ontology in Information Systems. In: Proceedings of FOIS 1998, pp. 3–15. IOS Press, Amsterdam (1998)

[G98b] Guarino, N.: Some Ontological Principles for Designing Upper Level Lexical Resources. In: Rubio, A., Gallardo, N., Castro, R., Tejada, A. (eds.) Proceedings of First International Conference on Language Resources and Evaluation. ELRA - European Language Resources Association, Granada, Spain (1998)

[GG95] Guarino, N., Giaretta, P.: Ontologies and Knowledge Bases: Towards a Terminological Clarification. In: Mars, N. (ed.) Towards Very Large Knowledge Bases: Knowledge Building and Knowledge Sharing, pp. 25–32. IOS Press, Amsterdam (1995)

[GGMO01] Gangemi, A., Guarino, N., Masolo, C., Oltramari, A.: Understanding toplevel ontological distinctions. In: Proceedings of IJCAI 2001 Workshop on Ontologies and Information Sharing, pp. 26–33. AAAI Press, Seattle (2001)

[GGO02] Gangemi, A., Guarino, N., Oltramari, A., Borgo, S.: Cleaning-up WordNet's toplevel. In: Proceedings of the 1st International WordNet Conference (January 2002)

[GHW02] Guizzardi, G., Herre, H., Wagner, G.: Towards Ontological Foundations for UML Conceptual Models. In: Meersman, R., Tari, Z., et al. (eds.) CoopIS 2002, DOA 2002, and ODBASE 2002. LNCS, vol. 2519, pp. 1100–1117. Springer, Heidelberg (2002)

[GN87] Genesereth, M.R., Nilsson, N.J.: Logical Foundation of Artificial Intelligence. Morgan Kaufmann, Los Altos (1987)

[GW02] Guarino, N., Welty, C.: Evaluating Ontological Decisions with OntoClean. Communications of the ACM 45(2), 61–65 (2002)

[HST99] Horrocks, I., Sattler, U., Tobies, S.: Practical reasoning for expressive description logics. In: Ganzinger, H., McAllester, D., Voronkov, A. (eds.) LPAR 1999. LNCS, vol. 1705, pp. 161–180. Springer, Heidelberg (1999)

[Inn+03] Persidis, A., Niederée, C., Muscogiuri, C., Bouquet, P., Wynants, M.: Innovation Engineering for the Support of Scientific Discovery. Innovanet Project (IST-2001-38422), deliverable D1 (2003)

[J05] Jarrar, M.: Towards Methodological Principles for Ontology Engineering. PhD thesis, Vrije Universiteit Brussel (May 2005)

[J05a] Jarrar, M.: Modularization and automatic composition of Object-Role Modeling (ORM) Schemes. In: Meersman, R., Tari, Z., Herrero, P. (eds.) OTM-WS 2005. LNCS, vol. 3762, pp. 613–625. Springer, Heidelberg (2005)

[J06] Jarrar, M.: Towards the notion of gloss, and the adoption of linguistic resources in formal ontology engineering. In: Proceedings of the 15th International World Wide Web Conference (WWW 2006), Scotland, pp. 497–503. ACM Press, New York (2006)

[J07c] Jarrar, M.: Towards Effectiveness and Transparency in e-Business Transactions, An Ontology for Customer Complaint Management, ch. 7. Semantic Web Methodologies for E-Business Applications. IGI Global (October 2008) ISBN: 978-1-60566-066-0

[J07] Jarrar, M.: Towards Automated Reasoning on ORM Schemes. -Mapping ORM into the DLR_idf description logic. In: Parent, C., Schewe, K.-D., Storey, V.C., Thalheim, B. (eds.) ER 2007. LNCS, vol. 4801, pp. 181–197. Springer, Heidelberg (2007)

[J07b] Jarrar, M.: Mapping ORM into the SHOIN/OWL Description Logic- Towards a Methodological and Expressive Graphical Notation for Ontology Engineering. In: Meersman, R., Tari, Z., Herrero, P. (eds.) OTM-WS 2007, Part I. LNCS, vol. 4805, pp. 729–741. Springer, Heidelberg (2007)

[JDM03] Jarrar, M., Demy, J., Meersman, R.: On Using Conceptual Data Modeling for Ontology Engineering. In: Spaccapietra, S., March, S., Aberer, K. (eds.) Journal on Data Semantics I. LNCS, vol. 2800, pp. 185–207. Springer, Heidelberg (2003)

[JE06] Jarrar, M., Eldammagh, M.: Reasoning on ORM using Racer. Technical report, Vrije Universiteit Brussel, Brussels, Belgium (August. 2006)

[JH07] Jarrar, M., Heymans, S.: Towards Pattern-based Reasoning for Friendly Ontology Debugging. Journal of Artificial Intelligence Tools 17(4) (August 2008)

[JM02a] Jarrar, M., Meersman, R.: Formal Ontology Engineering in the DOGMA Approach. In: Meersman, R., Tari, Z., et al. (eds.) CoopIS 2002, DOA 2002, and ODBASE 2002. LNCS, vol. 2519, pp. 1238–1254. Springer, Heidelberg (2002)

[KN03] Klein, M., Noy: A component-based framework for ontology evolution. Technical Report IR-504, Vrije Universiteit Amsterdam (March 2003)

[KTT03] Kerremans, K., Temmerman, R., Tummers, J.: Representing multilingual and culture-specific knowledge in a VAT regulatory ontology: support from the termontography approach. In: OTM 2003 Workshops. Springer, Tübingen (2003)

[M55] Martinet, A.: Economie des changements phonétiques, pp. 157–158. Francke, Berne (1955)

[M93] McCarthy, J.: Notes on Formalizing Context. In: Proceedings of IJCAI 1993. Morgan Kaufmann, San Francisco (1993)

[M96] Meersman, R.: An essay on the Role and Evolution of Data(base) Semantics. In: Meersman, R., Mark, L. (eds.) Proceeding of the IFIP WG 2.6 Working Conference on Database Applications Semantics (DS-6). CHAPMAN & HALL, Atlanta (1996)

[M99a] Meersman, R.: The Use of Lexicons and Other Computer-Linguistic Tools. In: Zhang, Y., Rusinkiewicz, M., Kambayashi, Y. (eds.) Semantics, Design and Cooperation of Database Systems, The International Symposium on Cooperative Database Systems for Advanced Applications (CODAS 1999), pp. 1–14. Springer, Heidelberg (1999)

[M99b] Meersman, R.: Semantic Ontology Tools in Information System Design. In: Ras, Z., Zemankova, M. (eds.) Proceedings of the ISMIS 1999 Conference. LNCS, vol. 1609, pp. 30–45. Springer, Heidelberg (1999)

[M01] Meersman, R.: Ontologies and Databases: More than a Fleeting Resemblance. In: d'Atri, A., Missikoff, M. (eds.) OES/SEO 2001 Rome Workshop. Luiss Publications (2001)

[MBFGM90] Miller, G., Beckwith, R., Fellbaum, F., Gross, D., Miller, K.: Introduction to wordnet: an on-line lexical database. International Journal of Lexicography 3(4), 235–244 (1990)

[N90] Nöth, W.: Handbook of Semiotics. Indiana University Press, Bloomington (1990)

[NM02] Nakhimovsky, A., Myers, T.: Web Services: Description, Interfaces and Ontology. In: Geroimenko, V., Chen, C. (eds.) Visualizing the Semantic Web, pp. 135–150. Springer, Heidelberg (2002)

[P05] Pretorius, A.J.: Visual Analysis for Ontology Engineering. Journal of Visual Languages and Computing 16(4), 359–381 (2005)

[PFP+92] Patil, R., Fikes, R., Patel-Schneider, P., McKay, D., Finin, T., Gruber, T., Neches, R.: The DARPA Knowledge Sharing Effort: Progress Report. In: Proceedings of Knowledge Representation and Reasoning, pp. 777–788 (1992)

[PS06] Pazienza, M., Stellato, A.: Linguistic Enrichment of Ontologies: a methodological framework. In: Second Workshop on Interfacing Ontologies and Lexical Resources for Semantic Web Technologies (OntoLex 2006), Italy, May 24-26 (2006)

[Q91] Qmair, Y.: Foundations of Arabic philosophy. Dar al-Shoroq, Beirut (1991) ISBN 2-7214-8024-3

[R88] Reiter, R.: Towards a Logical Reconstruction of Relational Database Theory. In: Readings in AI and Databases. Morgan Kaufmann, San Francisco (1988)

[S03a] Smith, B.: Ontology. In: Floridi, L. (ed.) Blackwell Guide to the Philosophy of Computing and Information, pp. 155–166. Blackwell, Oxford (2003)

[S95] Shapiro, S.: Propositional, First-Order And Higher-Order Logics: Basic Definitions, Rules of Inference, Examples. In: Iwanska, L., Stuart, S. (eds.) Natural Language Processing and Knowledge Representation: Language for Knowledge and Knowledge for Language. AAAI Press/The MIT Press, Menlo Park (1995)

[T00] Temmerman, T.: Towards New Ways of Terminology Description, the sociocognitive approach. John Benjamins Publishing Company, Amsterdam (2000) ISBN 9027223262

[TV06] Trog, D., Vereecken, J.: Context-driven Visualization For Ontology Engineering, p. 237. Computer Science, Brussels (2006)

[SMJ02] Spyns, P., Meersman, R., Jarrar, M.: Data modelling versus Ontology engineering. SIGMOD Record 31(4), 12–17 (2002)

[V82] Van Griethuysen, J.J. (ed.): Concepts and Terminology for the Conceptual Schema and Information Base. International Standardization Organization, Publication No. ISO/TC97/SC5- N695 (1982)

[VDM04] Verheyden, P., De Bo, J., Meersman, R.: Semantically unlocking database content through ontology-based mediation. In: Bussler, C.J., Tannen, V., Fundulaki, I. (eds.) SWDB 2004. LNCS, vol. 3372, pp. 109–126. Springer, Heidelberg (2005)

[WG01] Welty, C., Guarino, N.: Support for Ontological Analysis of Taxonomic Relationships. Journal of Data and Knowledge Engineering 39(1), 51–74 (2001)

[ZD04] Ziegler, P., Dittrich, K.: User-Specific Semantic Integration of Heterogeneous Data: The SIRUP Approach. In: Proceeding of the International Conference on Semantics of a Networked World. LNCS, pp. 14–44. Springer, Paris (2004)

Process Mining towards Semantics

A.K. Alves de Medeiros and W.M.P. van der Aalst

Eindhoven University of Technology,
P.O. Box 513, NL-5600 MB, Eindhoven, The Netherlands
{a.k.medeiros,w.m.p.v.d.aalst}@tue.nl

Abstract. Process mining techniques target the *automatic* discovery of information about process models in organizations. The discovery is based on the execution data registered in *event logs*. Current techniques support a variety of practical analysis, but they are somewhat limited because the labels in the log are not linked to any concepts. Thus, in this chapter we show how the analysis provided by current techniques can be improved by including *semantic* data in event logs. Our explanation is divided into two main parts. The first part illustrates the power of current process mining techniques by showing how to use the open source process mining tool ProM to answer concrete questions that managers typically have about business processes. The second part utilizes usage scenarios to motivate how process mining techniques could benefit from *semantic annotated* event logs and defines a concrete semantic log format for ProM. The ProM tool is available at www.processmining.org.

1 Introduction

Nowadays, most organizations use information systems to support the execution of their business processes [21]. Examples of information systems supporting operational processes are Workflow Management Systems (WMS) [10,15], Customer Relationship Management (CRM) systems, Enterprise Resource Planning (ERP) systems and so on. These information systems may contain an explicit model of the processes (for instance, workflow systems like Staffware [8], COSA [1], etc.), may support the tasks involved in the process without necessarily defining an explicit process model (for instance, ERP systems like SAP R/3 [6]), or may simply keep track (for auditing purposes) of the tasks that have been performed without providing any support for the actual execution of those tasks (for instance, custom-made information systems in hospitals). Either way, these information systems typically support logging capabilities that register what has been executed in the organization. These produced logs usually contain data about cases (i.e. process instances) that have been executed in the organization, the times at which the tasks were executed, the persons or systems that performed these tasks, and other kinds of data. These logs are the starting point for process mining, and are usually called *event logs*. For instance, consider the event log in Table 1. This log contains information about four process instances (cases) of a process that handles fines.

Process mining targets the *automatic* discovery of information from an event log. This discovered information can be used to deploy new systems that support

T.S. Dillon et al. (Eds.): Advances in Web Semantics I, LNCS 4891, pp. 35–80, 2008.

Table 1. Example of an event log

Case ID	Task Name	Event Type	Originator	Timestamp	Extra Data
1	File Fine	Completed	Anne	20-07-2004 14:00:00	...
2	File Fine	Completed	Anne	20-07-2004 15:00:00	...
1	Send Bill	Completed	system	20-07-2004 15:05:00	...
2	Send Bill	Completed	system	20-07-2004 15:07:00	...
3	File Fine	Completed	Anne	21-07-2004 10:00:00	...
3	Send Bill	Completed	system	21-07-2004 14:00:00	...
4	File Fine	Completed	Anne	22-07-2004 11:00:00	...
4	Send Bill	Completed	system	22-07-2004 11:10:00	...
1	Process Payment	Completed	system	24-07-2004 15:05:00	...
1	Close Case	Completed	system	24-07-2004 15:06:00	...
2	Send Reminder	Completed	Mary	20-08-2004 10:00:00	...
3	Send Reminder	Completed	John	21-08-2004 10:00:00	...
2	Process Payment	Completed	system	22-08-2004 09:05:00	...
2	Close case	Completed	system	22-08-2004 09:06:00	...
4	Send Reminder	Completed	John	22-08-2004 15:10:00	...
4	Send Reminder	Completed	Mary	22-08-2004 17:10:00	...
4	Process Payment	Completed	system	29-08-2004 14:01:00	...
4	Close Case	Completed	system	29-08-2004 17:30:00	...
3	Send Reminder	Completed	John	21-09-2004 10:00:00	...
3	Send Reminder	Completed	John	21-10-2004 10:00:00	...
3	Process Payment	Completed	system	25-10-2004 14:00:00	...
3	Close Case	Completed	system	25-10-2004 14:01:00	...

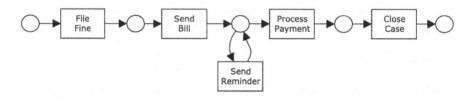

Fig. 1. Petri net illustrating the control-flow perspective that can be mined from the event log in Table 1

the execution of business processes or as a feedback tool that helps in auditing, analyzing and improving already enacted business processes. The main benefit of process mining techniques is that information is *objectively* compiled. In other words, process mining techniques are helpful because they gather information about what is *actually* happening according to an event log of an organization, and not what people *think* that is happening in this organization.

The type of data in an event log determines which *perspectives* of process mining can be discovered. If the log (i) provides the tasks that are executed in the process and (ii) it is possible to infer their order of execution and link these taks to individual cases (or process instances), then the *control-flow perspective* can be mined. The log in Table 1 has this data (cf. fields "Case ID", "Task Name" and "Timestamp"). So, for this log, mining algorithms could discover the process in Figure 1[1]. Basically, the process describes that after a fine is entered in the system, the bill is sent to the driver. If the driver does not pay the bill within

---

[1] The reader unfamiliar with Petri nets is referred to [17,30,32].

one month, a reminder is sent. When the bill is paid, the case is archived. If the log provides information about the persons/systems that executed the tasks, the *organizational perspective* can be discovered. The organizational perspective discovers information like the social network in a process, based on transfer of work, or allocation rules linked to organizational entities like roles and units. For instance, the log in Table 1 shows that "Anne" transfers work to both "Mary" (case 2) and "John" (cases 3 and 4), and "John" sometimes transfers work to "Mary" (case 4). Besides, by inspecting the log, the mining algorithm could discover that "Mary" never has to send a reminder more than once, while "John" does not seem to perform as good. The managers could talk to "Mary" and check if she has another approach to send reminders that "John" could benefit from. This can help in making good practices a common knowledge in the organization. When the log contains more details about the tasks, like the values of data fields that the execution of a task modifies, the *case perspective* (i.e. the perspective linking data to cases) can be discovered. So, for instance, a forecast for executing cases can be made based on already completed cases, exceptional situations can be discovered etc. In our particular example, logging information about the profiles of drivers (like age, gender, car etc.) could help in assessing the probability that they would pay their fines on time. Moreover, logging information about the places where the fines were applied could help in improving the traffic measures in these places. From this explanation, the reader may have already noticed that the control-flow perspective relates to the "How?" question, the organizational perspective to the "Who?" question, and the case perspective to the "What?" question. All these three perspectives are complementary and relevant for process mining.

Current process mining techniques can address all these three perspectives [9,11,12,16,20,22,28,29,31,33,34,35,38,40]. Actually, many of these techniques are implemented in the open-source tool ProM [19,39]. As we show in this chapter, the ProM tool can be used to answer common questions about business processes, like "How is the distribution of all cases over the different paths in through the processes?", "Where are the bottlenecks in the process?" or "Are all the defined rules indeed being obeyed?". Showing how to use the current process mining techniques implemented in ProM to answer these kinds of questions is the first contribution of this chapter.

However, although the process mining techniques implemented in ProM can answer many of the common questions, these techniques are somewhat limited because their analysis is purely based on the *labels* in the log. In other words, the techniques are unable to reason about *concepts* in an event log. For instance, if someone wants to get feedback about *billing* processes in a company or the webservices that provide a certain service, this person has to manually specify all the labels that map to billing processes or webservices providing the given service. Therefore, as supported by [27], we believe that the automatic discovery provided by process mining techniques can be augmented if we include *semantic information* about the elements in an event log. Note that semantic process mining techniques bring the discovery to the conceptual (or semantical) level. Furthermore, because semantic logs will link to concepts in ontologies, it is possible to embed ontology reasoning in the mining techniques. However, when supporting the links to the ontologies, it is important to make sure that the semantically annotated logs can also be mined by current process mining techniques. This way we avoid recoding of good existing solutions. Thus, the second contribution of this chapter consists of showing (i) how to extend our mining format to support links to ontologies and (ii) providing usage scenarios that illustrate the gains of using such semantic logs.

The remainder of this chapter is organized as follows. Section 2 shows how to use ProM plug-ins to answer common questions that managers have about business processes. Section 3 explains how to extend the current log format used by ProM to support the link to ontologies and discusses usage scenarios based on this format. Section 4 concludes this chapter.

2   Process Mining in Action

In this section we show how to use the ProM tool to answer the questions in Table 2. These are common questions that managers typically have about business processes. The ProM framework [19,39] is an open-source tool specially tailored to support the development of process mining plug-ins. This tool is currently at version 4.0 and contains a wide variety of plug-ins. Some of them go beyond process mining (like doing process verification, converting between different modelling notations etc). However, since in this chapter our aim is to show how to use ProM plug-ins to answer *common questions about processes in companies*, we focus on the plug-ins that use as input (i) an event log only or (ii) an event log and a process model. Figure 2 illustrates how we "categorize" these plug-ins. The plug-ins based on data in the event log only are called *discovery* plug-ins because they do not use any existing information about deployed models. The plug-ins that check how much the data in the event log matches the prescribed behavior in the deployed models are called *conformance* plug-ins. Finally, the plug-ins that need both a model and its logs to discover information that will enhance this model are called *extension* plug-ins. In the context of our common questions, we use (i) discovery plug-ins to answer questions like "How are the cases actually being executed? Are the rules indeed being obeyed?", (ii) conformance plug-ins to questions like "How compliant are the cases (i.e. process instances) with the deployed process models? Where are the problems? How frequent is the (non-)compliance?", and (iii) extension plug-ins to questions like "What are the business rules in the process model?"

The remainder of this section illustrates how to use ProM to answer the questions in Table 2. The provided explanations have a tutorial-like flavor because you should be able to reproduce the results when using ProM over the example in this section or while analyzing other logs. The explanation is supported by the running example in Subsection 2.1. Subsection 2.3 describes how you can inspect and clean an event log before performing any mining. Subsection 2.4 shows how to use *discovery* plug-ins (cf. Figure 2) and Subsection 2.5 describes

Table 2. Common questions that managers usually have about processes in organizations

---

1. What is the most frequent path for every process model?
2. How is the distribution of all cases over the different paths through the process?
3. How compliant are the cases (i.e. process instances) with the deployed process models? Where are the problems? How frequent is the (non-) compliance?
4. What are the routing probabilities for each split/join task?
5. What is the average/minimum/maximum throughput time of cases?
6. Which paths take too much time on average? How many cases follow these routings? What are the critical sub-paths for these paths?
7. What is the average service time for each task?
8. How much time was spent between any two tasks in the process model?
9. How are the cases actually being executed?
10. What are the business rules in the process model?
11. Are the process rules/constraints indeed being obeyed?
12. How many people are involved in a case?
13. What is the communication structure and dependencies among people?
14. How many transfers happen from one role to another role?
15. Who are important people in the communication flow?
16. Who subcontract work to whom?
17. Who work on the same tasks?

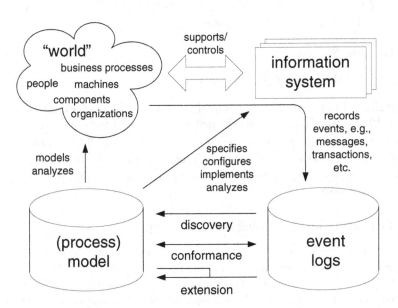

Fig. 2. Sources of information for process mining. The *discovery* plug-ins use only an event log as input, while the *conformance* and *extension* plug-ins also need a (process) model as input.

how to mine with *conformance* and *extension* plug-ins. Finally, we advice you to have the ProM tool at hand while reading this section. This way you can play with the tool while reading the explanations. Subsection 2.2 explains how to get started with ProM.

2.1   Running Example

The running example is about a *process to repair telephones in a company*. The company can fix 3 different types of phones ("T1", "T2" and "T3"). The process starts by registering a telephone device sent by a customer. After registration, the telephone is sent to the Problem Detection (PD) department. There it is analyzed and its defect is categorized. In total, there are 10 different categories of defects that the phones fixed by this company can have. Once the problem is identified, the telephone is sent to the Repair department and a letter is sent to the customer to inform him/her about the problem. The Repair (R) department has two teams. One of the teams can fix *simple* defects and the other team can repair *complex* defects. However, some of the defect categories can be repaired by both teams. Once a repair employee finishes working on a phone, this device is sent to the Quality Assurance (QA) department. There it is analyzed by an employee to check if the defect was indeed fixed or not. If the defect is not repaired, the telephone is again sent to the Repair department. If the telephone is indeed repaired, the case is archived and the telephone is sent to the customer. To save on throughput time, the company only tries to fix a defect a limited number of times. If the defect is not fixed, the case is archived anyway and a brand new device is sent to the customer.

2.2   Getting Started

To prepare for the next sections, you need to do:

1. Install the ProM tool. This tool is freely available at http:// prom.sourceforge. net. Please download and run the installation file for your operating system.
2. Download the two log files for the running example. These logs files are located at (i) http://www.processmining.org/_media/tutorial/repairExample.zip and (ii) http://www.processmining.org/_media/tutorial/repairExampleSample2.zip.

2.3   Inspecting and Cleaning an Event Log

Before applying any mining technique to an event log, we recommend you to first get an idea of the information in this event log. The main reason for this is that you can only answer certain questions if the data is in the log. For instance, you cannot calculate the throughput time of cases if the log does not contain information about the times (timestamp) in which tasks were executed. Additionally, you may want to remove unnecessary information from the log before you start the mining. For instance, you may be interested in mining only information about the cases that are completed. For our running example

Table 3. Log Pre-Processing: questions and pointers to answers

Question	Subsection
How many cases (or process instances) are in the log? How many tasks (or audit trail entries) are in the log? How many originators are in the log? Are there running cases in the log? Which originators work in which tasks?	2.3.1
How can I filter the log so that only completed cases are kept? How can I see the result of my filtering? How can I save the pre-processed log so that I do not have to redo work?	2.3.2

(cf. Section 2.1), all cases without an archiving task as the last one correspond to running cases and should not be considered. The cleaning step is usually a projection of the log to consider only the data you are interested in. Thus, in this section we show how you can inspect and clean (or pre-process) an event log in ProM. Furthermore, we show how you can save the results of the cleaned log, so that you avoid redoing work.

The questions answered in this section are summarized in Table 3. As you can see, Subsection 2.3.1 shows how to answer questions related to log inspection and Subsection 2.3.2 explains how to filter an event log and how to save your work. Note that the list of questions in Table 3 is not exhaustive, but they are enough to give you an idea of the features offered by ProM for log inspection and filtering.

2.3.1 Inspecting the Log

The first thing you need to do to inspect or mine a log is to load it into ProM. In this section we use the log at the location *http://www.processmining.org/_media/ tutorial/repairExample.zip*. This log has process instances of the running example described in Section 2.1.

To open this log, do the following:

1. Download the log for the running example and save it at your computer.
2. Start the ProM framework. You should get a screen like the one in Figure 3. Note that the ProM menus are context sensitive. For instance, since no log has been opened yet, no mining algorithm is available.
3. Open the log via clicking *File→Open MXML log*, and select your saved copy of the log file for the running example. Once your log is opened, you should get a screen like the one in Figure 4. Note that now more menu options are available.

Now that the log is opened, we can proceed with the actual log inspection. Recall that we want to answer the following questions:

1. How many cases (or process instances) are in the log?
2. How many tasks (or audit trail entries) are in the log?

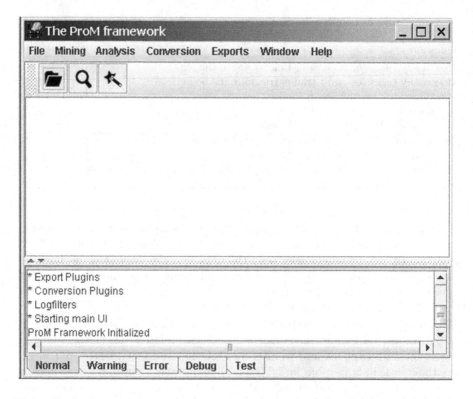

Fig. 3. Screenshot of the main interface of ProM. The menu *File* allows to open event logs and to import models into ProM. The menu *Mining* provides the mining plug-ins. These mining plug-ins mainly focus on discovering information about the control-flow perspective of process models or the social network in the log. The menu *Analysis* gives access to different kinds of analysis plug-ins for opened logs, imported models and/or mined models. The menu *Conversion* provides the plug-ins that translate between the different notations supported by ProM. The menu *Exports* has the plug-ins to export the mined results, filtered logs etc.

3. How many originators are in the log?

4. Are there running cases in the log?

5. Which originators work in which tasks?

The *first four questions* can be answered by the analysis plug-in *Log Summary*. To call this plug-in, choose *Analysis→[log name...]→Log Summary*. Can you now answer the first four questions of the list above? If so, you probably have noticed that this log has 104 *running* cases and 1000 *completed* cases. You see that from the information in the table "Ending Log Events" of the log summary (cf. Figure 5). Note that only 1000 cases end with the task "Archive Repair". The *last question* of the list above can be answered by the analysis plug-in *Originator by Task Matrix*. This plug-in can be started by clicking the menu *Analysis→[log*

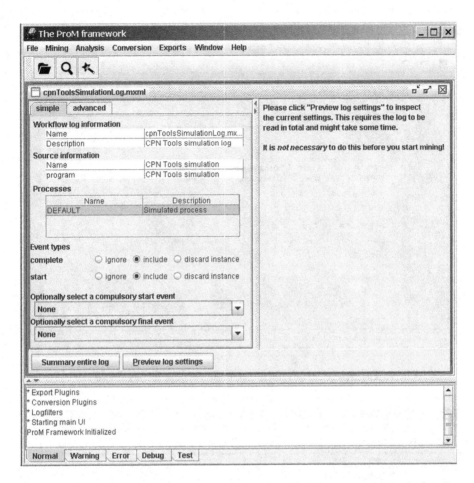

Fig. 4. Screenshot of ProM after opening the log of the running example (cf. Section 2.1)

name...]→Originator by Task Matrix. Can you identify which originators perform the same tasks for the running example log? If so, you probably have also noticed that there are 3 people in each of the teams in the Repair department (cf. Section 2.1) of the company [2]. The employees with login "SolverC..." deal with the complex defects, while the employees with login "SolverS..." handle the simple defects.

Take your time to inspect this log with these two analysis plug-ins and find out more information about it. If you like, you can also inspect the *individual* cases by clicking the button *Preview log settings* (cf. bottom of Figure 4) and then double-clicking on a specific process instance.

---

[2] See the originators working on the tasks "Repair (Complex)" and "Repair (Simple)" in Figure 6.

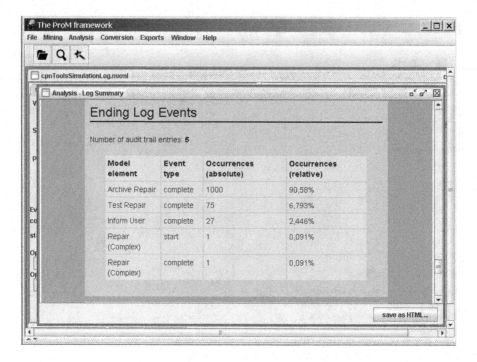

Fig. 5. Excerpt of the result of the analysis plug-in *Log Summary*

Fig. 6. Scheenshot with the result of the analysis plug-in *Originator by Task Matrix*

2.3.2 Cleaning the Log

In this chapter we use the process mining techniques to get insight about the process for repairing telephones (cf. Section 2.1). Since our focus in on the process *as a whole*, we will base our analysis on the *completed* process instances only. Note that it does not make much sense to talk about the most frequent path if it is not complete, or reason about throughput time of cases when some of them are still running. In short, we need to pre-process (or clean or filter) the logs.

In ProM, a log can be filtered by applying the provided *Log Filters*. In Figure 4 you can see three log filters (see bottom-left of the panel with the log): *Event Types*, *Start Event* and *End Event*. The *Event Types* log filter allows us to select the type of events (or tasks or audit trail entries) that we want to consider while mining the log. For our running example, the log has tasks with two event types: *complete* and *start*. If you want to (i) keep all tasks of a certain event, you should select the option "include" (as it is in Figure 4), (ii) omit the tasks with a certain event type from a trace, select the option "ignore", and (iii) discard all traces with a certain event type, select the option "discard instance". This last option may be useful when you have aborted cases etc. The *Start Event* filters the log so that only the traces (or cases) that start with the indicated task are kept. The *End Event* works in a similar way, but the filtering is done with respect to the final task in the log trace.

From the description of our running example, we know that the completed cases are the ones that start with a task to *register* the phone and end with a task to *archive* the instance. Thus, to filter the completed cases, you need to execute the following procedure:

1. Keep the event types selection as in Figure 4 (i.e., "include" all the *complete* and *start* event types);
2. Select the task "Register (complete)" as the *compulsory start event*;
3. Select the task "Archive Repair (complete)" as the *compulsory final event*.

If you now inspect the log (cf. Section 2.3.1), for instance, by calling the analysis plug-in *Log Summary*, you will notice that the log contains fewer cases (Can you say how many?) and all the cases indeed start with the task "Register (complete)" and finish with the task "Archive Repair (complete)".

Although the log filters we have presented so far are very useful, they have some limitations. For instance, you can only specify one task as the start task for cases. It would be handy to have more flexibility, like saying "Filter all the cases that start with task X or task Y". For reasons like that, the *advanced* tab of the panel with the log (cf. Figure 7) provides more powerful log filters. Each log filter has a *Help*, so we are not going into details about them. However, we strongly advise you to spend some time trying them out and getting more feeling about how they work. Our experience shows that the advanced log filters are especially useful when handling real-life logs. These filters not only allow for projecting data in the log, but also for adding data to the log. For instance, the log filters *Add Artificial Start Task* and *Add Artificial End Task* support the respective addition of tasks at the begin and end of traces. These two log filters are handy when applying process mining algorithms that assume the target model to have a single start/end point.

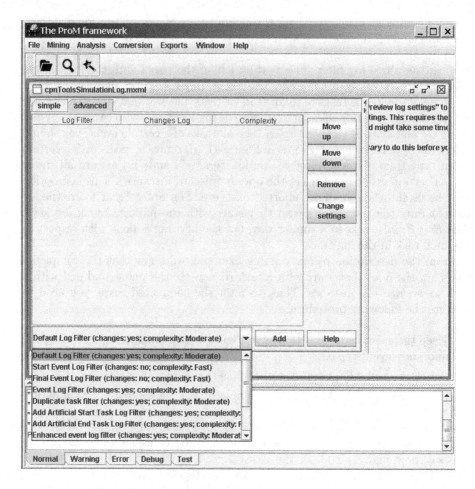

Fig. 7. Scheenshot of the *Advanced Log Filters* in ProM

Once you are done with the filtering, you can save your results in two ways:

1. Export the filtered log by choosing the export plug-in *XML log file*. This will save a copy of the log that contains all the changes made by the application of the log filters.

2. Export the configured log filters themselves by choosing the export plug-in *Log Filter (advanced)*. Exported log filters can be imported into ProM at a later moment and applied to a (same) log. You can import a log filter by selecting *File→[log name...]→Open Log Filter (advanced)*.

If you like, you can export the filtered log for our running example. Can you open this exported log into ProM? What do you notice by inspecting this log? Note that your log should only contain 1000 cases and they should all start and end with a single task.

Table 4. Discovery Plug-ins: questions and pointers to answers

Question	Subsection
How are the cases actually being executed?	2.4.1
What is the most frequent path for every process model? How is the distribution of all cases over the different paths through the process?	2.4.2
How many people are involved in a case? What is the communication structure and dependencies among people? How many transfers happen from one role to another role? Who are the important people in the communication flow? Who subcontract work to whom? Who work on the same tasks?	2.4.3
Are the process rules/constraints indeed being obeyed?	2.4.4

2.4 Questions Answered Based on an Event Log Only

Now that you know how to inspect and pre-process an event log (cf. Subsection 2.3), we proceed with showing how to answer the questions related to the *discovery* ProM plug-ins (cf. Figure 2). Recall that a log is the only input for these kinds of plug-ins.

The questions answered in this section are summarized in Table 3. Subsection 2.4.1 shows how to mine the *control-flow* perspective of process models. Subsection 2.4.2 explains how to mine information regarding certain aspects of cases. Subsection 2.4.3 describes how to mine information related to the roles/employees in the event log. Subsection 2.4.4 shows how to use temporal logic to verify if the cases in a log satisfy certain (required) properties.

2.4.1 Mining the Control-Flow Perspective of a Process

The control-flow perspective of a process establishes the dependencies among its tasks. Which tasks precede which other ones? Are there concurrent tasks? Are there loops? In short, what is the process model that summarizes the flow followed by most/all cases in the log? This information is important because it gives you feedback about how *cases are actually being executed* in the organization.

As shown in Figure 8, ProM supports various plug-ins to mine the control-flow perspective of process models. In this section, we will use the mining plug-in *Alpha algorithm plugin*. Thus, to mine the log of our running example, you should perform the following steps:

1. Open the filtered log that contains only the completed cases (cf. Section 2.3.2), or redo the filtering for the original log of the running example.
2. Verify with the analysis plug-in *Log Summary* if the log is correctly filtered. If so, this log should contain 1000 process instances, 12 audit trail entries, 1 start event ("Register"), 1 end event ("Archive Repair"), and 13 originators.
3. Run the *Alpha algorithm plugin* by choosing the menu *Mining→[log name...] →Alpha algorithm plugin* (cf. Figure 8).

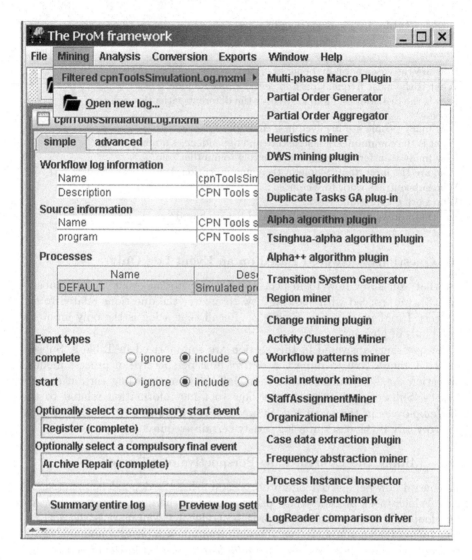

Fig. 8. Scheenshot of the mining plug-ins in ProM

4. Click the button *Start mining*. The resulting mined model should look like the one in Figure 9. Note that the *Alpha algorithm plugin* uses Petri nets[3] as its notation to represent process models. From this mined model, you can observe that:

---

[3] Different *mining* plug-ins can work with different notations, but the main idea is always the same: portray the dependencies between tasks in a model. Furthermore, the fact that different mining plug-ins may work with different notations does not prevent the interoperability between these representations because the ProM tool offers *conversion* plug-ins that translate models from one notation to another [39].

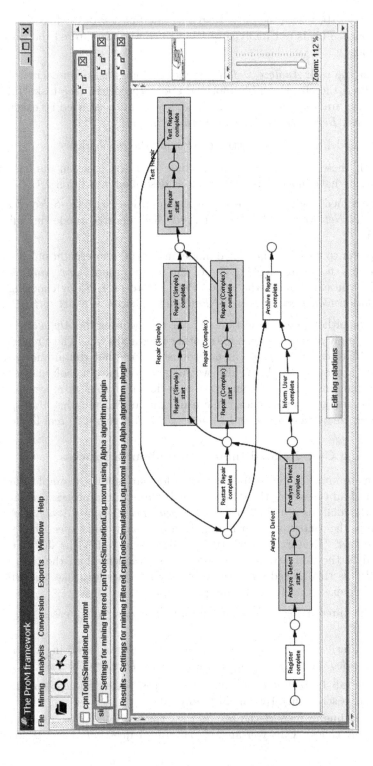

Fig. 9. Scheenshot of the mined model for the log of the running example

- All cases start with the task "Register" and finish with the task "Archive Repair". This is not really surprising since we have filtered the cases in the log.

- After the task *Analyze Defect* completes, some tasks can occur in parallel: (i) the client can be informed about the defect (see task "Inform User"), **and** (ii) the actual fix of the defect can be started by executing the task *Repair (Complete)* **or** *Repair (Simple)*.

- The model has a loop construct involving the repair tasks.

Based on these remarks, we can conclude that the cases in our running example log have indeed been executed as described in Section 2.1.

5. Save the mined model by choosing the menu option *Exports→Selected Petri net→Petri Net Kernel file*. We will need this exported model in Subsection 2.5.

6. If you prefer to visualize the mined model in another representation, you can convert this model by invoking one of the menu option *Conversion*. As an example, you can convert the mined Petri net to an EPC by choosing the menu option *Conversion→Selected Petri net→Labeled WF net to EPC*.

As a final note, although in this section we mine the log using the *Alpha algorithm plugin*, we strongly recommend you to try other plug-ins as well. The main reason is that the *Alpha algorithm plugin* is not robust to logs that contain noisy data (like real-life logs typically do). Thus, we suggest you have a look at the help of the other plug-ins before choosing for a specific one. In our case, we can hint that we have had good experience while using the mining plug-ins *Multi-phase Macro plugin*, *Heuristics miner* and *Genetic algorithm plugin* to real-life logs.

2.4.2   Mining Case-Related Information about a Process

Do you want to know the most frequent path for our running example? Or the distribution of all cases over the different paths through the process? Then you should use the analysis plug-in *Performance Sequence Diagram Analysis*. As an illustration, in the context of our running example, one would expect that paths without the task "Restart Repair" (i.e., situations in which the defect could not be fixed in the first attempt) should be less frequent than the ones with this task. But is this indeed the current situation? Questions like this will be answered while executing the following procedure:

1. Open the filtered log that contains only the completed cases (cf. Section 2.3.2).

2. Run the *Performance Sequence Diagram Analysis* by choosing the menu *Analysis→[log name...]→Performance Sequence Diagram Analysis*.

3. Select the tab *Pattern diagram* and click on the button *Show diagram*. You should get a screen like the one in Figure 10.
   Take your time to inspect the results (i.e., the sequence patterns and their throughput times). Can you answer our initial questions now? If so, you

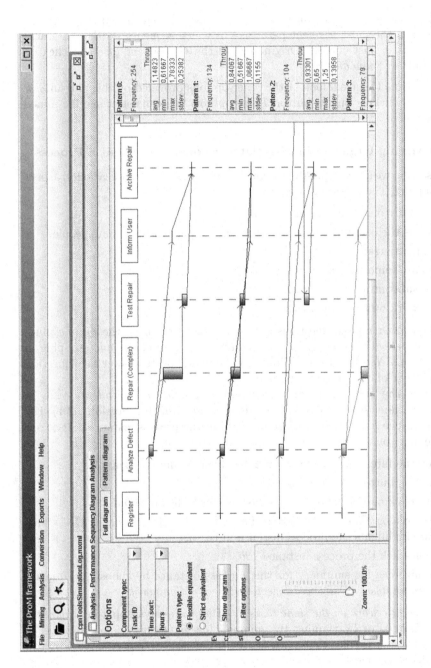

Fig. 10. Scheenshot of the analysis plug-in *Performance Sequence Diagram Analysis*. The configuration options are on the left side, the sequence diagrams are on the middle and the patterns frequence and throughput times are on the right side.

have probably notice that the 73,5% of the defects could be fixed in the first attempt[4].

4. Now, how about having a look at the resources? Which employees are involved in the most frequent patterns? In which sequence do they interact? To see that, just choose "Originator" as the *Component type* and click on the button *Show diagram*.

Take your time to have a look at the other options provided by this plug-in. For instance, by clicking on the button *Filter options* you can select specific mined patterns etc.

2.4.3   Mining Organizational-Related Information about a Process

In this section we answer questions regarding the social (or organizational) aspect of a company. The questions are:

- How many people are involved in a specific case?
- What is the communication structure and dependencies among people?
- How many transfers happen from one role to another role?
- Who are important people in the communication flow?
- Who subcontract work to whom?
- Who work on the same tasks?

These and other related questions can be answered by using the *mining* plug-ins *Social Network Miner* and *Organizational Miner*, and the *analysis* plug-in *Analyze Social Network*. In the following we explain how to answer each question in the context of our running example.

To know the *people that are involved in a specific case or all the cases in the log*, you can use the analysis plug-in *Log Summary* (cf. Section 2.3.1). For instance, to check which people are involved in the process instance *120* of our example log, you can do the following:

1. Open the filtered log (cf. Section 2.3.2) for the running example.
2. Click the button *Preview log settings*.
3. Right-click on the panel *Process Instance* and click on *Find....*
4. In the dialog *Find*, field "Text to find", type in *120* and click "OK". This option highlights the process instance in the list.
5. Double-click the process instance *120*.
6. Visualize the log summary for this process instance by choosing the menu option *Analysis→Previewed Selection...→Log Summary*.

You can see *who work on the same tasks* by using the analysis plug-in *Originator by Task Matrix*, or by running the mining plug-in *Organizational Miner*. For instance, to see the roles that work for the same tasks in our example log, you can do the following:

---

[4] See patterns 0 to 6, notice that the task "Restart Repair" does not belong to these patterns. Furthermore, the sum of the occurrences of these patterns is equal to 735.

1. Open the filtered log (cf. Section 2.3.2) for the running example.

2. Select the menu option *Mining→Filtered...→Organizational Miner*, choose the options "Doing Similar Task" and "Correlation Coefficient", and click on *Start mining*.

3. Select the tab *Organizational Model*. You should get a screen like the one in Figure 11. Take you time to inspect the information provided at the bottom of this screen. Noticed that the *automatically generated organizational model* shows that the people with the role "Tester..." work on the tasks "Analyze Defect" and "Test Repair", and so on. If you like, you can edit these automatically generated organizational model by using the functionality provided at the other two tabs *Tasks<->Org Entity* and *Org Entity<->Resource*. Note that organizational models can be exported and used as input for other organizational-related mining and analysis plug-ins.

The other remaining questions of the list on page 52 are answered by using the mining plug-in *Social Network* in combination with the analysis plug-in *Analyze Social Network*. For instance, in the context of our running example, we would like to check if there are employees that outperform others. By identifying these employees, one can try to make the good practices (or way of working) of these employees a common knowledge in the company, so that peer employees also benefit from that. In the context of our running example, we could find out which employees are better at fixing defects. From the process description (cf. Section 2.1) and from the mined model in Figure 9, we know that telephones which were not repaired are again sent to the Repair Department. So, we can have a look at the *handover of work* for the tasks performed by the people in this department. In other words, we can have a look at the handover of work for the tasks *Repair (Simple)* and *Repair (Complete)*. One possible way to do so is to perform the following steps:

1. Open the log for the running example.

2. Use the *advanced* log filter *Event Log Filter* (cf. Section 2.3.2) to filter the log so that only the four tasks "Repair (Simple) (start)", "Repair (Simple) (complete)", "Repair (Complex) (start)" and "Repair (Complex) (complete)" are kept. (Hint: Use the analysis plug-in *Log Summary* to check if the log is correctly filtered!).

3. Run the *Social Network Miner* by choosing the menu option *Mining→ Filtered...→Social network miner* (cf. Figure 8).

4. Select the tab *Handover of work*, and click the button *Start mining*. You should get a result like the one in Figure 12. We could already analyze this result, but we will use the analysis plug-in *Analyze Social Network* to do so. This analysis plug-in provides a more intuitive user interface. This is done on the next step.

5. Run the *Analyze Social Network* by choosing the menu option *Analysis→ SNA→Analyze Social Network*. Select the options "Vertex size", "Vertex degree ratio stretch" and set *Mouse Mode* to "Picking" (so you can use the

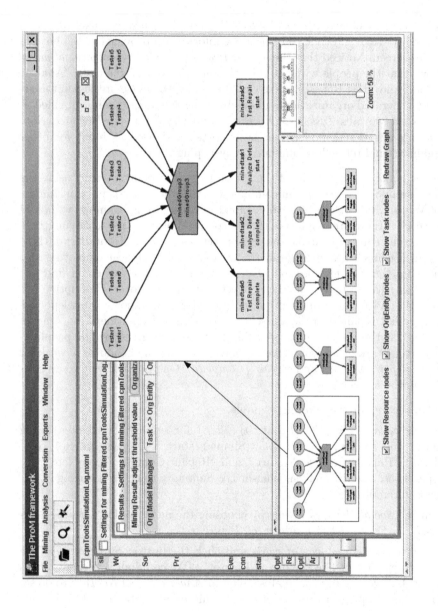

Fig. 11. Scheenshot of the mining plug-in *Organizational Miner*

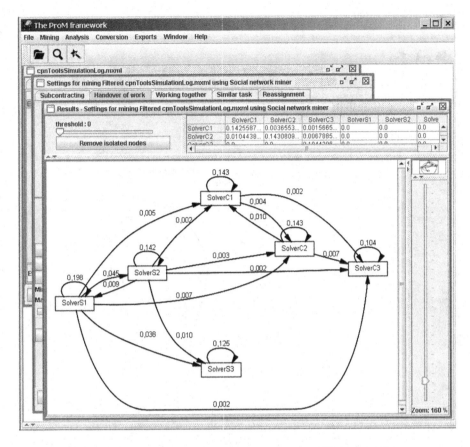

Fig. 12. Scheenshot of the mining plug-in *Social Network Miner*

mouse to re-arrange the nodes in the graph). The resulting graph (cf. Figure 13) shows which employees handed over work to other employees in the process instances of our running example. By looking at this graph, we can see that the employees with roles "SolverS3" and "SolverC3" outperform the other employees because the telephones these two employees fix always pass the test checks and, therefore, are not re-sent to the Repair Department (since no other employee has to work on the cases involving "SolverS3" and "SolverC3"). The oval shape of the nodes in the graph visually expresses the relation between the *in* and *out* degree of the connections (arrows) between these nodes. A higher proportion of ingoing arcs lead to more vertical oval shapes while higher proportions of outgoing arcs produce more horizontal oval shapes. From this remark, can you tell which employee has more problems to fix the defects?

Take you time to experiment with the plug-ins explained in the procedure above. Can you now answer the other remaining questions?

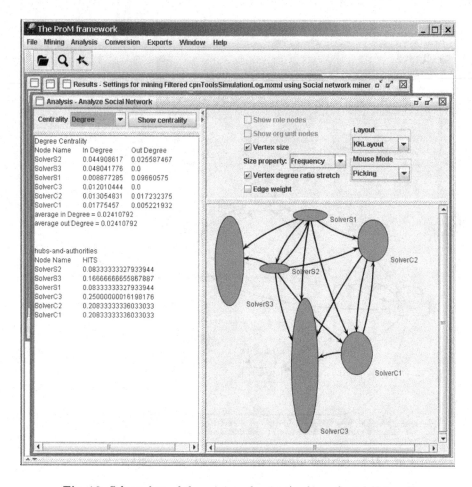

Fig. 13. Scheenshot of the mining plug-in *Analyzer Social Network*

As a final remark, we point out that the results produced by the *Social Network* mining plug-in can be exported to more powerful tools like AGNA[5] and NetMiner[6], which are especially tailored to analyze social networks and provide more powerful user interfaces.

2.4.4 Verifying Properties in an Event Log

It is often the case that processes in organizations should obey certain rules or principles. One common example is the "four-eyes principle" which determines that some tasks should not be executed by a same person within a process instance. These kinds of principles or rules are often used to ensure quality of the delivered products and/or to avoid frauds. One way to check if these rules

---

[5] http://agna.gq.nu
[6] http://www.netminer.com/

are indeed being obeyed is to audit the log with data about what has happened in an organization. In ProM, auditing is provided by the analysis plug-in *Default LTL Checker Plugin*[7].

From the description of our running example (cf. Section 2.1), we know that after a try to fix the defect, the telephone should be tested to check if it is indeed repaired. Thus, we could use the *Default LTL Checker Plugin* to verify the property: *Does the task "Test Repair" always happen* after *the tasks "Repair (Simple)" or "Repair (Complex)"* and before *the task "Archive Repair"*? We do so by executing the following procedure:

1. Open the filtered log (cf. Section 2.3.2) for the running example.

2. Run the *Default LTL Checker Plugin* by selecting the menu option *Analysis* →*[log name...]*→*Default LTL Checker Plugin*. You should get a screen like the one in Figure 14.

3. Select the formula "eventually_activity_A_then_B_then_C".

4. Give as values: (i) activity A = *Repair (Simple)*, (ii) activity B = *Test Repair* and (iii) activity C = *Archive Repair*. Note that the LTL plug-in is case sensitive. So, make sure you type in the task names as they appear in the log.

5. Click on the button *Check*. The resulting screen should show the log split into two parts: one with the cases that satisfy the property (or formula) and another with the cases that do not satisfy the property. Note that the menu options now also allow you to do mining, analysis etc. over the split log. For instance, you can apply again the LTL plug-in over the *incorrect process instances* to check if the remaining instances refer to situations in which the task "Repair (Complex)" was executed. Actually, this is what we do in the next step.

6. Run the *Default LTL Checker Plugin* over the *Incorrect Process Instances* by choosing *Analysis*→*Incorrect Instances (573)*→*Default LTL Checker Plugin*.

7. Select the same formula and give the same input as in steps 3 and 4 above. However, this time use activity A = *Repair (Complex)*.

8. Click on the button *Check*. Note that all 573 cases satisfy the formula. So, for this log, there are not situations in which a test does not occur after a repair.

Take your time to experiment with the LTL plug-in. Can you identify which pre-defined formula you could use to check for the "four-eyes principle"?

In this section we have shown how to use the pre-defined formulae of the LTL analysis plug-in to verify properties in a log. However, you can also *define your own formulae and import them into ProM*. The tutorial that explains how to do so is provided together with the documentation for the ProM tool[8].

---

[7] LTL stands for Linear Temporal Logic.

[8] For Windows users, please see *Start*→*Programs*→*Documentation*→*All Documentation*→*LTLChecker-Manual.pdf*.

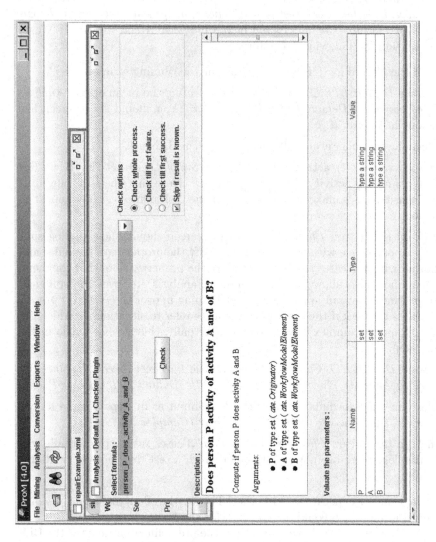

Fig. 14. Scheenshot of the analysis plug-in *Default LTL Checker Plugin*

Table 5. Conformance and Extension Plug-ins: questions and pointers to answers

Question	Subsection
How compliant are the cases (i.e. process instances) with the deployed process models? Where are the problems? How frequent is the (non-) compliance?	2.5.1
What are the routing probabilities for each slipt/join task? What is the average/minimum/maximum throughput time of cases? Which paths take too much time on average? How many cases follow these routings? What are the critical sub-paths for these routes? What is the average service time for each task? How much time was spent between any two tasks in the process model?	2.5.2
What are the business rules in the process model?	2.5.3

2.5 Questions Answered Based on a Process Model Plus an Event Log

In this section we explain the ProM analysis plug-ins that are used to answer the questions in Table 5. These plug-ins differ from the ones in Section 2.4 because they require a log *and* a (process) model as input (cf. Figure 2). Subsection 2.5.1 explains a *conformance* ProM plug-in that detects discrepancies between the flows prescribed in a model and the actual process instances (flows) in a log. Subsections 2.5.2 and 2.5.3 describe *extension* ProM plug-ins that respectively extend the models with performance characteristics and business rules.

2.5.1 Conformance Checking

Nowadays, companies usually have some process-aware information system [21] to support their business process. However, these process models may be incomplete because of reasons like: one could not think of all possible scenarios while deploying the model; the world is dynamic and the way employees work may change but the prescribed process models are not updated accordingly; and so on. Either way, it is always useful to have a tool that provides feedback about this.

The ProM analysis plug-in that checks *how much process instances in a log match a model and highlights discrepancies* is the *Conformance Checker*. As an illustration, we are going to check the exported mined model (cf. Section 2.4.1, page 50) for the log of the running example against a new log provided by the company. Our aim is to check how compliant this new log is with the prescribed model. The procedure is the following:

1. Open the log "repairExampleSample2.zip". This log can be downloaded from *http://www.processmining.org/_media/tutorial/repairExampleSample2.zip.*

2. Open the exported PNML model that you created while executing the procedure on page 50.

3. Check if the automatically suggested mapping from the tasks in the log to the tasks in the model is correct. If not, change the mapping accordingly.

4. Run the *Conformance Checker* plug-in by choosing the menu option *Analysis* →*Selected Petri net*→*Conformance Checker*.

5. Deselect the options "Precision" and "Structure"[9], and click the button *Start analysis*. You should get results like the ones shown in figures 15 and 16, which respectively show screenshots of the *model and log diagnostic perspective* of the Conformance Checker plug-in. These two perspectives provide detailed information about the problems encountered during the log replay. The *model perspective* diagnoses information about *token counter* (number of missing/left tokens), *failed tasks* (tasks that were not enabled), *remaining tasks* (tasks that remained enabled), *path coverage* (the tasks and arcs that were used during the log replay) and *passed edges* (how often every arc in the model was used during the log replay). The *log perspective* indicates the points of non-compliant behavior for every case in the log.

Take your time to have a look at the results. Can you tell how many traces are not compliant with the log? What are the problems? Have all the devices been tested after the repair took places? Is the client always being informed?

2.5.2   Performance Analysis

Like the *Conformance Checker* (cf. Section 2.5.1), the plug-in *Perfomance Analysis with Petri net* also requires a log and a Petri net as input[10]. The main difference is that this plug-in focuses on analyzing *time-related* aspects of the process instances. In other words, this plug-in can answer the questions:

– What are the routing probabilities for each slipt/join task?
– What is the average/minimum/maximum throughput time of cases?
– Which paths take too much time on average? How many cases follow these routings? What are the critical sub-paths for these routes?
– What is the average service time for each task?
– How much time was spent between any two tasks in the process model?

To execute the *Perfomance Analysis with Petri net* analysis plug-in over the log of our running example, perform the following steps:

1. Open the filtered log (cf. Section 2.3.2) for the running example.
2. Open the exported PNML model that you created while executing the procedure on page 50.
3. Run the *Perfomance Analysis with Petri net* analysis plug-in by selecting the menu option *Analysis*→*Selected Petri net*→*Performance Analysis with Petri net*.

---

[9] These are more advanced features that we do not need while checking for compliance. Thus, we will turn them off for now.

[10] If the model you have is not a Petri net but another one of the supported formats, you can always use one of the provided *conversion* plug-ins to translate your model to a Petri net.

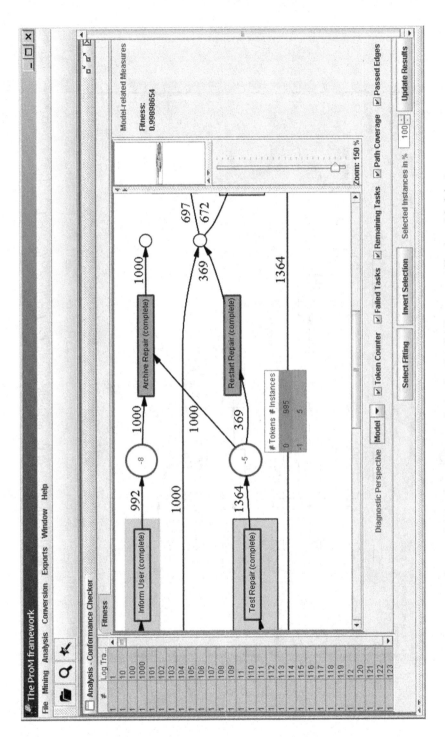

Fig. 15. Scheenshot of the analysis plug-in *Conformance Checker:* Model view

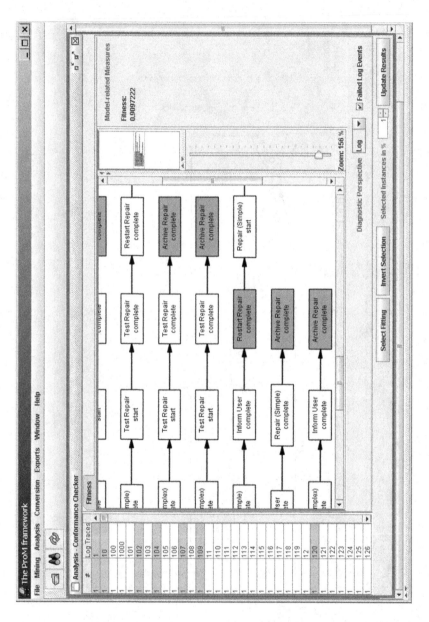

Fig. 16. Scheenshot of the analysis plug-in *Conformance Checker*: Log view

4. Set the field *Times measured in* to "hours" and click on the button *Start analysis*. The plug-in will start replaying the log and computing the time-related values. When it is ready, you should see a screen like the one in Figure 17.

5. Take your time to have a look at these results. Note that the right-side panel provides information about the *average/minimum/maximum throught-put* times. The central panel (the one with the Petri net model) shows (i) the *bottlenecks* (notice the different colors for the places) and (ii) the *routing probabilities* for each split/join tasks (for instance, note that only in 27% of the cases the defect could not the fixed on the first attempt). The bottom panel show information about the *waiting times* in the places. You can also select one or two tasks to respectively check for *average service times* and *the time spent between any two tasks in the process model*. If you like, you can also change the *settings* for the waiting time (small window at the bottom with *High, Medium* and *Low*).

The previous procedure showed how to answer all the questions listed in the beginning of this section, except for one: *Which paths take too much time on average? How many cases follow these routings? What are the critical sub-paths for these routes?* To answer this last question, we have to use the analysis plug-in *Performance Sequence Diagram Analysis* (cf. Section 2.4.2) in combination with the *Performance Analysis with Petri net*. In the context of our example, since the results in Figure 17 indicate that the cases take on average *1.11 hours* to be completed, it would be interesting to analyze what happens for the cases that take longer than that. The procedure to do so has the following steps:

1. If the screen with the results of the *Performance Analysis with Petri net* plug-in is still open, just choose the menu option *Analysis→ Whole Log→ Performance Sequence Diagram Analysis*. Otherwise, just open the filtered log for the running example and run Step 2 described on page 50.

2. In the resulting screen, select the tab *Pattern Diagram*, set *Time sort* to hours, and click on the button *Show diagram*.

3. Now, click on the button *Filter Options* to filter the log so that only cases with throughput time superior to 1.11 hours are kept.

4. Select the option *Sequences with throughput time*, choose "above" and type in "1.1" in the field "hours". Afterwards, click first on the button *Update* and then on button *Use Selected Instances*. Take your time to analyze the results. You can also use the *Log Summary* analysis plug-ins to inspect the *Log Selection*. Can you tell how many cases have throughput time superior to 1.1 hours? Note that the *Performance Sequence Diagram Analysis* plug-in shows how often each cases happened in the log. Try playing with the provided options. For instance, what happens if you now select the *Component Type* as "Originator"? Can you see how well the employees are doing? Once you have the log selection with the cases with throughput time superior to 1.1 hour, you can check for *critical sub-paths* by doing the remaining steps in this procedure.

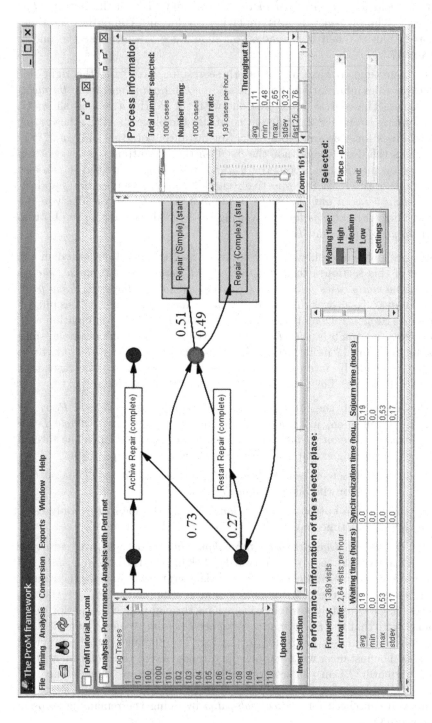

Fig. 17. Scheenshot of the analysis plug-in *Performance Analysis with Petri net*

5. Open exported PNML model that you created while executing the procedure on page 50, but this time link it to the *selected cases* by choosing the menu option *File→Open PNML file→With:Log Selection*. If necessary, change the automatically suggested mappings and click on the button *OK*. Now the imported PNML model is linked to the process instances with throughput times superior to 1.1 hours.

6. Run the analysis plug-in *Performance Analysis with Petri net* to discover the *critical sub-paths* for these cases. Take your time to analyze the results. For instance, can you see that now 43% of the defects could not be fixed on the first attempt?

Finally, we suggest you spend some time reading the Help documentation of this plug-in because it provides additional information to what we have explained in this section. Note that the results of this plug-in can also be exported.

2.5.3 Decision Point Analysis

To discover the *business rules* (i.e. the conditions) that influence the points of choice in a model, you can use the *Decision Point Analysis* plug-in. For instance, in the context of our running example, we could investigate which defect types (cf. Section 2.1) are fixed by which team. The procedure to do so has the following steps:

1. Open the filtered log (cf. Section 2.3.2) for the running example.

2. Open the exported PNML model that you created while executing the procedure on page 50.

3. Run the *Decision Point Analysis* plug-in by selecting the menu option *Analysis→Selected Petri net→Decision Point Analysis*.

4. Double-click the option "Choice 4 p2". This will select the point of choice between execution the task "Repair (Complex)" or "Repair (Simple)"[11].

5. Select the tab *Attributes* and set the options: (i) *Attribute selection scope* = "just before", (ii) change the *Attribute type* of the field *defectType* to "numeric". Afterwards, click on the button *Update results*. This analysis plug-in will now invoke a *data mining* algorithm (called J48) that will discover which fields in the log determine the choice between the different branches in the model.

6. Select the tab *Result* to visualize the mined rules (cf. Figure 18). Note that cases with a defect types[12] from 1 to 4 are routed to the task "Repair (Simple)" and the ones with defect type bigger than 4 are routed to the task "Repair (Complex)". This is the rule that covers the *majority* of the cases

---

[11] **Note:** If your option "Choice 4 p2" does not correspond to the point of choice we refer to in the model, please identify the correct option on your list. The important thing in this step is to select the correct point of choice in the model.

[12] From the problem description (cf. Section 2.1), we know that there are 10 types of defects.

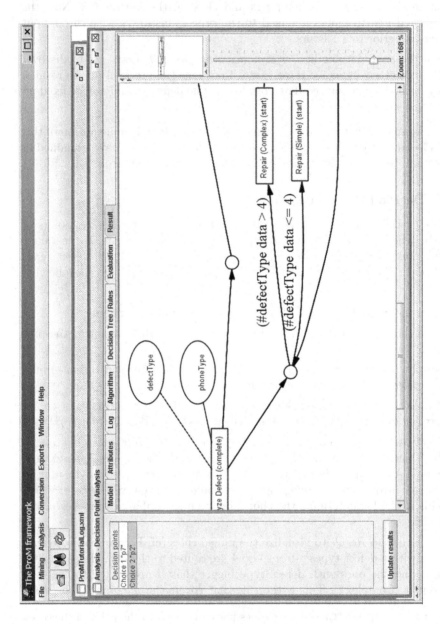

Fig. 18. Scheenshot of the analysis plug-in *Decision Point Analysis: Result* tab

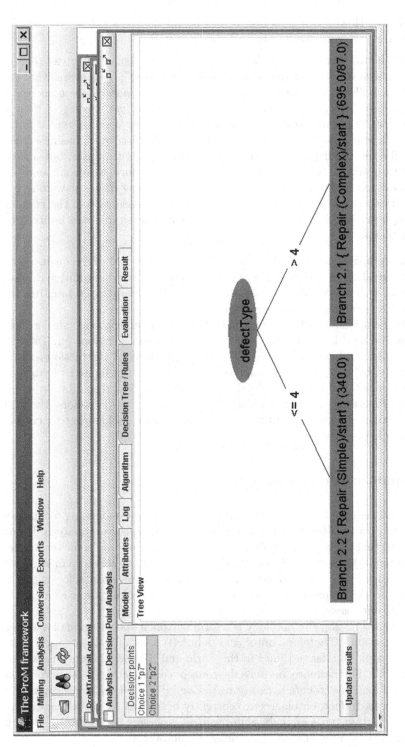

Fig. 19. Scheenshot of the analysis plug-in *Decision Point Analysis: Decision Tree/Rules* tab

in the log. However, it does not mean that all the cases follow this rule. To check for this, you have to perform the next step.

7. Select the tab *Decision Tree/Rules* (cf. Figure 19). Remark the text *(695.0/ 87.0)* inside the box for the task "Repair (Complex)". This means that *according to the discovered business rules, 695 cases should have been routed to the task "Repair (Complex)" because their defect type was bigger than 4. However, 87 of these cases are* misclassified *because they were routed to the task "Repair (Simple)"*. Thus, the automatically discovered business rules describe the conditions that apply to the majority of the cases, but it does not mean that all the cases will fit these rules. Therefore, we recommend you to always check for the results in the tab *Decision Tree/Rules* as well. In our case, these result makes sense because, from the description of the running example (cf. Section 2.1), we know that some defect types can be fixed by both teams.

To get more insight about the *Decision Point Analysis*, we suggest you spend some time reading its Help documentation because it provides additional information that was not covered in this section. As for the many ProM plug-ins, the mined results (the discovered rules) can also be exported.

The explanations in this section show that the current process mining techniques can be used to answer the set of typical common questions (cf. Table 2) for the analysis of business processes. However, the level of abstraction and re-use provided by these techniques for this analysis is quite limited because all the queries are based on a label-level. For instance, it is not possible to define a generic property that checks for the "four-eyes principle". Therefore, the next section motivates which benefits the use of semantics could bring to current process mining techniques.

3 Semantic Process Mining

Semantic process mining aims at bringing the current process mining techniques from the level of *label-based* analysis to the level of *concept-based* analysis. Recall that the starting point of any mining algorithm is a log and that some techniques also use a model as input (cf. Section 2, Figure 2). Thus, the core idea in semantic process mining is to explicitly relate (or annotate) elements in a log with the concepts that they represent. This can be achieved by linking these elements to concepts in *ontologies*.

As explained in [25], "an ontology is a formal explicit specification of a shared conceptualization". Therefore, ontologies define (i) a set of concepts used by (a group of) people to refer to things in the world and (ii) the relationships among these concepts. Furthermore, because the concepts and relationships are *formally* defined, it is possible to *automatically reason or infer other relationships* among these concepts. In fact, ontologies are currently been used for modelling and for performing certain types of analysis in business processes [14,23,24,37]. Thus, it is realistic to base semantic process mining on ontologies.

To illustrate how ontologies can enhance the analysis provided by process mining tools, let us return to our running example (cf. Section 2.1). For this example, one could use the ontologies in Figure 20 to express the concepts in the business process for repairing the telephones. This figure shows three ontologies: *TaskOntology*, *RoleOntology* and *PerformerOntology*. In this figure, the concepts are modelled by ellipses and the instances of the concepts by rectangles. Additionally, the arrows define the relationships between the concepts and the instances. The arrows go from a concept to a superconcept, or from an instance to a concept. By looking at these ontologies, it is possible to infer subsumption relations among the concepts and instances. For instance, it is possible to identify that the elements "Repair (Complex)" and "Repair (Simple)" are *tasks* for *repairing* purposes. The use of ontologies (and the automatic reasoning they support) enables process mining techniques to analyze at different abstraction levels (i.e., at the level of instances and/or concepts) and, therefore, promotes re-use. Remark that current process mining techniques only provide for analysis based on the "instance" level. As an illustration, consider the last procedure on page 53. In this procedure we had to filter the log so that only the tasks "Repair (Simple)" and "Repair (Complex)" would be kept in the log. Now, assume that the elements in the log would link to the ontologies as illustrated in Figure 20. In this setting, the filtering performed on page 53 (cf. Step 2) could be simplified **from** *keep all the tasks "Repair (Simple)" and "Repair (Complex)"* **to** *keep all the tasks linking to the concept "TaskOntology:Repair"* [13]. Note that the filtering now specifies a *set of concepts* to maintain in the log. Therefore, it is up to the process mining log filter to analyze the concepts in the log and automatically infer that the tasks "Repair (Simple)" and "Repair (Complex)" should be kept in the log because these two tasks link to *subconcepts* of the concept "TaskOntology:Repair".

The remainder of this section provides an outlook on the possibilities for semantic process mining. Our aim is not to describe concrete semantic process mining algorithms, but rather motivate the opportunities and identify the elements that are needed to proceed towards these semantic algorithms. This is done in Subsection 3.1. Additionally, as a first step for realizing semantic process mining, Subsection 3.2 explains a concrete semantically annotated extension of the input format for logs mined by the ProM tool. The elements in this semantically annotated format support the kind of analysis discussed in the usage scenarios.

3.1 Usage Scenarios

The usage scenarios presented in this section illustrate the benefits of bringing the analysis discussed in Section 2 to the *semantic* level. While describing these scenarios, we assume that (i) the elements in event logs and models given as input link to concepts in ontologies, and (ii) the mining algorithms are able to load and infer subsumption relationships about the concepts that are referenced

---

[13] In this section, we use the notation "<ontology name>:<ontology concept>" when referring to concepts of the ontologies illustrated in Figure 20.

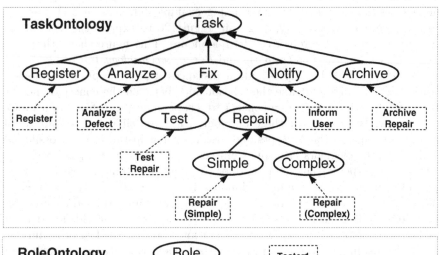

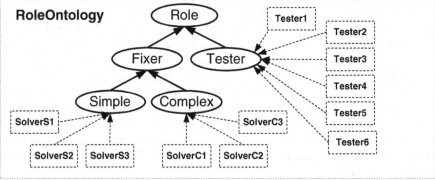

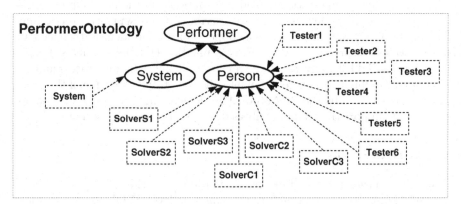

Fig. 20. Example of ontologies that could be defined for the running example in Section 2.1. Concepts are represented by ellipses and instances by rectangles. The root ellipses specify the most generic concepts and the leaf ellipses the most specific subconcepts.

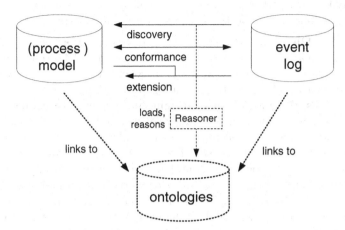

Fig. 21. Sources of information for *semantic* process mining. The additional elements (with dashed lines) are necessary to support the mining at the conceptual level(cf. Figure 2).

in these logs/models. Both the loading of the ontologies and the respective subsumption inferences are provided by ontology reasoners [3,4,5,26,36]. Figure 21 describes the sources of information for semantic process mining. In the following we elaborate on the usage scenarios. All the scenarios are based on the ontologies in Figure 20 and the running example in Subsection 2.1.

Scenarios for Log Inspection. Log inspection provides an overview of the elements in the log (cf. Subsection 2.3.1). Thus, when a log has links to concepts in ontologies, the log inspection techniques could also give an overview of different semantic perspectives. As a minimal requirement, these techniques should support the visualization of the concepts that are referenced in the log, their instances (i.e. the actual elements), and all the superconcepts of these concepts. This is necessary because the user needs to know which concepts he/she can use while performing the mining. Other possibilities would be to automatically find out (i) completed cases (or process instances) in the log (e.g., the ontologies could contain axioms that would define when instances of a given process were completed), (ii) which roles are in the log and which tasks were executed by originators with these roles, (iii) which concepts link to which labels, (iv) which tasks are executed by people and which by systems, and so on.

Scenarios for Cleaning the Log. Log cleaning allows for projecting and/or adding data to a log (cf. Subsection 2.3.2). Thus, a semantically annotated log could also be cleaned based on the concepts its elements link to. For instance, keep in the log only the tasks that are instances (i.e., link to) concepts X, W and Z or any of their subconcepts. Note that log filters defined over concepts are more generic because they can be reused over multiple logs involving the same concepts. At the current situation, the reuse of log filters only makes sense when the logs have elements with identical labels. Thus, the use of concepts would boost the (re-)use of log filters.

Scenarios for Discovery Plug-ins. The analysis performed by discovery plug-ins is based on the log only (cf. Subsection 2.4). The *control-flow mining plug-ins* (cf. Subsection 2.4.1) discover a process model by inferring ordering relations between tasks in a log. When these tasks link to concepts in ontologies, these algorithms can mine process models at different levels of abstraction by inferring ordering relations between these concepts. The higher the considered concepts are in the subsumption trees derived from ontologies, the higher the level of abstraction of the mined models. For instance, for the ontology "TaskOntology", if a control-flow mining algorithm would use only the concepts at level 1 of its tree (i.e., use only the concepts "Register", "Analyze", "Fix", "Notify" and "Archive"), a process model like the one in Figure 22 could be discovered. In a similar way, if this same algorithm would use the instances of the concepts in this ontology, the mined model could be just like the one in Figure 9. Note that the mined model in Figure 22 is more compact (i.e., has a higher abstraction level) than the one in Figure 9. In as similar way, the *case-related information* plug-ins (cf. Subsection 2.4.2) could show most frequent paths with respect to concepts in the log. For the plug-ins that mine *organizational-related information* (cf. Subsection 2.4.3), the ontological concepts linked to the originator and tasks in the log would allow for a more precise analysis of the organizational model expressed in the log. For instance, consider the automatically discovered organizational model for the running example (cf. Figure 11). In this case, the tasks "Analyze Defect..." and "Test Repair..." are grouped together because they are executed by the same originators. However, if the link to ontologies would be present in the log, the groups in Figure 23 could be automatically inferred. Note that the use of ontological concepts would make it possible to (i) distinguish between the tasks "Analyze Defect..." and "Test Repair..." and (ii) identify that all originator "Tester..." have two roles: "RoleOntology:Classifier" and "RoleOntology:Tester". Finally, mining plug-ins for *verification of properties in the log* (cf. Subsection 2.4.4) could also benefit from a reasoning at the concept level. For instance, in the procedure on page 57, we have checked if *all "repaired" devices would always be tested before archiving*. To do so, we had to explicitly inform the *Default LTL Checker Plugin* the labels for the two repair tasks "Repair (Simple)" and "Repair (Complex)". If there were concepts in the log, this verification could be simplified to the formula "eventually_concept_A_then_B_then_C", where A = "TaskOntology:Repair", B = "TaskOntology:Test" and C = "TaskOntology:Archive". By using ontology reasoners, the plug-in would automatically find out the appropriate task labels to verify. Furthermore, like it happens for the log filters, LTL formulae defined over concepts can be more easily reused than the ones defined over labels.

Scenarios for Conformance and Extension Plug-ins. These plug-in enhance existing models by adding to them extra information discovered from

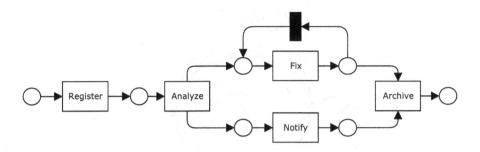

Fig. 22. Example of a mined model for the running example when only the concepts at the level 1 of the tree for the ontology "TaskOntology" (cf. Figure 20) are considered. Note that this mined model is more compact than the one in Figure 9.

logs. Because they need a log and a model while executing their analysis, these plug-ins would require the provided (process) models to also contain links to concepts in ontologies. This way they can find out relations between elements in models and logs the surpass the string matching level. Therefore: (i) the *conformance checking* plug-ins (cf. Subsection 2.5.1) would be performed at the conceptual level (with subsumption inferences taken into account), as well as the automatically suggested mapping between tasks in the log and in the model (cf. Step 3, on page 59); (ii) the *performance analysis* plug-ins (cf. Subsection 2.5.2) would be able to answer questions like "What are the routing probabilities of split/joint points of a *certain concept?*" or "Given tasks of a certain concept, which subconcepts outperform/underperform others in terms of service times?"; and (iii) *decision point analysis* plug-ins (cf. Subsection 2.5.3) would be able to automatically infer if a data value is *nominal* or *numeric*.

The starting point to realize the semantic process mining techniques illustrated in this subsection is to define a *semantic annotated format for event logs*. This format is the subject of the next subsection.

3.2   Semantic Annotated Mining XML Format

The Semantic Annotated Mining eXtensible Markup Language (SA-MXML) format is a *semantic annotated version* of the MXML format used by the ProM framework. In short, the SA-MXML incorporates the *model references* (between elements in logs and concepts in ontologies) that are necessary to implement our approach. However, before explaining the SA-MXML, let us first introduce the MXML format.

The Mining XML format (MXML) started as an initiative to share a common input format among different mining tools [11]. This way, event logs could be shared among different mining tools. The schema for the MXML format (depicted in Figure 24) is available at is.tm.tue.nl/research/processmining/Workflow-Log.xsd.

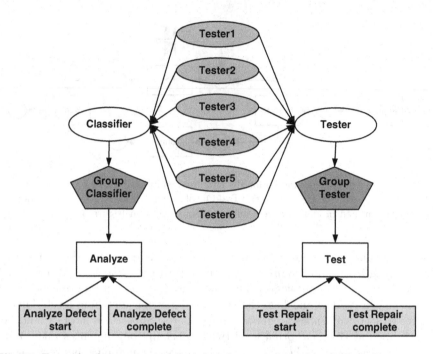

Fig. 23. Example of an automatically inferred organizational model based on a seman-tically annotated log for the running example. In this example we assume that in the log (i) the tasks "Analyze Defect..." link to the concept "TaskOntology:Analyze" and the tasks "Test Repair..." to "TaskOntology:Repair"; and (ii) the originators "Tester..." link to the concept "RoleOntology:Tester" when executing the tasks "Test Repair..." and to the concept "RoleOntology:Classifier" while performing the task "Analyze De-fect...". Note that this model more precisely identifies the groups in the log than the one in Figure 11.

As can be seen in Figure 24, an event log (element *WorkflowLog*) contains the execution of one or more processes (element *Process*), and optional information about the source program that generated the log (element *Source*) and additional data elements (element *Data*). Every process (element *Process*) has zero or more cases or process instances (element *ProcessInstance*). Similarly, every process instance has zero or more tasks (element *AuditTrailEntry*). Every task or audit trail entry (ATE) must *at least* have a name (element *WorkflowModelElement*) and an event type (element *EventType*). The event type determines the state of the tasks. There are 13 supported event types: schedule, assign, reassign, start, resume, suspend, autoskip, manualskip, withdraw, complete, ate_abort, pi_abort and unknown. The other task elements are optional. The *Timestamp* element supports the logging of time for the task. The *Originator* element records the person/system that performed the task. The *Data* element allows for the logging of additional information. Figure 25 shows an excerpt of the running example

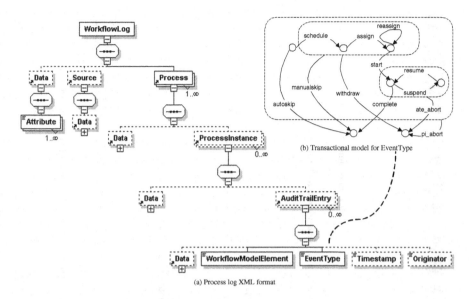

Fig. 24. The visual description of the schema for the Mining XML (MXML) format

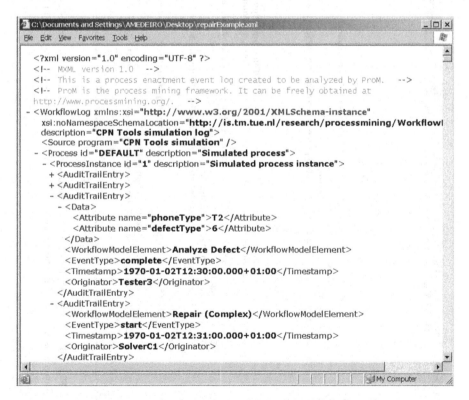

Fig. 25. Excerpt of the running example log (cf. Section 2.1) in the the MXML format

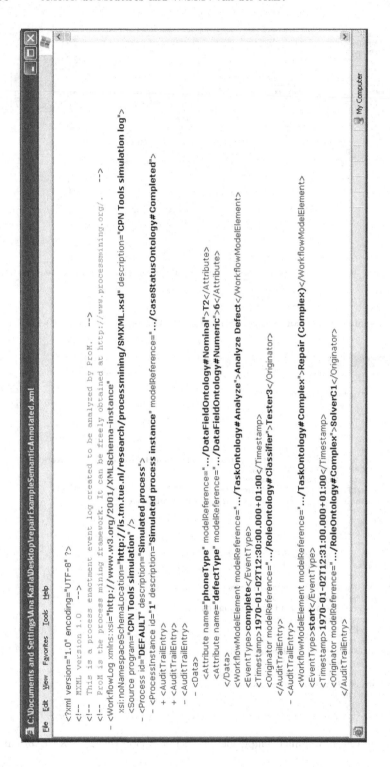

Fig. 26. Excerpt of the running example log (cf. Section 2.1) in the SA-MXML format. Note that the optional field *modelRef-erence* is used to link the elements of the MXML format to concepts in ontologies.

(cf. Subsection 2.1) log in the MXML format. More details about the MXML format can be found in [18,19].

The SA-MXML format is just like the MXML format plus the addition that *all elements (except for AuditTrailEntry and Timestamp) have an optional extra attribute* called *modelReference*. This attribute links to a *list of concepts* in ontologies and, therefore, support the necessary *model references* for our approach. The concepts are expressed as URIs and the elements in the list are separated by blank spaces. Actually, the use of *modelReference* in the SA-MXML format is based on the work for the semantic annotations provided by SAWSDL (Semantically Annotated Web Service Definition Language) [7]. The schema for the SA-MXML format is available at is.tm.tue.nl/research/processmining/SAMXML.xsd. Figure 3.2 shows an example of a log with semantic annotations for the log of our running example with respect to the ontologies in Figure 20. Note that the fields "ProcessInstance", "Data", "WorkflowModelElement" and "Originator" link to concepts in the ontologies. Furthermore, note that the SA-MXML format is *backwards compatible* with MXML format. This way the process mining techniques that do not support a semantic treatment can also be directly applied to logs in SA-MXML.

4   Conclusions

In this chapter we have shown (i) how to use the open source process mining tool ProM to get useful feedback about processes, (ii) how the inclusion of semantic information in event logs can empower process mining techniques, and (iii) we have defined a concrete semantic log format, the SA-MXML format.

Since our focus when illustrating the power of current process mining techniques was on answering a set of common questions that managers usually have about processes, we have not covered many of the other plug-ins that are in ProM. We hope that the subset we have shown in this chapter will help you in finding your way in ProM. However, if you are interested, you could have a further look in plug-ins to *verify (process) models and detect potential problems* (by using the analysis plug-ins *Check correctness of EPC, Woflan Analysis* or *Petri net Analysis*), *quantify (from 0% until 100%) how much behavior two process models have in common with respect to a given even log* (by using the analysis plug-in *Behavioral Precision/Recall*), *create simulation models with the different mined perspectives* (by using the export plug-in *CPN Tools*) etc. The ProM tool can be downloaded at www.processmining.org.

Embedding semantic information in event logs brings the process mining techniques from the level of label-based analysis to the concept-based one. This allows for working with different levels of abstractions while getting feedback about processes and properties in a log. Furthermore, it also supports easier reuse of queries defined over logs.

The SA-MXML format is the first step towards creating semantic process mining algorithms. This format extends the current MXML format by specifying that any element present in the MXML format may link to a set of concepts in ontologies. Following steps will focus on developing semantic process mining algorithms to implement the usage scenarios described in this chapter.

Acknowledgements

The authors would like to thank Ton Weijters, Boudewijn van Dongen, Anne Rozinat, Christian Günther, Minseok Song, Ronny Mans, Huub de Beer, Laura Maruster and Peter van den Brand for their on-going work on process mining techniques and tools at Eindhoven University of Technology. Additionally, the authors would like to thank Monique Jansen-Vullers, Mariska Netjes, Irene Vanderfeesten and Hajo Reijers for their input while selecting the common questions used in Section 2. The work presented in this chapter was partially funded by the European Commission under the project SUPER (FP6-026850).

References

1. COSA Business Process Management, http://www.cosa-bpm.com/
2. Extensible Markup Language (XML), http://www.w3.org/XML/
3. KAON2, http://kaon2.semanticweb.org/
4. Pellet, http://pellet.owldl.com/
5. Racer, http://www.racer-systems.com/
6. SAP, http://www.sap.com/
7. Semantic Annotations for Web Service Description Language (SA-WSDL), http://www.w3.org/TR/2006/WD-sawsdl-20060630/
8. Staffware Process Suite, http://www.staffware.com/
9. van der Aalst, W.M.P., de Beer, H.T., van Dongen, B.F.: Process Mining and Verification of Properties: An Approach Based on Temporal Logic. In: Meersman, R., Tari, Z. (eds.) OTM 2005. LNCS, vol. 3760, pp. 130–147. Springer, Heidelberg (2005)
10. van der Aalst, W.M.P., van Hee, K.M.: Workflow Management: Models, Methods, and Systems. MIT Press, Cambridge (2002)
11. van der Aalst, W.M.P., van Dongen, B.F., Herbst, J., Maruster, L., Schimm, G., Weijters, A.J.M.M.: Workflow Mining: A Survey of Issues and Approaches. Data and Knowledge Engineering 47(2), 237–267 (2003)
12. van der Aalst, W.M.P., Weijters, A.J.M.M., Maruster, L.: Workflow Mining: Discovering Process Models from Event Logs. IEEE Transactions on Knowledge and Data Engineering 16(9), 1128–1142 (2004)
13. Bussler, C., Haller, A. (eds.): BPM 2005. LNCS, vol. 3812. Springer, Heidelberg (2006)
14. Casati, F., Shan, M.C.: Semantic Analysis of Business Process Executions. In: Jensen, C.S., Jeffery, K., Pokorný, J., Šaltenis, S., Bertino, E., Böhm, K., Jarke, M. (eds.) EDBT 2002. LNCS, vol. 2287, pp. 287–296. Springer, Heidelberg (2002)
15. Workflow Management Coalition. WFMC Home Page, http://www.wfmc.org
16. Cook, J.E., Du, Z., Liu, C., Wolf, A.L.: Discovering Models of Behavior for Concurrent Workflows. Computers in Industry 53(3), 297–319 (2004)
17. Desel, J., Esparza, J.: Free Choice Petri Nets. Cambridge Tracts in Theoretical Computer Science, vol. 40. Cambridge University Press, Cambridge (1995)

18. van Dongen, B.F., van der Aalst, W.M.P.: A Meta Model for Process Mining Data. In: Proceedings of the CAiSE 2005 WORKSHOPS, vol. 2. FEUP (2005)
19. van Dongen, B.F., Alves de Medeiros, A.K., Verbeek, H.M.W., Weijters, A.J.M.M., van der Aalst, W.M.P.: The ProM Framework: A New Era in Process Mining Tool Support. In: Ciardo, G., Darondeau, P. (eds.) ICATPN 2005. LNCS, vol. 3536, pp. 444–454. Springer, Heidelberg (2005)
20. van Dongen, B.F., van der Aalst, W.M.P.: Multi-phase Process mining: Aggregating Instance Graphs into EPCs and Petri Nets. In: Proceedings of the Second International Workshop on Applications of Petri Nets to Coordination, Workflow and Business Process Management (PNCWB) (2005)
21. Dumas, M., van der Aalst, W.M.P., ter Hofstede, A.H. (eds.): Process-Aware Information Systems: Bridging People and Software Through Process Technology. John Wiley & Sons Inc., Chichester (2005)
22. Greco, G., Guzzo, A., Pontieri, L.: Mining Hierarchies of Models: From Abstract Views to Concrete Specifications. In: van der Aalst, W.M.P., Benatallah, B., Casati, F., Curbera, F. (eds.) BPM 2005. LNCS, vol. 3649, pp. 32–47. Springer, Heidelberg (2005)
23. Green, P., Rosemann, M.: Business Systems Analysis with Ontologies. Idea Group Publishing (2005)
24. Grigori, D., Casati, F., Castellanos, M., Dayal, U., Sayal, M., Shan, M.C.: Mining Exact Models of Concurrent Workflows 53(3), 321–343 (2004)
25. Gruninger, M., Lee, J.: Special Issue on Ontology applications and design - Introduction. Communications of ACM 45(2), 39–41 (2002)
26. Haarslev, V., Möller, R.: Racer: A core inference engine for the semantic web. In: Sure, Y., Corcho, Ó. (eds.) EON. CEUR Workshop Proceedings, vol. 87 (2003) CEUR-WS.org
27. Hepp, M., Leymann, F., Domingue, J., Wahler, A., Fensel, D.: Semantic Business Process Management: a Vision Towards Using Semantic Web services for Business Process Management. In: IEEE International Conference on e-Business Engineering (ICEBE 2005), pp. 535–540 (2005)
28. Herbst, J., Karagiannis, D.: Workflow Mining with InWoLvE. Computers in Industry 53(3), 245–264 (2004)
29. Thao Ly, L., Rinderle, S., Dadam, P., Reichert, M.: Mining staff assignment rules from event-based data. In: Bussler and Haller [13], pp. 177–190
30. Murata, T.: Petri Nets: Properties, Analysis and Applications. Proceedings of the IEEE 77(4), 541–580 (1989)
31. Pinter, S.S., Golani, M.: Discovering Workflow Models from Activities Lifespans. Computers in Industry 53(3), 283–296 (2004)
32. Reisig, W., Rozenberg, G. (eds.): APN 1998. LNCS, vol. 1491. Springer, Heidelberg (1998)
33. Rozinat, A.: Decision mining in prom. In: Dustdar, S., Fiadeiro, J.L., Sheth, A.P. (eds.) BPM 2006. LNCS, vol. 4102, pp. 420–425. Springer, Heidelberg (2006)
34. Rozinat, A., van der Aalst, W.M.P.: Conformance Testing: Measuring the Fit and Appropriateness of Event Logs and Process Models. In: Bussler and Haller [13], pp. 163–176
35. Schimm, G.: Mining Exact Models of Concurrent Workflows. Computers in Industry 53(3), 265–281 (2004)
36. Sirin, E., Parsia, B., Grau, B.C., Kalyanpur, A., Katz, Y.: Pellet: A practical OWL-DL reasoner. Web Semantics: Science, Services and Agents on the World Wide Web 5(2), 51–53 (2007)

37. Thomas, M., Redmond, R., Yoon, V., Singh, R.: A semantic approach to monitor business process. Communications of ACM 48(12), 55–59 (2005)
38. van der Aalst, W.M.P., Reijers, H.A., Song, M.: Discovering Social Networks from Event Logs. Computer Supported Cooperative Work 14(6), 549–593 (2005)
39. Verbeek, H.M.W., van Dongen, B.F., Mendling, J., van der Aalst, W.M.P.: Interoperability in the ProM Framework. In: Latour, T., Petit, M. (eds.) Proceedings of the CAiSE 2006 Workshops and Doctoral Consortium, pp. 619–630. Presses Universitaires de Namur (June 2006)
40. Wen, L., Wang, J., Sun, J.: Detecting Implicit Dependencies Between Tasks from Event Logs. In: Zhou, X., Li, J., Shen, H.T., Kitsuregawa, M., Zhang, Y. (eds.) APWeb 2006. LNCS, vol. 3841, pp. 297–306. Springer, Heidelberg (2006)

Models of Interaction as a Grounding for Peer to Peer Knowledge Sharing

David Robertson[1], Adam Barker[1], Paolo Besana[1], Alan Bundy[1],
Yun Heh Chen-Burger[1], David Dupplaw[2], Fausto Giunchiglia[3], Frank van Harmelen[4],
Fadzil Hassan[1], Spyros Kotoulas[4], David Lambert[1], Guo Li[1], Jarred McGinnis[1],
Fiona McNeill[1], Nardine Osman[1], Adrian Perreau de Pinninck[5], Ronny Siebes[4],
Carles Sierra[5], and Chris Walton[1]

[1] Informatics, University of Edinburgh, UK
[2] Electronics and Computer Science, University of Southampton, UK
[3] Information and Communication Technology, University of Trento, Italy
[4] Mathematics and Computer Science, Free University, Amsterdam, Netherlands
[5] Artificial Intelligence Research Institute, Barcelona, Spain

Abstract. Most current attempts to achieve reliable knowledge sharing on a large scale have relied on pre-engineering of content and supply services. This, like traditional knowledge engineering, does not by itself scale to large, open, peer to peer systems because the cost of being precise about the absolute semantics of services and their knowledge rises rapidly as more services participate. We describe how to break out of this deadlock by focusing on semantics related to interaction and using this to avoid dependency on *a priori* semantic agreement; instead making semantic commitments incrementally at run time. Our method is based on interaction models that are mobile in the sense that they may be transferred to other components, this being a mechanism for service composition and for coalition formation. By shifting the emphasis to interaction (the details of which may be hidden from users) we can obtain knowledge sharing of sufficient quality for sustainable communities of practice without the barrier of complex meta-data provision prior to community formation.

1 Introduction

At the core of this paper is an unusual view of the semantics of Web service coordination. When discussing semantics it is necessary to ground our definitions in some domain in order to decide whether our formal machinery performs appropriate inference. Normally this grounding is assumed to be in the Web services themselves, so formal specification focuses on individual services. It seems natural then to assume that having defined the semantics of services precisely we can combine them freely as long as our means of combination preserves the local semantics of those services. This assumption is ill founded for large scale systems because, when we combine services, we normally share information (by connecting inputs to outputs) and this raises the issue of whether the semantics of information provided by a service is preserved by another service obtaining that information. Universal standardisation of semantics across services appears impractical on a large scale; partly because broad ontological consensus

T.S. Dillon et al. (Eds.): Advances in Web Semantics I, LNCS 4891, pp. 81–129, 2008.

is difficult to achieve but also because the semantics of service interfaces derives from the complex semantics of the programs providing those services.

We explore an alternative approach, where services share explicit knowledge of the interactions in which they are engaged and these models of interaction are used operationally as the anchor for describing the semantics of the interaction. By shifting our view in this way we change the boundaries of the semantic problem. Instead of requiring a universal semantics across services we require only that semantics is consistent (separately) for each instance of an interaction. This is analogous to the use in human affairs of contracts, which are devices for standardising and sharing just those aspects of semantics necessary for the integrity of specific interactions.

In what follows we focus on interaction and we use models of the salient features of required interactions in order to provide a context for knowledge sharing. We are able to exchange interaction models (and associated contextual knowledge) between peers[1] that may not have collaborated before, with the context being extended and adapted as interaction proceeds. This changes the way in which key elements of distributed knowledge sharing, such as ontology alignment, are approached because semantic commitments made for the purposes of interaction are not necessarily commitments to which individual peers must adhere beyond the confines of a specific interaction. Different types of interaction require different commitments, and different levels of integrity in maintaining these.

1.1 A Scenario

To ground our discussion, we give a scenario to which we shall return throughout this paper:

> Sarah works for a high-precision camera manufacturing company and is responsible for periodic procurement for market research. Her job includes identifying competitors' newest products as they come onto the market and purchasing them for internal analysis. She knows which types of camera she needs, but, to save money, may vary the ways she purchases them. She isn't always sure what is the best way to purchase these cameras, but via recommendation service she learns of an Internet shop; an auction service (where she sets initial and maximum prices and the auction service finds a supplier by competitive bidding) and a purchasing service direct from the manufacturer (which allows some configuration of the order to take place on-line). Each of these three services has a different way of describing what they do but the system she uses to access the services can supply the translation necessary to see each of the three interactions through. She tries all three automatically; compares the prices offered; checks that she is comfortable with the way in which the interaction was performed for her; then buys from the one she prefers.

Sarah will have encountered, in the scenario above, several issues that will be explored later in this paper. Her ontology for describing a camera purchase had to be

---

[1] We use the term "peer" above to emphasise the independence of the services involved, rather than to suggest any specific peer-to-peer architecture.

matched to those of available services (Section 3). Her recommendation service had to know which services might best be able to interact and enable them to do so (Sections 2 and 4). When they interact the contextual knowledge needed to interact reliably should propagate to the appropriate services as part of the interaction (Section 5). When the services are being coordinated then it might be necessary to reconcile their various constraints in order to avoid breaking the collaboration (Section 6). Before she started, Sarah might have wanted to be reassured that the interaction conforms to the requirements of her business process (Section 7) and that her interaction was reliable (Section 8).

1.2   Structure of This Paper and Its Link to Computation

Central to this paper is the idea that models of interaction can be specified independently of services but used operationally to coordinate specific services. In Section 2 we describe a compact language for this purpose. The mechanisms needed to make the language operational in a peer to peer setting are sufficiently compact that they can specified in totality in this paper; and they are sufficiently close to an encoding in a declarative language that a Prolog interpreter can be obtained by a simple syntactic translation of the specification in Section 2. Similarly, the interaction model examples of Figures 2, 3, 4, 9 and 13 translate directly to their operational versions. Together, this makes it practical for the reader to understand and test at coding level the core mechanisms specified in this paper. This style of compact, complete, executable specification (made easy for programmers to pick up) is in the tradition of lightweight formal methods [26, 48].

Our lightweight language allows us to demonstrate how an interaction oriented view of semantics allows us to tackle traditional semantic web problems in unusual ways. Space prohibits us from presenting these in complete, executable detail but for each method there exists a detailed paper (see below), so it is sufficient to provide a compact formal reconstruction of these in a uniform style. Then, in Section 9, we summarise an implemented system for peer to peer knowledge sharing that embodies many of these ideas. The issues addressed are:

Dynamic ontology mapping (Section 3): necessary because we cannot rely solely on *a priori* ontology mapping. Details in [8].

Coalition formation (Section 4): necessary because the semantics of an interaction is sensitive to the choice of which peers participate in the interaction. Details in [46].

Maintaining interaction context (Section 5): necessary because interaction models represent contracts between peers and the semantics of the interaction depends on contextual information accompanying these contracts. Details in [47].

Making interactions less brittle (Section 6): necessary because peers are not as predictable or stable as subroutines in a programming language so it is useful to have mechanisms to avoid this or reduce its impact. Details in [23, 36].

Satisfying requirements on the interaction process (Section 7): necessary because on the Internet languages for relating requirements (particularly business requirements) to services are becoming established, so interaction models should connect to these rather than compete with them. Details in [32].

Building interaction models more reliably (Section 8): necessary because the design of interaction models is of similar engineering complexity to the design of programs, hence we need analogous robust mechanisms to reduce error in design. Details in [42, 55].

The aim of this paper is to describe an alternative view of interaction between peers that share knowledge. What a "peer" might be is understood differently in different communities (in multi-agent systems a peer is an agent; in a semantic web a peer is a program supplying a service) and is supported on different infrastructures (in multi-agent systems via performative based message passing; in Web service architectures by connecting to WSDL interfaces). Our specification language though computational, is more compact than specification languages that have grown within those communities and infrastructures but it can be related back to them, as we discuss in Section 10. We have also applied our methods direclty in business modelling [33] and in e-Science [5, 57].

2 Interaction Modelling

In this section we describe a basic language for modelling interactions. Our use of this language in the current section will be for specification of interactions but in the sections that follow it will be used for executing interactions. We do not claim that this language, as it stands, is ideally suited to deployment in the current Web services arena - on the contrary, we would expect it to be adapted to whatever specification standards emerge (the most stable currently being RDF and OWL, see Section 10) and linked to appropriate forms of service invocation (for example, we have used WSDL). The aim of this paper is to present the essentials of our interaction oriented method in as compact a form as possible.

2.1 A Lightweight Coordination Calculus

Our aim in this section is to define a language that is as simple as possible while also being able to describe interactions like the one in our scenario of Section 1.1. It is built upon a few simple principles:

- Interactions can be defined as a collection of (separate) definitions for the roles of each peer in the interaction.
- To undertake their roles, peers follow a sequence of activities.
- The most primitive activity is to send or receive a message.
- Peers may change role (recursively) as an activity.
- Constraints may be defined on activities or role changes.

Figure 1 defines the syntax of the Lightweight Coordination Calculus (LCC). An interaction model in LCC is a set of clauses, each of which defines how a role in the interaction must be performed. Roles are described by the type of role and an identifier

$$
\begin{aligned}
Model &:= \{Clause, \ldots\} \\
Clause &:= Role :: Def \\
Role &:= a(Type, Id) \\
Def &:= Role \mid Message \mid Def \ then \ Def \mid Def \ or \ Def \\
Message &:= M \Rightarrow Role \mid M \Rightarrow Role \leftarrow C \mid M \Leftarrow Role \mid C \leftarrow M \Leftarrow Role \\
C &:= Constant \mid P(Term, \ldots) \mid \neg C \mid C \wedge C \mid C \vee C \\
Type &:= Term \\
Id &:= Constant \mid Variable \\
M &:= Term \\
Term &:= Constant \mid Variable \mid P(Term, \ldots) \\
Constant &:= \text{lower case character sequence or number} \\
Variable &:= \text{upper case character sequence or number}
\end{aligned}
$$

Fig. 1. LCC syntax

for the individual peer undertaking that role. The definition of performance of a role is constructed using combinations of the sequence operator ('*then*') or choice operator ('*or*') to connect messages and changes of role. Messages are either outgoing to another peer in a given role ('$\Rightarrow$') or incoming from another peer in a given role ('$\Leftarrow$'). Message input/output or change of role can be governed by a constraint defined using the normal logical operators for conjunction, disjunction and negation. Notice that there is no commitment to the system of logic through which constraints are solved - so different peers might operate different constraint solvers (including human intervention).

2.2 Return to Scenario

To demonstrate how LCC is used, we describe the three interaction models of our scenario from Section 1.1. Figure 2 gives the first of these: a basic shopping service. This contains two clauses: the first defining the interaction from the viewpoint of the buyer; the second from the role of the shopkeeper. Only two roles are involved in this interaction so it is easy to see the symmetry between messages sent by one peer and received by the other. The interaction simply involves the buyer asking the shopkeeper if it has the item, X, then the shopkeeper sending the price, P, then the buyer offering to buy at that price and the shopkeeper confirming the sale.

The constraints in the interaction model of Figure 2 - $need(X)$, $shop(S)$, $afford(X, P)$ and $in\_stock(X, P)$ - must be satisfied by the peers in the role to which the constraint is attached (for example the buyer must satisfy the $afford$ constraint). We write $known(A, C)$ to denote that the peer with identifier A knows the axiom C. LCC is not predicated on a specific constraint language (in fact we shall encounter two constraint languages in this paper) but a common choice of constraint representation in programming is Horn clauses, so we follow this conventional path. By supplying Horn clause axioms in this way we can describe peer knowledge sufficient to complete the

$a(buyer, B) ::$
$\quad ask(X) \Rightarrow a(shopkeeper, S) \leftarrow need(X) \ and \ shop(S) \ then$
$\quad price(X, P) \Leftarrow a(shopkeeper, S) \ then$
$\quad buy(X, P) \Rightarrow a(shopkeeper, S) \leftarrow afford(X, P) \ then$
$\quad sold(X, P) \Leftarrow a(shopkeeper, S)$

$a(shopkeeper, S) ::$
$\quad ask(X) \Leftarrow a(buyer, B) \ then$
$\quad price(X, P) \Rightarrow a(buyer, B) \leftarrow in\_stock(X, P) \ then$
$\quad buy(X, P) \Leftarrow a(buyer, B) \ then$
$\quad sold(X, P) \Rightarrow a(buyer, B)$

Fig. 2. Shop interaction model

interaction model. For instance, if we have a buyer, b, and a shopkeeper, s, that know the following:

$known(b, need(canonS500))$
$known(b, shop(s))$
$known(b, afford(canonS500, P) \leftarrow P \leq 250)$
$known(s, in\_stock(canonS500, 249))$

then the sequence of messages in Table 1 satisfies the interaction model.

Table 1. Message sequence satisfying interaction model of Figure 2

Recipient	Message	Sender
a(shopkeeper,s)	ask(canonS500)	a(buyer,b))
a(buyer,b)	price(canonS500,249)	a(shopkeeper,s))
a(shopkeeper,s)	buy(canonS500,249)	a(buyer,b))
a(buyer,b)	sold(canonS500,249)	a(shopkeeper,s))

Figure 3 gives the second scenario in which a peer, S, seeking a vendor for an item, X, sends a message to an auctioneer, A, stating that S requires X and wants the auction for it to start at purchase value I and stop if the purchase value exceeds maximum value M with the bid value increasing in increments of I. On receiving this requirement the auctioneer assumes the role of a caller for bids from the set of vendors, Vs, that it recognises and, if the call results in a bid to sell the item at some price, P, then the auctioneer offers that price to the seeker who clinches the deal with the vendor and gets its agreement - otherwise the auctioneer signals that no offer was obtained. The role of caller (assumed by the auctioneer) involves two recursions. The first recursion is over the value set by the seeker: the caller starts with the initial value, L, and changes role to a notifier for the vendor peers in Vs that they have a potential sale of an item of type X at value L. If a vendor, V, is obtained by the notifier then the offered price, P, is set to L; if not the price is incremented by the given amount, I, and the role recurses. The

$a(seeker, S)$ ::

$\quad require(X, L, M, I) \Rightarrow a(auctioneer, A) \leftarrow need(X, L, M, I) \; and \; auction\_house(A) \; then$

$$\left(\begin{array}{l} offer(V, X, P) \Leftarrow a(auctioneer, A) \; then \\ clinch(X, P) \Rightarrow a(vendor, V) \; then \\ agreed(X, P) \Leftarrow a(vendor, V) \end{array} \right) \quad or$$

$\quad no\_offer(X) \Leftarrow a(auctioneer, A)$

$a(auctioneer, A)$ ::

$\quad require(X, L, M, I) \Leftarrow a(seeker, S) \; then$

$\quad a(caller(Vs, X, L, M, I, V, P), A) \leftarrow vendors(Vs) \; then$

$$\left(\begin{array}{l} offer(V, X, P) \Rightarrow a(seeker, S) \leftarrow not(P = failed) \; or \\ no\_offer(X) \Rightarrow a(seeker, S) \leftarrow P = failed \end{array} \right)$$

$a(caller(Vs, X, L, M, I, V, P), A)$ ::

$\quad a(notifier(Vs, X, L, Ps), A) \; then$

$$\left(\begin{array}{l} null \leftarrow s(V) \in Ps \; and \; P = L \; or \\ null \leftarrow L > M \; and \; P = failed \; or \\ a(caller(Vs, X, Ln, M, I, V, P), A) \leftarrow not(s(V) \in Ps) \; and \; Ln = L + I \; and \; Ln \le M \end{array} \right)$$

$a(notifier(Vs, X, C, Ps), A)$ ::

$$\left(\begin{array}{l} need(X, C) \Rightarrow a(vendor, V) \leftarrow Vs = [V|Vr] \; then \\ \left(\begin{array}{l} Ps = [s(V)|Pr] \leftarrow supply(X, C) \Leftarrow a(vendor, V) \; or \\ Ps = Pr \leftarrow decline(X, C) \Leftarrow a(vendor, V) \end{array} \right) \; then \\ a(notifier(Vr, X, C, Pr), A) \end{array} \right) \quad or$$

$\quad null \leftarrow Vs = [] \; and \; Ps = []$

$a(vendor, V)$ ::

$$\left(\begin{array}{l} need(X, C) \Leftarrow a(notifier(Vs, X, C, Ps), A) \; then \\ \left(\begin{array}{l} supply(X, C) \Rightarrow a(notifier(Vs, X, C, Ps), A) \leftarrow sell(X, C) \; or \\ decline(X, C) \Rightarrow a(notifier(Vs, X, C, Ps), A) \leftarrow not(sell(X, C)) \end{array} \right) \; then \\ a(vendor, V) \end{array} \right) \quad or$$

$\quad (clinch(X, P) \Leftarrow a(seeker, S) \; then$

$\quad agreed(X, P) \Rightarrow a(seeker, S))$

Fig. 3. Auction interaction model

second recursion is within the notifier which tells each vendor, V, in Vs that item X is needed at current offer price, C; then receives a message from V either offering to supply or declining.

Now in order to satisfy the interaction model we define the following example knowledge possessed by seeker, b, auctioneer, a, and two vendors, $v1$ and $v2$:

$$known(b, need(canonS500, 100, 200, 10))$$
$$known(b, auction\_house(a))$$
$$known(a, vendors([v1, v2]))$$
$$known(v1, sell(canonS500, 110))$$
$$known(v2, sell(canonS500, 170))$$

and then the sequence of messages in Table 2 satisfies the interaction model.

Table 2. Message sequence satisfying interaction model of Figure 3

Recipient	Message	Sender
$a(auctioneer, a)$	$require(canonS500, 100, 200, 10)$	$a(seeker, b)$
$a(vendor, v1)$	$need(canonS500, 100)$	$a(notifier([v1, v2], canonS500, 100, Ps1), a)$
$a(notifier([v1, v2], canonS500, 100, Ps1), a)$	$decline(canonS500, 100)$	$a(vendor, v1)$
$a(vendor, v2)$	$need(canonS500, 100)$	$a(notifier([v2], canonS500, 100, Ps1), a)$
$a(notifier([v2], canonS500, 100, Ps1), a)$	$decline(canonS500, 100)$	$a(vendor, v2))$
$a(vendor, v1)$	$need(canonS500, 110)$	$a(notifier([v1, v2], canonS500, 110, Ps2), a)$
$a(notifier([v1, v2], canonS500, 110, Ps2), a)$	$supply(canonS500, 110)$	$a(vendor, v1)$
$a(vendor, v2)$	$need(canonS500, 110)$	$a(notifier([v2], canonS500, 110, Ps3), a)$
$a(notifier([v2], canonS500, 110, Ps3), a)$	$decline(canonS500, 110)$	$a(vendor, v2)$
$a(seeker, b)$	$offer(v1, canonS500, 110)$	$a(auctioneer, a)$
$a(vendor, v1)$	$clinch(canonS500, 110)$	$a(seeker, b)$
$a(seeker, b)$	$agreed(canonS500, 110)$	$a(vendor, v1)$

Figure 4 gives the third scenario in which a peer that wants to be a customer of a manufacturer asks to buy an item of type X from the manufacturer, then enters into a negotiation with the manufacturer about the attributes required to configure the item to the customer's requirements. The negotiation is simply a recursive dialogue between manufacturer and customer with, for each attribute (A) in the set of attributes (As), the manufacturer offering the available attribute and the customer accepting it. When all the attributes have been accepted in this way, there is a final interchange committing the customer to the accepted attribute set, Aa, for X.

$a(customer, C) ::$
$\quad ask(buy(X)) \Rightarrow a(manufacturer, M) \leftarrow need(X) \text{ and } sells(X, M) \text{ then}$
$\quad a(n\_cus(X, M, []), C)$

$a(n\_cus(X, M, Aa), C) ::$
$$\left(\begin{array}{l} offer(A) \Leftarrow a(n\_man(X, C, \_), M) \text{ then} \\ accept(A) \Rightarrow a(n\_man(X, C, \_), M) \leftarrow acceptable(A) \text{ then} \\ a(n\_cus(X, M, [att(A)|Aa]), C) \end{array} \right) \text{ or}$$
$$\left(\begin{array}{l} ask(commit) \Leftarrow a(n\_man(X, C, \_), M) \text{ then} \\ tell(commit(Aa)) \Rightarrow a(n\_man(X, C, \_), M) \text{ then} \\ tell(sold(Aa)) \Leftarrow a(n\_man(X, C, \_), M) \end{array} \right)$$

$a(manufacturer, M) ::$
$\quad ask(buy(X)) \Leftarrow a(customer, C) \text{ then}$
$\quad a(n\_man(X, C, As), M) \leftarrow attributes(X, As)$

$a(n\_man(X, C, As), M) ::$
$$\left(\begin{array}{l} offer(A) \Rightarrow a(n\_cus(X, M, \_), C) \leftarrow As = [A|T] \text{ and } available(A) \text{ then} \\ accept(A) \Leftarrow a(n\_cus(X, M, \_), C) \text{ then} \\ a(n\_man(X, C, T), M) \end{array} \right) \text{ or}$$
$$\left(\begin{array}{l} ask(commit) \Rightarrow a(n\_cus(X, M, \_), C) \leftarrow As = [] \text{ then} \\ tell(commit(As)) \Leftarrow a(n\_cus(X, M, \_), C) \text{ then} \\ tell(sold(As)) \Rightarrow a(n\_cus(X, M, \_), C) \end{array} \right)$$

Fig. 4. Manufacturer interaction model

In order to satisfy this interaction model we define the following example of knowledge possessed by customer, b, and manufacturer, m:

$$known(b, need(canonS500))$$
$$known(b, sells(canonS500, m))$$
$$known(b, acceptable(memory(M)) \leftarrow M \geq 32)$$
$$known(b, acceptable(price(M, P)) \leftarrow P \leq 250)$$
$$known(m, attributes(canonS500, [memory(M), price(M, P)]))$$
$$known(m, available(memory(32)))$$
$$known(m, available(memory(64)))$$
$$known(m, available(memory(128)))$$
$$known(m, available(price(M, P)) \leftarrow P = 180 + M)$$

and then the sequence of messages in Table 3 satisfies the interaction model.

Table 3. Message sequence satisfying interaction model of Figure 4

Recipient	Message	Sender
$a(manufacturer, m)$	$ask(buy(canonS500))$	$a(customer, b)$
$a(n\_cus(canonS500, m, Aa1), b)$	$offer(memory(32))$	$a(n\_man(canonS500, b, \left[\begin{smallmatrix} memory(32), \\ price(32, P) \end{smallmatrix}\right]), m)$
$a(n\_man(canonS500, b, As1), m)$	$accept(memory(32))$	$a(n\_cus(canonS500, m, []), b)$
$a(n\_cus(canonS500, m, Aa2), b)$	$offer(price(32, 212))$	$a(n\_man(canonS500, b, [price(32, 212)]), m)$
$a(n\_man(canonS500, b, As2), m)$	$accept(price(32, 212))$	$a(n\_cus(canonS500, m, [att(memory(32))]), b)$
$a(n\_cus(canonS500, m, Aa3), b)$	$ask(commit)$	$a(n\_man(canonS500, b, []), m)$
$a(n\_man(canonS500, b, As3), m)$	$tell(commit(\left[\begin{smallmatrix} att(price(32, 212)), \\ att(memory(32)) \end{smallmatrix}\right]))$	$a(n\_cus(canonS500, m, \left[\begin{smallmatrix} att(price(32, 212)), \\ att(memory(32)) \end{smallmatrix}\right]), b)$

Note the duality in our understanding of the interaction models we have described in this section. The interaction models of figures 2, 3 and 4 are programs because they use the data structures and recursive computation of traditional (logic) programming languages. They also are distributed process descriptions because their purpose is to constrain the sequences of messages passed between peers and the clauses of interaction models constrain processes on (possibly) different physical machines.

2.3 Executing Interaction Models

LCC is a specification language but it is also executable and, as is normal for declarative languages, it admits many different models of computation. Our choice of computation method, however, has important engineering implications. To demonstrate this, consider the following three computation methods:

Interaction model run on a single peer: With this method there is no distribution of the model to peers. Instead, the model is run on a single peer (acting as a server). This is the style of execution often used with, for example, executable business process modelling languages such as BPEL4WS. It is not peer to peer because it is rigidly centralised but we have used LCC in this way to coordinate systems composed from traditional Web services offering only WSDL interfaces ([57]).

Interaction model clauses distributed across peers: Each clause in an LCC interaction model is independent of the others so as peers assume different roles in an interaction they can choose the appropriate clause from the model and run with it. Synchronisation is through message passing only, so clauses can be chosen and used by peers independently as long as there is a mechanism for knowing from which interaction model each clause has been derived. This is a peer to peer solution because all peers have the same clause interpretation abilities. It allows the interaction model to be distributed across peers but, since it is distributed by clause, it is not always possible to reconstruct the global state of the interaction model (as we can when confined to a single peer), since this would mean synchronising the union of its distributed clauses. Reconstructing the global state is not necessary for many applications but, where it is, there is another peer to peer option.

Interaction model transported with messages: In this style of deployment we distribute the interaction model clauses as above but each peer that receives a message containing an interaction model, having selected and used an appropriate clause, replaces it with the interaction model and transmits it with the message to the next peer in the interaction. This keeps the current state of the interaction together but assumes a linear style of interaction in which exactly one message is in transit at any instant - in other words the interaction consists of a chain of messages totally ordered over time. Many common service interactions are of this form, or can be constructed from chains of this form.

In the remainder of this paper we shall adopt the linear model of computation in which global state is transmitted with messages, because this is simpler to discuss. Many of the concepts we introduce also apply to non-linear interactions without global state.

Figure 5 describes informally the main components of interaction between peers. Later, Figure 6 gives a precise definition, using a linear computation model, that is consistent with the informal description. Ellipses in Figure 5 are processes; rectangles are data; and arcs denote the main inputs and outputs of processes. The large circle in the diagram encloses the components effecting local state of the peer with respect to the interaction, which interacts with the internal state of the peer via constraints specified in the interaction model. The only means of peer communication is by message passing and we assume a mechanism (specific to the message passing infrastructure) for decoding from any message appropriate information describing the interaction model associated with that message (see Section 2.4 for an example of this sort of mechanism). To know its obligations within the interaction the peer must identify the role it is to perform and (by choosing the appropriate clause) find the current model for that role. It then must attempt to discharge any obligations set by that model, which it does by identifying those constraints the model places at the current time and, if possible, satisfying them. In the process it may accept messages sent to it by other peers and send messages out to peers. Each message sent out must be routed to an appropriate peer, and the choice of recipient peer may be determined by the sender directly (if our infrastructure requires strictly point to point communication between peers) or entrusted to a message router (if a routing infrastructure such as JXTA is used).

The overview of Figure 5 is high level, so it makes no commitment to how messages are structured or how the obligations of interactions are discharged. To make these

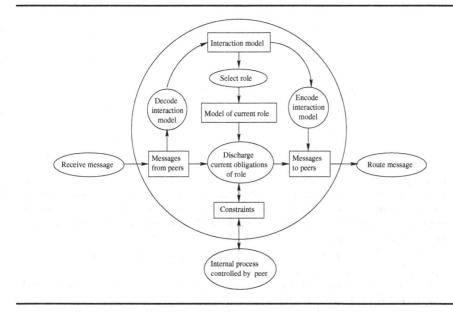

Fig. 5. Conceptual model of local state change of a peer

commitments we must be more specific about computation, which is the topic of our next section.

2.4 A Basic, Linear Computation Method for Interaction Models

To be precise in our analysis of the interaction between peers we introduce, in Figure 6, a formal definition of the linear computation introduced in the previous section. Although this assumes a shared interaction model (transmitted via messages), each transition is performed by a specific peer using only the part of the interaction visible to that peer. A sequence of transitions is initiated by a single peer with some goal to achieve and the sequence terminates successfully only when that peer can infer from its local knowledge of the interaction that it has obtained that goal.

Expressions 1 and 2 in Figure 6 generate transition sequences. Expression 3 in Figure 6 describes the general form of any such sequence. Following through expression 3 for a peer $p1$ attempting to establish goal G_{p1}, it begins with the selection by $p1$ of an interaction model ($\Omega \overset{p1}{\ni} S_1$); then selection of the local state pertaining to $p1$ from the shared model ($S_1 \overset{s}{\supseteq} S_{p1}$); then a transition of the local state for $p1$ according to the model ($S_{p1} \xrightarrow{M_1, S_1, M_2} S'_{p1}$); then merging of the ensuing local state with the shared state to produce a new shared interaction state ($S'_{p1} \overset{s}{\cup} S_1 = S_2$); then repeating the transitions until reaching a final state in which the required goal can be derived using $p1$'s local knowledge ($k_{p1}(S_f) \vdash G_{p1}$).

$$\sigma(p, G_p) \quad \leftrightarrow \quad \Omega \overset{p}{\ni} \mathcal{S} \wedge i(\mathcal{S}, \phi, \mathcal{S}_f) \wedge k_p(\mathcal{S}_f) \vdash G_p \tag{1}$$

$$i(\mathcal{S}, M_i, \mathcal{S}_f) \quad \leftrightarrow \quad \mathcal{S} = \mathcal{S}_f \vee \begin{pmatrix} \mathcal{S} \overset{s}{\supseteq} \mathcal{S}_p \wedge \\ \mathcal{S}_p \xrightarrow{M_i, S, M_n} \mathcal{S}'_p \wedge \\ \mathcal{S}'_p \overset{s}{\cup} \mathcal{S} = \mathcal{S}' \wedge \\ i(\mathcal{S}', M_n, \mathcal{S}_f) \end{pmatrix} \tag{2}$$

where:

- p is a unique identifier for a peer.
- G_p is a goal that peer P wants to achieve.
- $\mathcal{S}$, is a state of interaction which contains the interaction model used to coordinate peers; the knowledge shared by peers participating in the interaction; and a description of their current progress in pursuing the interaction. Ω is the set of all the available initial interaction states. $\Omega \overset{p}{\ni} \mathcal{S}$ selects an initial interaction state, $\mathcal{S}$, for peer P from Ω.
- M is the current set of messages sent by peers. The empty set of messages is ϕ.
- $\sigma(P, G_p)$ is true when goal G_p is attained by peer P.
- $i(\mathcal{S}, M_1, \mathcal{S}_f)$ is true when a sequence of interactions allows state $\mathcal{S}_f$ to be derived from $\mathcal{S}$ given an initial set of messages M_1.
- $k_p(\mathcal{S})$ is a function giving the knowledge visible to peer P contained in state $\mathcal{S}$.
- $\mathcal{S} \overset{s}{\supseteq} \mathcal{S}_p$ selects the state, $\mathcal{S}_p$, pertaining specifically to peer P from the interaction state $\mathcal{S}$.
- $\mathcal{S}_p \xrightarrow{M_i, S, M_n} \mathcal{S}'_p$ is a transition of the state of peer P to a new state, $\mathcal{S}'_p$, given the current set of inter-peer messages, M_i, and producing the new set of messages M_n.
- $\mathcal{S}_p \overset{s}{\cup} \mathcal{S}$ is a function that merges the state, $\mathcal{S}_p$, specific to peer P with interaction state $\mathcal{S}$ (replacing any earlier interaction state for peer P).

Every successful, terminating interaction satisfying $\sigma(p1, G_{p1})$ can then be described by the following sequence of relations (obtained by expanding the 'i' relation within expression 1 using expression 2):

$$\Omega \overset{p1}{\ni} \mathcal{S}_1 \overset{s}{\supseteq} \mathcal{S}_{p1} \xrightarrow{M_1, S_1, M_2} \mathcal{S}'_{p1} \overset{s}{\cup} \mathcal{S}_1 = \mathcal{S}_2 \overset{s}{\supseteq} \mathcal{S}_{p2} \xrightarrow{M_2, S_2, M_3} \mathcal{S}'_{p2} \overset{s}{\cup} \mathcal{S}_2 = \mathcal{S}_3 \ldots k_{p1}(\mathcal{S}_f) \vdash G_{p1} \tag{3}$$

Fig. 6. Formal model of linearised peer interaction

Having defined, in Figure 6, a formal model of interaction, we describe in Sections 2.5 to 2.8 how this connects to the LCC language that we introduces in Section 2.1. The operations that our language must support in order to conform to Figure 6 are given as part of each section heading.

2.5   Initial Interaction State: $\Omega \overset{p}{\ni} \mathcal{S}$

For a peer, p, to initiate an interaction it must select an appropriate initial state, S, from the set of possible such initial states, Ω. In Section 2.1 we have given a way to describe S, based on interaction models. This, however, allows infinitely many possible elements of Ω to be constructed in theory (through manual coding, synthesis, *etc.* In practise, these interaction models are (like traditional programs) built or borrowed for specific tasks of use to p, so Ω might come from a local library. In an open system, where knowing of the existence of helpful interaction models is an issue, then the contents of Ω may not fully be known to p and mechanisms of discovery are required, as we discuss in Section 4. Automated synthesis of some elements of Ω is discussed in Section 6.

Given some way of (partially) populating Ω, there remains the issue for p of choosing which initial state to use. This is determined by some choice made by p based on the interaction model, $\mathcal{P}_g$, and shared knowledge, K, components of the initial states (recall that in LCC the initial state is a term of the form $m(\phi, \mathcal{P}_g, K)$). We denote this choice as $c(p, \mathcal{P}_g, K)$ in the expression below but do not define the mechanism by which that choice is made, since it varies according to application - anything from a fully automated choice to a decision made by a human operator.

$$\Omega \overset{p}{\ni} \mathcal{S} \;\leftrightarrow\; \mathcal{S} \in \Omega \,\wedge\, \mathcal{S} = m(\phi, \mathcal{P}_g, K) \,\wedge\, c(p, \mathcal{P}_g, K) \tag{4}$$

2.6 State Selection by a Peer: $\mathcal{S} \overset{s}{\supseteq} \mathcal{S}_p$

Given that in LCC the state of interaction always is expressed as a term of the form $m(\mathcal{P}_s, \mathcal{P}_g, K)$, the selection of the current state for a peer, p, simply requires the selection of the appropriate clause, $a(R, p) :: D$, defining (in D) the interaction stare for p when performing role R.

$$\mathcal{S} \overset{s}{\supseteq} \mathcal{S}_p \;\leftrightarrow\; \exists R, D.(\mathcal{S}_p \in \mathcal{S} \,\wedge\, \mathcal{S}_p = a(R, p) :: D) \tag{5}$$

2.7 State Transition by a Peer: $\mathcal{S}_p \xrightarrow{M_i, \mathcal{S}, M_n} \mathcal{S}_p'$

Recall that, from Section 2.6 we can know the state of a specific role in the interaction by selecting the appropriate clause. This clause gives us $\mathcal{S}_p$ and we now explain how to advance the state associated with this role to the new version of that clause, $\mathcal{S}_p'$, given an input message set, M_i, and producing a new message set, M_n, which contains those messages from M_i that have not been processed plus additional messages added by the state transition. Since we shall need a sequence of transitions to the clause for $\mathcal{S}_p$ we use C_i to denote the start of that sequence and C_j the end. The rewrite rules of Figure 7 are applied to give the transition sequence of expression 6.

$$C_i \xrightarrow{M_i, \mathcal{S}, M_n} C_j \;\leftrightarrow\; \exists R, D.(C_i = a(R, p) :: D) \,\wedge \tag{6}$$
$$\begin{pmatrix} C_i \xrightarrow{R_i, M_i, M_{i+1}, \mathcal{S}, O_i} C_{i+1} \,\wedge \\ C_{i+1} \xrightarrow{R_i, M_{i+1}, M_{i+2}, \mathcal{S}, O_{i+1}} C_{i+2} \,\wedge \\ \cdots \\ C_{j-1} \xrightarrow{R_i, M_{j-1}, M_j, \mathcal{S}, O_j} C_j \end{pmatrix} \wedge$$
$$M_n = M_j \cup O_j$$

$$R :: B \xrightarrow{R_i,M_i,M_o,S,O} A :: E \qquad\qquad\qquad\qquad if\ B \xrightarrow{R,M_i,M_o,S,O} E$$

$$A_1\ or\ A_2 \xrightarrow{R_i,M_i,M_o,S,O} E \qquad\qquad\qquad if\ \neg closed(A_2)\ \wedge$$
$$A_1 \xrightarrow{R_i,M_i,M_o,S,O} E$$

$$A_1\ or\ A_2 \xrightarrow{R_i,M_i,M_o,S,O} E \qquad\qquad\qquad if\ \neg closed(A_1)\ \wedge$$
$$A_2 \xrightarrow{R_i,M_i,M_o,S,O} E$$

$$A_1\ then\ A_2 \xrightarrow{R_i,M_i,M_o,S,O} E\ then\ A_2 \qquad\quad if\ A_1 \xrightarrow{R_i,M_i,M_o,S,O} E$$
$$A_1\ then\ A_2 \xrightarrow{R_i,M_i,M_o,S,O} A_1\ then\ E \qquad\quad if\ closed(A_1)\ \wedge$$
$$A_2 \xrightarrow{R_i,M_i,M_o,S,O} E$$

$$C \leftarrow M \Leftarrow A \xrightarrow{R_i,M_i,M_i-\{m(R_i,M \Leftarrow A)\},S,\emptyset} c(M \Leftarrow A)\ if\ m(R_i,M \Leftarrow A) \in M_i\ \wedge$$
$$satisfy(C)$$

$$M \Rightarrow A \leftarrow C \xrightarrow{R_i,M_i,M_o,S,\{m(R_i,M \Rightarrow A)\}} c(M \Rightarrow A) \quad if\ satisfied(S,C)$$

$$null \leftarrow C \xrightarrow{R_i,M_i,M_o,S,\emptyset} c(null) \qquad\qquad\qquad if\ satisfied(S,C)$$

$$a(R,I) \leftarrow C \xrightarrow{R_i,M_i,M_o,S,\emptyset} a(R,I) :: B \qquad\qquad if\ clause(S,a(R,I)::B)\ \wedge$$
$$satisfied(S,C)$$

An interaction model term is decided to be closed as follows:

$$closed(c(X))$$
$$closed(A\ then\ B)\ \leftarrow\ closed(A)\ \wedge\ closed(B) \qquad\qquad (7)$$
$$closed(X :: D)\ \leftarrow\ closed(D)$$

$satisfied(S,C)$ is true if constraint C is satisfiable given the peer's current state of knowledge. $clause(S,X)$ is true if clause X appears in the interaction model S, as defined in Figure 1.

Fig. 7. Rewrite rules for expansion of an interaction model clause

2.8   Merging Interaction State: $S_p \overset{s}{\cup} S = S'$

The interaction state, S, is a term of the form $m(P_s, P_g, K)$ and the state relevant to an individual peer, S_p, always is a LCC clause of the form $a(R,p) :: D$. Merging S_p with S therefore is done simply by replacing in S the (now obsolete) clause in which p plays role R with its extended version S_p.

$$(a(R,p) :: D)\ \overset{s}{\cup}\ S\ =\ (S \overset{s}{-} \{a(R,p) :: D'\}) \cup \{a(R,p) :: D)\} \qquad (8)$$

2.9   Interaction-Specific Knowledge: $k_p(S) \vdash G_p$

Shared knowledge in LCC is maintained in the set of axioms, K, in the interaction state $m(P_s, P_g, K)$ so a peer's goal, G_p, can be satisfied if it is satisfiable from K or through the peer's own internal satisfiability mechanisms. This corresponds to the $satisfied$ relation introduced with the rewrite rules of Figure 7.

$$k_p(S) \vdash G_p\ \leftrightarrow\ satisfied(S, G_p) \qquad\qquad (9)$$

This completes our operational specification for LCC combined with the style of linear deployment given in Figure 6. In an ideal world (in which all peers were aware of each other; conformed to the same ontology and cooperated perfectly to obtain desired interactions) we would now simply deploy this system by implementing it for an appropriate message passing infrastructure. The Internet, however, presents many obstacles to so doing. In the remainder of this paper we consider how some of these obstacles may be overcome. An important point throughout is that, by anchoring our solutions in a peer to peer architecture with a strong notion of modelling of interaction, it is technically possible to tackle these obstacles using well known methods that do not necessarily require huge investments up-front from the Web community.

3   Dynamic Ontology Matching

We defined, in Section 2.4, a mechanism by which a peer may make a state transition, $\mathcal{S}_p \xrightarrow{M_i, S, M_n} \mathcal{S}_p'$ but we assumed when doing this that the terminology used in the message set, M_i, is consistent with the terminology actually used by the host peer, p. In an open system (where there is no restriction on the terminology used by each peer) this need not be true. When a message is created concepts in the sender's representation of the domain are mapped to the terms that compose the message, conforming to the syntax of that language. Then the receiver must map the terms in the message to the concepts in its own representation, helped by the syntax rules that structure the message. If the terms are mapped to equivalent concepts by the sender and by the receiver peers, then the understanding is correct. A misunderstanding happens if a term is mapped to different concepts by the sender and the receiver, while the interaction may fail spuriously if the receiver does not succeed in mapping a term that should correspond.

To avoid such misunderstandings we have two means of control: adapt the peer's state, $\mathcal{S}_p$ or map between terms in $\mathcal{S}_p$ and M_i. Mappings are normally viewed as a more modular form of control because they allow the ontological alignment of interaction to remain distinct from whatever inference is performed locally by a peer. Indeed, much of traditional ontology mapping is supposed to be done on ontology representations independent of interactions and prior to those ontologies being used [19, 28]. The problem with this *a priori* approach is that mapping cannot be determined separately from interaction state unless it is possible to predict all the states of all the peers in all potential interactions. This is demonstrated by examining our example interaction models in Figures 2 to 4, all of which contain constraints that an appropriate peer must satisfy in order to change its state in the interaction. For example, the peer adopting the buyer role in Figure 2 must satisfy the constraint $need(X)$ and $shop(S)$ in order to send the first message, $ask(X)$ to shopkeeper S. The identity of X is determined by the current state of the buyer but the ontology used has to match that of the shopkeeper. We can only judge whether a mapping of terms was needed for X if we know the buyer's and seller's constraint solving choices, which normally are part of the private, internal state of each peer. Therefore, we cannot expect reliable ontological mapping in open systems (in which peer state is not known and interaction models are not fixed) without some form of dynamic mapping to keep ontology alignment on track in situations that were not predicted prior to interaction.

It is straightforward to insert an ontology mapping step into our conceptual model of interaction from Section 2.4 by adapting expressions 1 and 2 to give expressions 10 and 11 respectively. Expression 10 is obtained by defining interaction models in the set Ω with an accompanying set, O, of ontological constraints (which may be empty). These constraints are carried along with S into the interaction, producing as a consequence some final set of constraints, O_f. Expression 11 applies state transitions for each peer, S_p, but requires that a mapping relation, $map(M_i, O_i, S_p, M_i', O_n)$, applies between the current message set, M_i, and the message set, M_i', used in applying the appropriate state transition. This mapping can adapt the ontological constraints from the current set, O_i, to the new set, O_n.

$$\sigma(p, G_p) \quad \leftrightarrow \quad \Omega \overset{p}{\ni} \langle S, O \rangle \wedge i(S, \phi, O, S_f, O_f) \wedge k_p(S_f) \vdash G_p \qquad (10)$$

$$i(S, M_i, O_i, S_f, O_f) \quad \leftrightarrow \quad S = S_f \vee \begin{pmatrix} S \overset{s}{\supseteq} S_p \wedge \\ map(M_i, O_i, S_p, M_i', O_n) \wedge \\ S_p \xrightarrow{M_i', S, M_n} S_p' \wedge \\ S_p' \overset{s}{\cup} S = S' \wedge \\ i(S', M_n, O_n, S_f, O_f) \end{pmatrix} \qquad (11)$$

The purpose of the map relation is to define a set of axioms that enable the ontology used in the messages, M_i, to connect to the ontology used in S_p. For example, in Section 2.2 the shopkeeper, s, using the interaction model of Figure 2 must receive from the buyer, b, a message of the form $ask(X)$. Suppose that the message actually sent by b was in fact $require(canonS500)$ because b used a different ontology. We then want the map relation at to add information sufficient for S to conclude $require(canonS500) \rightarrow ask(canonS500)$. Although this might seem an elementary example it introduces a number of key points:

- There is no need, as far as a specific interaction is concerned, to provide a mapping any more general than for a specific object. We do not care whether $require(X) \rightarrow ask(X)$ holds for any object other than $canonS500$ or for any peer other than s because only that object matters for this part of the interaction. This form of mapping will therefore tend, on each occasion, to be much more specific than generalised mappings between ontologies.
- We would not want to insist that s accept a general $\forall X.require(X) \rightarrow ask(X)$ axiom because, in general, we don't necessarily ask for all the things we require (nor, incidentally, do we always require the things we ask about).
- The word "require" is used in many different ways in English depending on the interaction. For instance a message of the form $require(faster\_service)$ places a very different meaning (one that is not anticipated by our example interaction model) on $require$ than our earlier requirement for a specific type of camera. We could attempt to accommodate such distinctions in a generic ontology mapping but this would require intricate forms of engineering (and social consensus) to produce "the correct" disambiguation. Wordnet defines seven senses of the word "ask" and four senses of the word "require" so we we would need to consider at least all the

combinations of these senses in order to ensure disambiguation when the words interact; then follow these through into the ontological definitions. To address just one of these combinations we would have to extend the mapping definitions to have at least two different classes of *require*, each applying to different classes of object - one a form of merchandise; the other a from of functional requirement - then we would need to specify (disjoint) classes of merchandise and functional requirements. This becomes prohibitively expense when we have to, simultaneously, consider all the other word-sense combinations and we can never exclude the possibility that someone invents a valid form of interaction that breaches our "consensus".

The above are reasons why the strong notion of global sets of pre-defined ontology mappings do not appear attractive for open peer to peer systems, except in limited cases where there is a stable consensus on ontology (for example when we are federating a collection of well known databases using database schema). What, then, is the least that our interaction models require?

Returning to our example, we want the *map* relation when applied by seller, s, interacting with buyer, b, as follows:

$$map(\ \{m(a(shopkeeper, s), require(canonS500) \ \Leftarrow \ a(buyer, b))\},$$
$$\{\},$$
$$a(shopkeeper, s) ::$$
$$ask(X) \ \Leftarrow \ a(buyer, B) \ then$$
$$price(X, P) \ \Rightarrow \ a(buyer, B) \ \leftarrow \ in\_stock(X, P) \ then$$
$$buy(X, P) \ \Leftarrow \ a(buyer, B) \ then$$
$$sold(X, P) \ \Rightarrow \ a(buyer, B)$$
$$M'_i,$$
$$O_n)$$

to give the bindings:

$$M'_i = \{m(a(shopkeeper, s), ask(canonS500) \ \Leftarrow \ a(buyer, b))\}$$
$$O_n = \{require(canonS500)@a(buyer, b) \ \rightarrow \ ask(canonS500)@a(shopkeeper, s)\}$$

where the expression $T@A$ denotes that proposition T is true for the peer A. Obtaining this result needs at least three steps of reasoning:

Detection of mapping need: It is necessary to identify in the LCC clause describing the current state of the seller, s, the transition step for which mapping may immediately be required. In the example above this is the first step in enactment of the *seller* role, since no part of the clause has been closed. Targets for mapping are then any terms in this step which do not have matching terms in the message set. The mapping need in the example is thus for $ask(X)$.

Hypothesis of mappings: Knowing where mappings are immediately needed, the issue then is whether plausible mappings may be hypothesised. In our example (since there is only one message available to the *seller*) we need to decide whether $require(canonS500)$ should map to $ask(X)$. There is no unique, optimal algorithm for this; we present an evidence-based approach below but other methods (such as statistical methods) are possible.

Description of mappings: Mappings may, in general, be in different forms. For instance, two terms (T_1 and T_2) may map via equivalence ($T_1 \leftrightarrow T_2$ or subsumption ($T_1 \rightarrow T_2$ or $T_1 \leftarrow T_2$.) This becomes a complex issue when attempting to map two ontologies exhaustively and definitively but for the purposes of a single interaction we are content with the simplest mapping that allows the interaction to continue. The simplest hypothesis is that the term in the message received allows us to imply the term we need - in our running example $require(canonS500) \rightarrow ask(canonS500)$.

The most difficult step of the three above is hypothesising a mapping. For this we need background information upon which to judge the plausibility of our hypothesis. Several sources of such information commonly are available:

Standard word usage: We could use a reference such as Wordnet to detect similar words (such as $require$ and ask in our example). A similarity detected via Wordnet would raise our confidence in a mapping.

Past experience: If we have interactions that have been successful in the past using particular sets of mappings we may be inclined to use these again (more on this subject in Section 4). In this way earlier successes raise our confidence in a mapping while failures reduce it.

Type hierarchies for peers and interactions: Any message passing event involves three elements: the peer sending the message; the peer receiving it and the interaction model in which they are involved. Any or all of these elements may possess ontological information influencing a mapping hypothesis. A mapping might be implied by one or more of the ontologies, raising our confidence in it, or it might be inconsistent with the ontologies, lowering our confidence.

Human operators: In some circumstances it may be necessary for a human to decide whether a hypothesised mapping is valid. This choice might be informed by evidence from any or all of the sources above.

Notice that all of the sources of evidence above are unreliable: standard word usage doesn't always cover a specific circumstance; past experience may no longer apply; type hierarchies aren't necessarily complete or compatible between peers; human operators make errors. Our method does not depend on the usefulness of any of these methods, however. One can program interactions without ontological mapping but those interactions then will need perfect matches between terms (like normal programs). Ontological mapping permits more terms to match during interaction and the major programming concern is whether this "looseness" can be controlled in appropriate programming settings. Some settings require no looseness - only a perfect match will do. Other settings, in which we know in advance the ontologies used in an application but do not know which peers will be involved, allow us to define a sufficient set of mappings (O in expression 10) along with the initial interaction model. More open settings require the mapping relation (map in expression 11) to hypothesise mappings that extend the ontological "safe envelope" maintained around the interaction (relating O_i to O_n in expression 11).

4   Coalition Formation

Interaction states change via the state changes of individual peers - giving the S_{pN} sequence in expression 3 of Figure 6. Crucial to the success of the interaction is the choice of pN at each step. For interactions involving finite numbers of peers for which the identity is known in advance there is no coalition formation problem: the LCC interaction model simply is executed with the given peers. Notice that the examples of Figures 2, 3 and 4 are like this - we were careful to define constraints that required the peers contacting others to determine precisely which those are (for example in Figure 2 the $shop(S)$ constraint determines which shop is contacted). It is more common, however, for the choice of individual peers not to be prescribed by the interaction model - for example in the interaction model of Figure 2 what happens if the buyer doesn't know which shop might be appropriate? In open systems, a peer is often unaware of the existence and capabilities of other peers in its world. When one peer must collaborate with another to achieve some goal, a mechanism must exist to enable the discovery of other peers and their abilities.

This problem is well known, an early instance appearing in the Contract Net system [51]. It remains a crucial issue in the deployment of agent-like systems [15, 30, 59], and is resurfacing as a fundamental problem in newer endeavours like the Semantic Web and Grid computing, a recent example being the Web Services Choreography Definition Language (WSCDL) [29]. The most popular response to this problem has been to focus on specialised agents, often called "middle agents" [15, 30] or "matchmakers". The first multi-agent systems to offer explicit matchmakers were ABSI, COINS, and SHADE. These set the mould for the majority of subsequent work on matchmaking, by defining two common features: matching based on similarity measures between atomic client requests and advertised provider services; and a consideration of purely two-party interaction. OWL-S [35] and many other matchmaking architectures presume a universe where a client wishes to fulfil some goal that can be achieved by a single service provider (which may interact with other services at its own discretion). Finding collaborators for multi-party web-service interactions is discussed in [62]. Our use of performance histories is predated by a similar approach found in [63], although that only examines the case of two-party interactions.

Our aim in this section is to show how the interaction models used in LCC support matchmaking based on data from previous interactions. Suppose that in our running example we decide to automate the purchase of a list of products. We define a LCC interaction model consisting of expressions 12 and 13 below plus the original clauses from Figures 2, 3 and 4.

$$a(purchaser(L), A) :: (a(buy\_item(X), A) \leftarrow L = [H|T] \text{ then } a(purchaser(T), A)) \text{ or}$$
$$null \leftarrow L = [] \tag{12}$$

$$a(buy\_item(X), A) :: (a(buyer(X), A) \text{ or } a(seeker(X), A) \text{ or } a(customer(X), A)) \tag{13}$$

Let us further suppose that the constraints used to identify the vendors in each of the original interaction models ($shop(S)$ in Figure 2, $auction\_house(A)$ in Figure 3 and $sells(X, M)$ in Figure 4) are removed. If we now wish to buy three different types of camera by performing the role of $a(purchaser([canonS500, olympusE300,$ $canonEOS1]), b)$ then we have three purchases to make and it will be necessary, when performing the role of *buyer* for each item, to choose one of the three available forms of buying model with appropriate choices of vendors. For example, we might satisfy our interaction model with the sequence of roles given below:

$a(purchaser([canonS500, olympusE300, canonEOS1]), b)$
$a(buy\_item(canonS500), b)$
$a(buyer(canonS500), b))$
Message sequence given in Table 1 when interacting with $a(shopkeeper, s)$
$a(purchaser([olympusE300, canonEOS1]), b)$
$a(buy\_item(olympusE300), b)$
$a(seeker(olympusE300), b))$
Message sequence given in Table 2 when interacting with $a(auctioneer, a)$
$a(purchaser([canonEOS1]), b)$
$a(buy\_item(canonEOS1), b)$
$a(customer(canonEOS1), b))$
Message sequence given in Table 3 when interacting with $a(manufacturer, m)$
$a(purchaser([]), b)$
---

in which case our interaction will have involved the set of peers $\{b, s, a, m\}$, but the sequence above is only one of many sequences we might have chosen for this set of peers. We might have chosen different roles (*e.g.* by buying the $canonS500$ at auction rather than at a shop) or different peers for the same roles (*e.g.* maybe peer m could take the role of a shopkeeper as well as or instead of its role as a manufacturer). We might have chosen to interact with only one peer in the same role each time (*e.g.* by shopping for all the cameras with peer s). The best choices of roles and peers are likely to depend on factors not expressed in the interaction model - for example, peer a might be unreliable; or peer s might give better service to some using it more frequently; or peers s and m may conspire (through separate communication channels) not to supply the same sources.

The task of a matchmaker is, by tackling problems like those above, to make the right choices of peer identifiers and roles as an interaction model unfolds over time. Given that the things that make or break an interaction often are task/domain specific in ways that cannot be analysed in detail, matchmaking algorithms may have to rely a great deal on empirical data describing successes or failures of previous interactions. This is analogous to the situation on the conventional Worldwide Web, where mass browsing behaviours continually influence the ranking of pages. Imagine, instead, that we want to rank choices of peers to involve in appropriate roles at a given stage of an interaction model's execution. Figure 8 defines a basic matchmaker capable of performing this task. A more extensive discussion (and more sophisticated matchmaking based on this principle) appears in [31].

The matchmaker of Figure 8 is an extension of the clause expansion rewrite rules of Figure 7. To each rewrite rule is added a parameter, Δ, that contains the set of peers that have been involved in closed parts of the clause, prior to the rewrite currently being applied (the predicate *closed* of arity 2 collects the appropriate peers, following the closed part of a clause similarly to *closed* of arity 1 which we defined in

$$R :: B \xrightarrow{R_i, M_i, M_o, S, O, \Delta} A :: E \qquad if \ B \xrightarrow{R, M_i, M_o, S, O, \{R\} \cup \Delta} E$$

$$A_1 \ or \ A_2 \xrightarrow{R_i, M_i, M_o, S, O, \Delta} E \qquad if \ \neg closed(A_2) \land$$
$$A_1 \xrightarrow{R_i, M_i, M_o, S, O, \Delta} E$$

$$A_1 \ or \ A_2 \xrightarrow{R_i, M_i, M_o, S, O, \Delta} E \qquad if \ \neg closed(A_1) \land$$
$$A_2 \xrightarrow{R_i, M_i, M_o, S, O, \Delta} E$$

$$A_1 \ then \ A_2 \xrightarrow{R_i, M_i, M_o, S, O, \Delta} E \ then \ A_2 \qquad if \ A_1 \xrightarrow{R_i, M_i, M_o, S, O, \Delta} E$$

$$A_1 \ then \ A_2 \xrightarrow{R_i, M_i, M_o, S, O, \Delta} A_1 \ then \ E \qquad if \ closed(A_1, \Delta_1) \land$$
$$A_2 \xrightarrow{R_i, M_i, M_o, S, O, \Delta \cup \Delta_1} E$$

$$C \leftarrow M \Leftarrow A \xrightarrow{R_i, M_i, M_i - \{m(R_i, M \Leftarrow A)\}, S, \emptyset, \Delta} c(M \Leftarrow A) \ if \ m(R_i, M \Leftarrow A) \in M_i \land$$
$$satisfy(C)$$

$$M \Rightarrow A \leftarrow C \xrightarrow{R_i, M_i, M_o, S, \{m(R_i, M \Rightarrow A)\}, \Delta} c(M \Rightarrow A) \quad if \ satisfied(S, C) \land$$
$$coalesce(\Delta, A)$$

$$null \leftarrow C \xrightarrow{R_i, M_i, M_o, S, \emptyset, \Delta} c(null) \qquad if \ satisfied(S, C)$$

$$a(R, I) \leftarrow C \xrightarrow{R_i, M_i, M_o, S, \emptyset, \Delta} a(R, I) :: B \qquad if \ clause(S, a(R, I) :: B) \land$$
$$satisfied(S, C)$$

$$\begin{aligned}
&closed(c(M \Leftarrow A), \{A\}) \\
&closed(c(M \Rightarrow A), \{A\}) \\
&closed(A \ then \ B, \Delta_1 \cup \Delta_2) \leftarrow closed(A, \Delta_1) \land closed(B, \Delta_2) \\
&closed(X :: D, \Delta) \leftarrow closed(D, \Delta)
\end{aligned} \qquad (14)$$

$$coalesce(\Delta, a(R, X)) \leftarrow \neg var(X) \lor$$
$$(var(X) \land X = sel(\{(X', P_p, P_n, N) | coalition(\Delta, a(R, X'), P_p, P_n, N)\})) \qquad (15)$$

$$coalition(\Delta, A, P_p, P_n, N) \leftarrow P_p = \frac{card(\{E | (event(A, E) \land success(E) \land co(\Delta, E)\}))}{card(\{E | co(\Delta, E)\})} \qquad (16)$$

$$P_n = \frac{card(\{E | (event(A, E) \land failure(E) \land co(\Delta, E)\}))}{card(\{E | co(\Delta, E)\})}$$
$$N = card(\{E | co(\Delta, E)\})$$

$$co(\Delta, E) \leftarrow (\exists A.A \in \Delta \land event(A, E) \land \neg(\exists R, X, X'.a(R, X) \in \Delta \land event(a(R, X'), E) \land X \neq X') \qquad (17)$$

Where: Δ is a set of the peers $(a(R, X))$ appearing in the clause along the path of rewrites (above).

$var(X)$ is true when X is a variable.

$card(S)$ returns the cardinality of set S.

$sel(S_x)$ returns a peer identifier, X, from an element, (X, P_p, P_n), of S_x selected according to the values of P_p and P_n

See Figure 7 for definitions of other terms.

Fig. 8. A basic event-based matchmaker

Figure 7). The set, Δ, is needed in the seventh rewrite rule which deals with sending a message out from the peer. At this point the identity of the peer, A, may not be known so the predicate $coalesce(\Delta, A)$ ensures that an identifier is assigned. Expression 15 attempts to find an identifier for the peer if its identifier, X, is a variable. It does this by selecting the best option from a set of candidates, each of the form (X', P_p, P_n, N) where: X' is an identifier; P_p is the proportion of previous interactions in which X' was part of a successful coalition with at least some of the peers in Δ; P_n is the proportion of such interactions where the coalition was unsuccessful; and N is the total number of appropriate coalitions. The selection function, sel, could take different forms depending on the application but typically would attempt to maximise P_p while minimising P_n. Expression 16 generates values for P_p and P_n for each appropriate instance, A of a peer, based on cached records of interaction events. An interaction event is recorded in the form $event(a(R, X), E)$ where $a(R, X)$ records the role and identifier of the peer and E is a unique identifier for the event. For instance, in our earlier shopping example there would be a unique event identifier for the sequence of roles undertaken and an $event$ definition for each role associated with that event (so if the event identifier was $e243$ then there would be an $event(a(purchaser([canonS500, olympusE300, canonEOS1]), b), e243)$ and so on).

To demonstrate matchmaking in this event-driven style, suppose that our automated camera purchaser is following the interaction model given in expression 12 and has already performed the part of the interaction needed to buy the first camera in our list (the $canonS500$) from a peer, s, using the interaction model of Figure 2. The state of the interaction (described as a partially expanded model in the LCC style) is given in expression 18 below.

$$
\begin{aligned}
&a(purchaser([canonS500, olympusE300, canonEOS1]), b) :: \\
&\quad a(buy\_item(canonS500), b) :: \\
&\quad\quad a(buyer(canonS500), b) \; then \\
&\quad\quad\quad ask(canonS500) \;\Rightarrow\; a(shopkeeper, s) \; then \\
&\quad\quad\quad price(canonS500, 249) \;\Leftarrow\; a(shopkeeper, s) \; then \\
&\quad\quad\quad buy(canonS500, 249) \;\Rightarrow\; a(shopkeeper, s) \; then \\
&\quad\quad\quad sold(canonS500, 249) \;\Leftarrow\; a(shopkeeper, s) \\
&\quad a(purchaser([olympusE300, canonEOS1]), b) :: \\
&\quad\quad a(buy\_item(olympusE300), b) :: \\
&\quad\quad\quad (a(buyer(olympusE300), b) \; or \; a(seeker(olympusE300), b) \; or \\
&\quad\quad\quad\quad a(customer(olympusE300), b))
\end{aligned}
$$

$$(18)$$

The choice at the end of expression 18 means that our *purchaser* peer now has to choose whether to become a *buyer* peer again or to be a *seeker* or a *customer*. This will require it to choose a model from either Figure 2, Figure 3 or Figure 4. This, inturn, will require it to identify either a *shopkeeper*, an *auctioneer* or a *manufacturer* (respectively) with which to interact when following its chosen interaction model. Suppose that our purchaser has access to the following results of previous interactions:

$$event(a(buyer(olympus E300), b), e1) \quad event(a(shopkeeper, s1), e1) \qquad failure(e1)$$
$$event(a(seeker(olympus E300), b), e2) \quad event(a(auctioneer, a1), e2) \qquad success(e2)$$
$$event(a(customer(olympus E300), b), e3) \quad event(a(manufacturer, m1), e3) \; success(e3)$$
$$event(a(seeker(olympus E300), b), e4) \quad event(a(auctioneer, a1), e4) \qquad failure(e4)$$
$$event(a(buyer(canon EOS1), b), e5) \quad event(a(shopkeeper, s1), e5) \qquad success(e5)$$

$$(19)$$

Applying the method described in Figure 8, the contextual set of peers, Δ, from expression 18 is:

$$\{ \; a(purchaser([canon S500, olympus E300, canon EOS1]), b)$$
$$a(buy\_item(canon S500), b)$$
$$a(buyer(canon S500), b)$$
$$a(shopkeeper, s)$$
$$a(purchaser([olympus E300, canon EOS1]), b)$$
$$a(buy\_item(olympus E300), b) \quad \}$$

and we can generate the following instances for $coalition(\Delta, A, P_p, P_n)$ via expression 16 of Figure 8:

$$For \;\; a(buyer(olympus E300), b) \qquad\quad : coalition(\Delta, a(shopkeeper, s1), e1, 0, 1, 1)$$
$$For \;\; a(seeker(olympus E300), b) \qquad\quad : coalition(\Delta, a(auctioneer, a1), 0.5, 0.5, 2)$$
$$For \;\; a(customer(olympus E300), b) : coalition(\Delta, a(manufacturer, m1), 1, 0, 1)$$

Our selection function (*sel* in expression 15 of Figure 8) must then choose which of the three options above is more likely to give a successful outcome. This is not clear cut because sample sizes vary as well as the proportion of successes to failures. It is possible, however, to rate the auctioneer or the manufacturer as the most likely to succeed, given the earlier events.

5   Maintaining an Interaction Context

When many peers interact we must make sure that the knowledge they share is consistent to the extent necessary for reliable interaction. This does not of course, require consistency across the totality of knowledge possessed by the peers - only the knowledge germane to the interaction. The general problem of attaining consistent common knowledge is known to be intractable (see for example [22]) so the engineering aim is to avoid, reduce or tolerate this theoretical worst case. The interaction model used in LCC identifies the points of contact between peers' knowledge and the interaction - these are the constraints associated with messages and roles. The knowledge to which these connections are made can be from two sources:

Devolved to the appropriate peers: so that the choice of which axioms and inference procedures are used to satisfy a constraint is an issue that is private and internal to the peer concerned. In this case there is no way of knowing whether one peer's constraint solving knowledge is consistent with another peer's internal knowledge.

Retained with the LCC interaction model: so the axioms used to satisfy a constraint
are visible at the same level as the interaction model and the inference procedures
may also be standardised and retained with the model. In this case we can identify
the knowledge relevant to the interaction and, if an appropriate consistency checking mechanism is available, we can apply it as we would to a traditional knowledge
base.

In practise it is necessary to balance retention of knowledge with an interaction
model (and the control that permits) against devolution to private peers (with the autonomy that allows). The way an engineer decides on this balance is, as usual, by
studying the domain of application. Where it is essential that constraints are satisfied
in a standardised way then axioms and inference methods are retained with the interaction model. Where it is essential that peers autonomously satisfy constraints then they
must be given that responsibility. What makes this more subtle than traditional software engineering is that axioms retained by the interaction model can be used to supply
knowledge hitherto unavailable to the (otherwise autonomous) peers using the model.
The remainder of this section demonstrates this using a standard example.

A classic logical puzzle involves a standardised form of interaction between a group
of people, each of which has an attribute which cannot be determined except by observ-
ing the behaviour of the others. This puzzle appears in different variants (including the
"cheating husbands", "cheating wives" and "wise men" puzzles) but here we use the
"muddy children" variant attributed to [6]. A paraphrased version of the puzzle is this:

> A number of daughters have got mud on their foreheads. No child can see the
> mud on her own forehead but each can see all the others' foreheads. Their
> father tells them that at least one of them is muddy. He then asks them, repeat-
> edly, whether any of them (without conversing) can prove that she is muddy.
> Assuming these children are clever logical reasoners, what happens?

The answer is that the children who are muddy will be able to prove this is so after the
father has repeated the question $n - 1$ times, where n is the number of muddy children.
The proof of this is inductive: for $n = 1$ the muddy child sees everyone else is clean so
knows she is muddy; for $n = 2$ the first time the question is asked the muddy children
can see $(n - 1) = 1$ other muddy child and, from the fact that no other child answered
"yes", knows that she also must be muddy so answers "yes" next time; similarly for
each $n > 2$.

The important features of this example for our purposes are that: the interaction
is essential to the peers acquiring the appropriate knowledge; the interaction must be
synchronised (otherwise the inductive proof doesn't hold); and, once the "trick" of in-
duction on the number of cycles of questioning is understood, the mechanism allowing
each peer to decide depends only on remembering the number of cycles. Figure 9 gives
a LCC interaction model that allows a group of peers to solve the muddy children puzzle. Notice that it is not our intention to unravel the semantics of such problems (as has
been done in, for example, [22]). Our aim is to show how to solve this problem simply.

To demonstrate how the interaction model of Figure 9 works, suppose that we have
two peers, $a1$ and $a2$, and that $a1$ knows $is\_muddy(a2)$ while $a2$ knows $is\_muddy(a1)$.
This knowledge is private to the peers concerned, and neither peer knows that it is itself

$a(coordinator(Cs, N), X) ::$
$$\left(\begin{array}{l} a(collector(Cs, Cs, N, Cs'), X) \leftarrow not(all\_known(Cs')) \; then \\ a(coordinator(Cs', N1), X) \leftarrow N1 = N + 1 \end{array} \right) \; or \quad (20)$$
$$null \leftarrow all\_known(Cs')$$

$a(collector(Cs, Rs, N, Cs'), X) ::$
$$\left(\begin{array}{l} poll(N, Rs) \Rightarrow a(child, Y) \leftarrow select(k(Y, Rp), Cs, Cr) \; then \\ Cs' = \{k(Y, R)\} \cup Cr' \leftarrow reply(R) \Leftarrow a(child, Y) \; then \\ a(collector(Cr, Rs, N, Cr'), X) \end{array} \right) \; or \quad (21)$$
$$null \leftarrow Cs = [] \; and \; Cs' = []$$

$a(child, X) ::$
$$poll(N, Rs) \Leftarrow a(collector(Cs, Rs, N, Cs'), X) \; then$$
$$\left(\begin{array}{l} reply(muddy) \Rightarrow a(collector(Cs, Rs, N, Cs'), X) \leftarrow muddy(N, Rs) \; or \\ reply(clean) \Rightarrow a(collector(Cs, Rs, N, Cs'), X) \leftarrow clean(N, Rs) \; or \\ reply(unknown) \Rightarrow a(collector(Cs, Rs, N, Cs'), X) \leftarrow unknown(N, Rs) \end{array} \right) \; then$$
$$a(child, X)$$
$$(22)$$

$$muddy(N, Rs) \leftarrow \begin{array}{l} Nk = card(\{Y | k(Y, muddy) \in Rs\}) \; and \\ Nm = card(\{Y' | is\_muddy(Y')\}) \; and \\ Nk = 0 \; and \; N \geq Nm \end{array} \quad (23)$$

$$clean(N, Rs) \leftarrow \begin{array}{l} Nk = card(\{Y | k(Y, muddy) \in Rs\}) \; and \\ Nk > 0 \end{array} \quad (24)$$

$$unknown(N, Rs) \leftarrow \begin{array}{l} Nk = card(\{Y | k(Y, muddy) \in Rs\}) \; and \\ Nm = card(\{Y' | is\_muddy(Y')\}) \; and \\ Nk = 0 \; and \; N < Nm \end{array} \quad (25)$$

Where: $all\_known(Cs)$ denotes that each element of Cs is known to be either muddy or clean.
$muddy(N, Rs)$ is true if the peer is muddy at cycle N given the previous response set Rs.
$clean(N, Rs)$ is true if the peer is clean.
$unknown(N, Rs)$ is true if the peer can't yet decide whether it is muddy or clean.
$is\_muddy(Y)$ denotes that the peer knows (another) peer Y to be muddy.
$card(S)$ returns the cardinality of set S.

Fig. 9. A muddy children LCC model

muddy. In order to work out whether they are muddy, one of the peers (let us choose $a1$) must assume the role of $coordinator(Cs, N)$ where $Cs = \{k(a1, unknown), k(a2, unknown)\}$ is the set of peers with their initial knowledge about their muddiness and $N = 1$ is the cycle number. This is analogous to the role of the father in the original puzzle and it could be performed by a third peer rather than by $a1$ or $a2$ but here we choose to let $a1$ be both coordinator and child. The coordinator role is recursive over N; on each cycle performing a round of polling for each child to find out its current answer. Each child has the obligation to reply when polled - its answer being either *muddy, clean* or *unknown* depending on whether it can satisfy axiom 23, 24 or 25. These are examples of axioms that it makes sense to retain with the interaction model because it is critical that all peers make this calculation in the same way and it is not guaranteed that each peer would possess the appropriate knowledge. By contrast the knowledge about which peers are known by a given peer to be muddy is assumed to be private to that peer so that is not retained with the model.

Interaction models do not solve the general problem of attaining common knowledge in a distributed system - no method appears capable of that. They do, however, give a way of identifying knowledge that must be shared and provide a basis for partitioning shared and private constraint solving knowledge.

6   Making Interactions Less Brittle

One of the ways in which our method differs from traditional programming is that execution of the clauses of an interaction model can happen across different machines, therefore satisfaction of constraints associated with a particular role in an interaction model is done in ignorance of constraints imposed by other peers in the interaction. Often a peer has a choice about how it satisfies constraints (binding interaction model variables in so doing) and if it makes the wrong choice relative to other peers' constraints then the interaction as a whole may fail. In this sense, interaction models can be more "brittle" than conventional programs. Since the messages physically sent during the interaction cannot be un-sent, the only way of bringing the interaction back on track is to reason about the interaction model. It is essential, however, that such reasoning does not invalidate the interaction model - in other words it should only make a given interaction model more likely to succeed as intended, not change the declarative meaning of the model. We shall consider two ways of doing this:

Allowing peers to set constraint ranges, rather than specific values, for variables shared in the interaction model - thus avoiding unnecessary early commitment.

Task/domain-specific adaptation of the interaction model during the process of interaction, making limited forms of "model patching" possible.

Before discussing these we study in a little more depth the programming issues raised by committed choice in LCC interaction models. Our interaction models allow choice via the *or* operators in model clauses (see Figure 1). For example, the clause:

$$a(r1, X) :: (m1 \Rightarrow a(r2, Y) \text{ then } D_1) \text{ or } (m2 \Rightarrow a(r2, Y)) \text{ then } D_2)$$

where D_1 and D_2 are some further definitions necessary for X to complete role $r1$, allows X to choose the state sequence commencing $m1 \Rightarrow a(r2, Y)$ or the sequence commencing $m2 \Rightarrow a(r2, Y)$. In some conventional logic programming languages (such as Prolog) this choice is not a commitment - if one of the options chosen fails then we may still complete the role by backtracking; rolling back the state of the clause to where it was before the choice was made, and choosing the alternative option. This

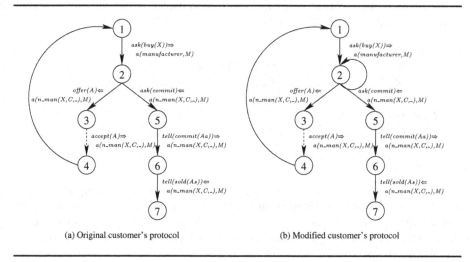

(a) Original customer's protocol   (b) Modified customer's protocol

Fig. 10. Customer's dialogue tree of the manufacturer scenario

is not possible in our situation because interacting peers may have (privately) altered their internal state, so for example $a(r2, Y)$ might have made some private internal commitments in response to message $m1$ that are not observable by $a(r1, X)$, which means that simply rolling back the state of the interaction model does not necessarily bring the interaction back to an identical state. Our choices are therefore committed choices. To make this issue concrete, let us return to our running example.

We can visualise a the possible sequences of messages sent or received when performing a role in interaction as a graph, an example of which is given in Figure 10 for the manufacturer's interaction model from Figure 4. Two peers are involved in this model, the customer and manufacturer. The dialogue tree of the customer is given in Figure 10(a). Nodes in the figure are states in the interaction (from the perspective of the customer). Solid arcs in the figure represent successful steps in the execution of the interaction model while the dashed arrow (from node 3) represents a possible failure that could occur if the customer could not accept the offer A proposed by the manufacturer. If that happens then our basic interaction model expansion mechanism, defined in Figure 7, will be unable to complete the interaction because from node 2 the only state other than the failing state (3) is node 5 but to reach it requires a different message to be received than the one actually received.

One proposed solution is to allow the peers to backtrack and try different offers. The interaction model should then be modified to allow the action of sending/receiving different offers possible. On the customer's side, the modification is represented in Figure 10(b). When the customer's model is at state 2, the customer can either receive an offer, or receive a message asking it to commit, or it can remain at state 2. This third and final option is added so that if any constraint failure occurs anywhere along the first two paths and the customer backtracks, then it will be able to receive another *offer* message.

The following is the modified customer's interaction model. A parallel modification would be required to the manufacturer's model.

$$a(n\_cus(X, M, Aa), C) ::$$
$$\begin{pmatrix} offer(A) &\Leftarrow a(n\_man(X, C, \_), M) \ then \\ accept(A) &\Rightarrow a(n\_man(X, C, \_), M) \leftarrow acceptable(A) \ then \\ a(n\_cus(X, M, [att(A)|Aa]), C) & \end{pmatrix} \ or$$
$$\begin{pmatrix} ask(commit) &\Leftarrow a(n\_man(X, C, \_), M) \ then \\ tell(commit(Aa)) &\Rightarrow a(n\_man(X, C, \_), M) \leftarrow acceptable(A) \ then \\ tell(sold(Aa)) &\Leftarrow a(n\_man(X, C, \_), M) & \end{pmatrix} \ or$$
$$a(n\_cus(X, M, Aa), C)$$

This is a pragmatic fix but it makes the interaction model more complex (hence harder to read and design) and it also changes the semantics of the model, since we have introduced an additional recursive option for the n_cus role that requires no interaction. Our only reason for changing the model in this way is to re-gain flexibility lost because of committed choice.

One way of regaining a form of backtracking for our interaction models is described in [41]. This involves an extension to the labelling used to denote closure of explored parts of the model, so that we can label parts of it as failed, then extending the model expansion rules (from Figure 7) to force (via failure messages) re-opening of previously closed sequences in the model when failure is detected. Although this gives a partial solution to the problem of backtracking, it does not address the problem that (invisible from the level of the interaction model) individual peers may have made internal commitments that prevent them "rolling back" their state to be consistent with backtracking at the model level. For this reason, it is interesting to explore (in Sections 6.1 and 6.2) ways of adding flexibility to interaction models without backtracking.

6.1   Brittleness through Variable Binding: Constraint Relaxation

The LCC language ensures coherent interaction between peers by imposing constraints relating to the message they send and receive in their chosen roles. The clauses of an interaction model are arranged so that, although the constraints on each role are independent of others, the ensemble of clauses operates to give the desired overall behaviour. For instance, the manufacturer interaction model of Figure 4 places two constraints on each attribute A in the set of attributes As: the first ($available(A)$) is a condition on the peer in the role of negotiating manufacturer sending the message $offer(A)$, and second ($acceptable(A)$) is a condition on the peer in the role of negotiating customer sending the message $accept(A)$ in reply. By (separately) satisfying $available(A)$ and $acceptable(A)$ the peers mutually constrain the attribute A.

In [23] we described how the basic clause expansion mechanism of LCC has been extended to preserve the ranges of finite-domain on variables. This allows peers to restrict rather than simply instantiate these constraints when interacting, thus allowing a less rigid interaction. For instance, applying this to our initial example of mutual finite-domain constraints in Figure 4, if the range of values permitted by the manufacturer for A by $available(A)$ is $\{32, 64, 128\}$, while the range of values permitted by the customer for A by $acceptable(A)$ is $\{greater\ than\ 32\}$, then were we to use finite-domain constraint solver, a constraint space of $\{64, 128\}$ is obtained – a range that would be attached to the variable returned in the $accept(X)$ message.

An important aspect of the interaction model between the manufacturer and customer roles defined in Figure 4 is the message passing that communicates the attributes of the digital camera to be purchased. This dialogue can continue only as long as there exists a match between the finite-domain ranges of attribute values offered by the negotiating manufacturer with those required by the negotiating customer. To illustrate this point, consider the following example.

Assuming that the customer agreed to accept a memory size of 64, then the following statements describe the knowledge and constraints private to the manufacturer and customer respectively, concerning the price of the digital camera to be negotiated:

Manufacturer: $available(price(P)) \leftarrow P = 244$
Customer: $acceptable(price(P)) \leftarrow P \leq 250$

Upon negotiating these mutual constraints via the defined interaction model, the value for price that meets the manufacturer's offer, and also the customer's requirement will be in the range: $244 \leq price(P) \leq 250$. Depending on the peer's strategies (e.g. choosing the maximum value within the agreed range, *etc.*), the final price can be assigned to a value within this agreed range. To support this we need a means of propagating constraint ranges across the interaction.

Similarly to our construction of expressions 1 and 2, for ontology mapping in Section 3, it is straightforward to propagate range constraints through our state transitions by (in expression 26) identifying the initial set, V, of variables (each with its range constraint) in the initial state, S and then threading this set of variable ranges through the state transition sequence (expression 27). Prior to each transition step the relation $apply\_ranges(V_i, S_p, S'_p)$ applies the range constraints, V_i, to the corresponding variables in the peer state S_p to give the range restricted state S'_p. After each transition step the relation $update\_ranges(S''_p, V_i, V_n)$ identifies each variable in V_i that has been restricted in the new peer state S''_p and adds the new range restrictions to produce V_n.

$$\sigma(p, G_p) \quad \leftrightarrow \quad \Omega \overset{p}{\ni} \langle S, V \rangle \wedge i(S, \phi, V, S_f, V_f) \wedge k_p(S_f) \vdash G_p \qquad (26)$$

$$i(S, M_i, V_i, S_f, V_f) \quad \leftrightarrow \quad S = S_f \vee \begin{pmatrix} S \overset{s}{\supseteq} S_p \wedge \\ apply\_ranges(V_i, S_p, S'_p) \wedge \\ S'_p \xrightarrow{M'_i, S, M_n} S''_p \wedge \\ update\_ranges(S''_p, V_i, V_n) \wedge \\ S''_p \overset{s}{\cup} S = S' \wedge \\ i(S', M_n, V_n, S_f, V_f) \end{pmatrix} \qquad (27)$$

Simply propagating variable range constraints defends against only limited forms of brittleness, however. Suppose that, instead of allowing memory size to be in the range

$\{64, 128\}$, the customer required a memory size of 128, then the following, consequent local constraints on price would break the interaction model:

Manufacturer: $available(price(P)) \leftarrow P = 308$
Customer: $acceptable(price(P)) \leftarrow P \leq 250$

In this situation, no match is found between the customer's expected price and the one that can be offered by the manufacturer. Rather than terminating the dialogue at this stage, we might reduce brittleness of this nature by including a constraint relaxation mechanism that allows manufacturer and customer to negotiate further by relaxing the specified mutual constraint on the value of the attribute. This issue is explored more fully in [24] so we do not expand on the theme here. Constraint relaxation only is possible, however, if the peers participating in the interaction are cognitively and socially flexible to the degree they can identify and fully or partially satisfy the constraints with which they are confronted. Such peers must be able to reason about their constraints and involve other peers in this reasoning process. Interaction models (like those of LCC) provide a framework for individual peers to analyse the constraints pertinent to them and to propagate constraints across the interaction.

6.2 Brittleness through Inflexible Sequencing: Interaction Model Adaptation

When tackling the brittleness problem in Section 6.1 the issue is to ensure that a given, fixed interaction model has more chance of concluding successfully. LCC models also break, however, when their designers have not predicted an interaction that needs to occur. Consider for example the model of Figure 2 in which a buyer asks for some item; receives a price; offers to buy the item; and its sale is confirmed. This is one way of purchasing but we could imagine more complex situations - for instance when the shopkeeper does not have exactly the requested item for sale but offers an alternative product. We could, of course, write a new model to cover this eventuality but this will lead us either to write many models (and still have the task of predicting the right one to use) or to write complex models that are costly to design and hard to test for errors. One way to reduce this problem is to allow limited forms of task/domain-specific synthesis of models, with the synthesised components being used according to need in the interaction.

Automated synthesis of LCC models has many similarities to traditional synthesis of relational/functional programs and to process/plan synthesis. It is not possible here to survey this large and diverse area. Instead we use an important lesson from traditional methods: that the problem of synthesis is greatly simplified when we limit ourselves to specific tasks and domains. This narrowing of focus is natural for interaction models which are devised with a task and domain in mind. To demonstrate this we develop an example based on the interaction model of Figure 2.

In order to synthesise models similar to those of Figure 2 we need to have some specification of our functional requirements. To make synthesis straightforward, we shall describe these in a domain-specific language close to the original model (remember we

$$\longrightarrow t(request(X), A1 \overset{*}{\Rightarrow} A2)$$

$$t(request(X), A1 \overset{*}{\Rightarrow} A2) \longrightarrow t(request(X), A1 \overset{*}{\Rightarrow} A2) \wedge$$
$$\Diamond t(describe(X, D), A1 \overset{*}{\Leftarrow} A2)$$

$$t(request(X), A1 \overset{*}{\Rightarrow} A2) \longrightarrow t(request(X), A1 \overset{*}{\Rightarrow} A2) \wedge$$
$$\Diamond t(alternative(X, Y), A1 \overset{*}{\Leftarrow} A2)$$

$$t(alternative(X, Y), A1 \overset{*}{\Leftarrow} A2) \longrightarrow t(alternative(X, Y), A1 \overset{*}{\Leftarrow} A2) \wedge$$
$$\Diamond t(request(Y), A1 \overset{*}{\Rightarrow} A2)$$

$$t(describe(X, D), A1 \overset{*}{\Leftarrow} A2) \longrightarrow t(describe(X, D), A1 \overset{*}{\Leftarrow} A2) \wedge$$
$$\Diamond t(propose(X, P), A1 \overset{*}{\Rightarrow} A2)$$

$$t(describe(X, D), A1 \overset{*}{\Leftarrow} A2) \longrightarrow t(describe(X, D), A1 \overset{*}{\Leftarrow} A2) \wedge$$
$$\Diamond t(describe(X, D'), A1 \overset{*}{\Leftarrow} A2)$$

$$t(propose(X, P), A1 \overset{*}{\Rightarrow} A2) \longrightarrow t(propose(X, P), A1 \overset{*}{\Rightarrow} A2) \wedge$$
$$\Diamond t(confirm(X, P), A1 \overset{*}{\Leftarrow} A2)$$

$$t(T, A1 \overset{*}{\Rightarrow} A2) \longrightarrow (m(A1, T \Rightarrow A2) \wedge m(A2, T \Leftarrow A1))$$
$$t(T, A1 \overset{*}{\Leftarrow} A2) \longrightarrow (m(A1, T \Leftarrow A2 \wedge m(A2, T \Rightarrow A1))$$

Where: $T1 \longrightarrow T2$ gives a permitted refinement of term $T1$ to term $T2$.

$\Diamond P$ denotes that expression P will be true at some future time.

$t(E, D)$ specifies a message interchange of type E between two peers, where D is either of the form $A1 \overset{*}{\Rightarrow} A2$, denoting a message sent from $A1$ to $A2$ or of the form $A1 \overset{*}{\Leftarrow} A2$, denoting a message sent from $A2$ to $A1$.

Domain terms: $alternative(X, Y)$ when Y is a product offered as a substitute for X.
$confirm(X, P)$ when the transaction is confirmed for X at price P.
$describe(X, D)$ when D describes X.
$propose(X, P)$ when price P is suggested for X.
$request(X)$ when a product of type X is requested.

Fig. 11. Rewrites for synthesis of a restricted, task-specific specification

only want to add some limited flexibility, not solve the general problem of model synthesis). In our language we have chosen to use five domain-specific terms that describe the type of interaction of message passing events in the model (listed at bottom of Figure 11). We then specify how these may be combined, starting from an initial *request* and describing how this can be extended via rewrites that add additional message passing behaviours. The syntax of the rewrites is described in Figure 11 but the intuition is that each rewrite extends some existing segment of message passing specification with an additional behaviour that may occur in future from the existing segment (the normal $\Diamond$ modal operator is used to denote an expression that must be true at some future time). For example, the second rewrite of Figure 11 says that a request from peer $A1$ to peer $A2$ can be extended with a subsequent description of X sent by $A2$ to $A1$. The operators $\overset{*}{\Rightarrow}$ and $\overset{*}{\Leftarrow}$, used to indicate the direction of message passing between peers, can be refined into the LCC message passing operators for individual peers using the two rules in the centre of Figure 11. Using all of these refinement rules, we can synthesise specifications of interaction model behaviour such as:

$$\diamond \left(\begin{array}{l} t(request(X), A1 \stackrel{*}{\Rightarrow} A2) \wedge \\ \left(\begin{array}{l} t(alternative(X,Y), A1 \stackrel{*}{\Leftarrow} A2) \wedge \\ \left(\begin{array}{l} t(request(Y), A1 \stackrel{*}{\Rightarrow} A2) \wedge \\ \left(\begin{array}{l} t(describe(Y,D), A1 \stackrel{*}{\Leftarrow} A2) \wedge \\ \left(\begin{array}{l} t(propose(Y,P), A1 \stackrel{*}{\Rightarrow} A2) \wedge \\ \diamond \left(t(confirm(Y,P), A1 \stackrel{*}{\Leftarrow} A2) \right) \end{array} \right) \end{array} \right) \end{array} \right) \end{array} \right) \end{array} \right) \quad (28)$$

If we then add additional refinements from the general domain terms of Figure 11 to the more specific terms used in the interaction model of Figure 2 as follows:

$$request(X) \longrightarrow ask(X)$$
$$describe(X,D) \longrightarrow price(X,D)$$
$$propose(X,P) \longrightarrow buy(X,P)$$
$$confirm(X,P) \longrightarrow sold(X,P)$$
$$alternative(X,Y) \longrightarrow similar\_product(X,Y)$$

then we can further refine the specification of expression 28 using these and the refinements message passing from Figure 11 to give the specification:

$$\diamond \left(\begin{array}{l} (m(A1, ask(X) \Rightarrow A2) \wedge m(A2, ask(X) \Leftarrow A1)) \wedge \\ \left(\begin{array}{l} (m(A1, similar\_product(X,Y) \Leftarrow A2) \wedge m(A2, similar\_product(X,Y) \Rightarrow A1)) \wedge \\ \left(\begin{array}{l} (m(A1, ask(Y) \Rightarrow A2) \wedge m(A2, ask(Y) \Leftarrow A1)) \wedge \\ \left(\begin{array}{l} (m(A1, price(Y,D) \Leftarrow A2) \wedge m(A2, price(Y,D) \Rightarrow A1)) \wedge \\ \left(\begin{array}{l} (m(A1, buy(Y,P) \Rightarrow A2) \wedge m(A2, buy(Y,P) \Leftarrow A1)) \wedge \\ \diamond ((m(A1, sold(Y,P) \Leftarrow A2) \wedge m(A2, sold(Y,P) \Rightarrow A1))) \end{array} \right) \end{array} \right) \end{array} \right) \end{array} \right) \end{array} \right)$$

$$(29)$$

Since the specification above describes a sequence of pairs of messages exchanged between peers $A1$ and $A2$ it is straightforward to "unzip" this sequence into the send and receive components of each exchange, providing the definition for a LCC model.

$A1 ::$

$$ask(X) \Rightarrow A2 \; then$$
$$similar\_product(X,Y) \Leftarrow A2 \; then$$
$$ask(Y) \Rightarrow A2 \; then$$
$$price(Y,D) \Leftarrow A2 \; then$$
$$buy(Y,P) \Rightarrow A2 \; then$$
$$sold(Y,P) \Leftarrow A2$$

$A2 ::$

$$ask(X) \Leftarrow A1 \; then$$
$$similar\_product(X,Y) \Rightarrow A1 \; then$$
$$ask(Y) \Leftarrow A1 \; then$$
$$price(Y,D) \Rightarrow A1 \; then$$
$$buy(Y,P) \Leftarrow A1 \; then$$
$$sold(Y,P) \Rightarrow A1$$

To arrive at this interaction model we have made numerous assumptions in order to make the task of synthesis easy. We assumed a narrow domain so that we had a small range of specifications to deal with, thus simplifying the task of writing the message passing specification. We assumed that only two peers were involved, simplifying both the specification and its refinement. We assumed a total ordering on the messages interchanged (with no interleaving between interchanges) giving an easy translation from temporal message sequence to interaction model. We would not, of course, expect these assumptions to hold for every domain - they are merely an example of assumptions that can usefully be made to simplify engineering in one specific domain. The hope for practical application of synthesis methods is that domains amenable to similar treatments occur commonly in practise.

7   Satisfying Requirements on the Interaction Process

Interaction models, like traditional programs, are written in order to control computations that satisfy requirements in some domain of application. In our scenario of Section 1.1, for example, Sarah has various interaction requirements: she wants by the end to purchase a camera; before then she wants to know the price of that camera; and ideally she would like the lowest price possible. These are all functional requirements (they can be judged with respect to the computation itself) and we shall confine ourselves to this class of requirements although we recognise that, as for traditional programs, non-functional requirements also are important (for example, Sarah might want to increase her stature in her company by buying cheap cameras but our interaction model says nothing about company politics).

There are many ways in which to express requirements but we shall focus on process-based requirements modelling, in particular business process modelling, because this is (arguably) the most popular style of requirements description for Web services. There are two ways of viewing these sorts of models:

- As specifications for the temporal requirements that must be satisfied by a business process when it is performed. In this case the business process model need not be executable.
- As a structure that may be interpreted in order to perform the business process. In this case there must exist an interpretation mechanism to drive the necessary computations from the business process model.

We demonstrate these views using an elementary process modelling language. Real process modelling languages (for example BPEL4WS [3]) are more complex but similar principles apply to them (see [32] for details of our methods applied to BPEL4WS). In our elementary language we shall assume that only two (given) peers interact to implement a process and that the process is described using a set of actions of the form $activity(I, X, Cp, Cr)$, where I identifies the peer that performs the activity; X is a descriptive name for the activity; Cp is the set of axioms forming a precondition for the activity; and Cr is the set of axioms established as a consequence of the activity. The expression $process(\{I, I'\}, A)$ relates the two peer identifiers I and I' with the set of activities A. Figure 12 gives a process model which one might write for the shopping

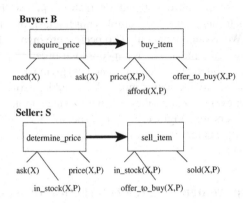

$$process(\{B, S\},$$
$$\{\, activity(B, enquire\_price, \quad \{need(X)\}, \qquad\qquad\qquad\qquad\qquad\qquad \{ask(X)\}),$$
$$activity(B, buy\_item, \quad\quad \{price(X, P), afford(X, P)\}, \qquad\qquad \{offer\_to\_buy(X, P)\}),$$
$$activity(S, determine\_price, \{ask(X), in\_stock(X, P)\}, \qquad\qquad\quad \{price(X, P)\}),$$
$$activity(S, sell\_item, \quad\quad\; \{offer\_to\_buy(X, P), in\_stock(X, P)\}, \{sold(X, P)\}) \,\} \,)$$

(30)

Fig. 12. Elementary process model for the shopping service

model that we introduced in Figure 2 of Section 2.2. The diagram at the top of the figure depicts the process: boxes are activities with preconditions to the left of each box and postconditions to the right. The *process* definition appears underneath, with the meaning of terms contained in activities being understood analogously to the meaning of messages and constraints in the model of Figure 2.

From the process model in Figure 12 we can infer some of the temporal constraints on interaction models enacting the business process - for instance: that for items needed that are in stock we eventually know a price; or that if an item can be afforded then it is sold.

$$need(X) \;\wedge\; shop(S) \;\wedge\; in\_stock(X, P) \;\rightarrow\; \Diamond price(X, P)$$
$$afford(X, P) \;\rightarrow\; \Diamond sold(X, P)$$

Perhaps the most obvious engineering strategy is now to test interaction models to see if they satisfy temporal requirements like those above - raising confidence that those models passing such tests are compliant with the required business process. For example, we could test whether the model of Figure 2 satisfies the temporal requirements given above. This is traditional requirements engineering re-applied so we shall not dwell on it here. There is, however, a more novel approach to ensuring compliance with business process models: write an interpreter for the models in the LCC language itself.

Figure 13 gives an interpreter for our elementary process modelling language. The key feature of this interaction model is that it takes as an argument (in $initiator(P)$ of expression 31) an entire process model, P (in our running example, this is the process

$$a(initiator(P), I) :: interpret(P, K) \Rightarrow A \leftarrow step(I, P, \phi, K)) \ then \qquad (31)$$
$$a(interpreter, I)$$

$$a(interpreter, I) :: step(I, P, K, Kn) \leftarrow interpret(P, K) \Leftarrow a(R, I') \ then \qquad (32)$$
$$interpret(P, Kn) \Rightarrow a(interpreter, I') \leftarrow then$$
$$a(interpreter, I)$$

$$step(I, process(S, A), K, K \cup Cr) \leftarrow I \in S \land activity(I, X, Cp, Cr) \in A \land satisfy(Cp, K)$$
$$(33)$$

Where: P is a process definition of the form $process(S, A)$.

$\quad$ S is a set of peer identifiers.

$\quad$ A is a set of activity definitions of the form $activity(I, X, Cp, Cr)$.

$\quad$ I is a peer identifier.

$\quad$ X is the name of the activity.

$\quad$ Cp is the set of axioms forming a precondition for the activity.

$\quad$ Cr is the set of axioms established as a consequence of the activity.

$\quad$ $step(I, P, K, Kn)$ defines a single step in the execution of a process, P, by peer I given shared knowledge K and generating the extended knowledge set Kn.

$\quad$ ϕ is the empty set of initial shared knowledge.

$\quad$ $satisfy(Cp, K)$ is true when the conjunctive set of propositions, Cp is satisfiable from shared knowledge K.

Fig. 13. A LCC interaction model used to define an interpreter for process models in the notation of Figure 12

definition of Figure 12). The model of expressions 31 to 33 then refers to P in order to determine the messages sent between the two peers involved. This approach is similar to meta-interpretation in logic programming languages, except in this case the meta-interpretation is being done by the interaction model.

By approaching the coordination problem in this way we allow languages used in domains of application to be interpreted directly, rather than having to translate them into equivalent LCC code. This makes it easier to gain confidence (through proof, testing or inspection) that the requirements associated with the domain model are satisfied. It also makes it easier to trace problems in interaction back to the domain model.

8 Building Interaction Models More Reliably

Normally people do not want, having chosen an interaction model, to find at some subsequent point in the interaction that the goal it was supposed to achieve is not attainable, or the way in which it is obtained is inappropriate to the problem in hand. To prevent this we need some explicit description of what the interaction model is (and that we have with LCC) combined with analytical tools that check whether it has the properties we desire. Put more formally, if we have an interaction model (such as the one in Figure 2) we wish to know whether a given form of enactment (such as the one defined in expressions 1 and 2) will generate at least one sequence (for example in the form shown in expression 3) such that each desired property holds and no sequence in which an undesirable property holds.

There are many different ways in which the correct outcome may fail to be reached. For convenience, we will overlook the possibility of the participants failing to obey the interaction model. We will also ignore issues of fault-tolerance, such as the failure of participants, network failures, or lost messages. These issues are outside the scope of the paper. Instead, we will focus on the case where the interaction model is correctly followed, but the desired outcome is not reached. In this case, the problem lies in specification of the interaction model, and not in the implementation of the external peers.

The design of an interaction model to solve a particular goal is a non-trivial task. The key difficulty lies in the nature of the peers being coordinated. The process of coordinating the external peers requires us to specify a complex concurrent system, involving the interactions of asynchronous entities. Concurrency introduces *non-determinism* into the system which gives rise to a large number of potential problems, such as synchronisation, fairness, and deadlocks. It is difficult, even for an experienced designer, to obtain a good intuition for the behaviour of a concurrent interaction model, primarily due to the large number of possible interleavings which can occur. Traditional debugging and simulation techniques cannot readily explore all of the possible behaviours of such systems, and therefore significant problems can remain undiscovered.

The prediction of the outcome in the presence of concurrency is typically accomplished by the application of formal verification techniques to the specification. In particular, the use of formal *model-checking* techniques [12] appears to hold a great deal of promise in the verification of interaction models. The appeal of model-checking is that it is an automated process, unlike theorem proving, though most model-checking is limited to finite-state systems. A model checker normally performs an exhaustive search of the state space of a system to determine if a particular property holds. Given sufficient resources, the procedure will always terminate with a yes/no answer. Model checking has been applied with considerable success in the verification of concurrent hardware systems, and it is increasingly being used as a tool for verifying concurrent software systems, including multi-agent systems [7, 9, 60].

It should be noted that the issues of outcome prediction are not simply an artifact of the use of complex interaction specifications, such as LCC models. Rather, the source of these problems is the need to coordinate complex asynchronous entities. In particular, the issues that we have highlighted occur even when our interactions are specified by simple linear plans. With asynchronous processes, the linear plans will be evaluated concurrently, and the individual plan actions will be interleaved. We will now sketch how the problems which arise in such concurrent systems can be detected by the use of model checking techniques.

We have used the SPIN model checker [25] as the basis for our verification. This model checker has been in development for many years and includes a large number of techniques for improving the efficiency of the model checking, e.g. partial-order reduction, and state-space compression. SPIN accepts design specifications in its own language PROMELA (PROcess MEta-LAnguage), and verifies correctness claims specified as Linear Temporal Logic (LTL) formula.

To perform model checking on the specification, we require an encoding of the specification into a form suitable for model checking. In [54] we define an encoding of the MAP agent interaction language, which is similar to LCC, into PROMELA. A similar technique has been defined for the AgentSpeak language [9]. In AgentSpeak, coordination is specified using the Belief-Desire-Intention (BDI) model. To illustrate the encoding process here, we sketch a translation from LCC into a state-based model in Figure 14. This figure illustrates the encoding of the main language constructs of LCC. The *e* label signifies an empty state, and *!* denotes logical negation. The encoding process is applied recursively to an LCC model. The outcome will be a state graph for each role in the model.

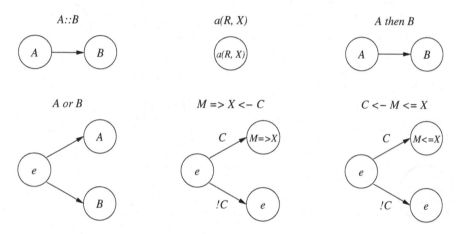

Fig. 14. State Space Encoding

The key feature of the encoding process for LCC is the treatment of the constraints. We make the observation that the purpose of a constraint is to impose a true/false decision on a model, and the purpose of the model checking process is to detect errors in the model and not in the constraints. Thus, based on these observations we can replace a constraint with a pair of states, one of which signifies that the constraint is true, and the other false. The exhaustive nature of the model checking process will mean that all possible behaviours of the interaction model will be explored. In other words, the model checker will explore all consequences for the model where the constraint was true, and all consequences where the constraint was false. Thus, we do not need to evaluate the actual constraints during the model checking process.

To illustrate the model checking process, we present a state-space encoding of the shop model in Figure 15. We have removed the redundant *e* states from the graphs, and we have abbreviated *shopkeeper* to *shop*. The state space defines all possible behaviours of the model. In this case, the model is linear, and we can examine the state space by hand. However, models that contain iteration and choice will rapidly expand the state space, and formal model checking is required. We can clearly see that the model leaves some behaviours undefined, indicated by the remaining *e* states in the graphs. These states occur as the shop model does not define what should be done when a constraint

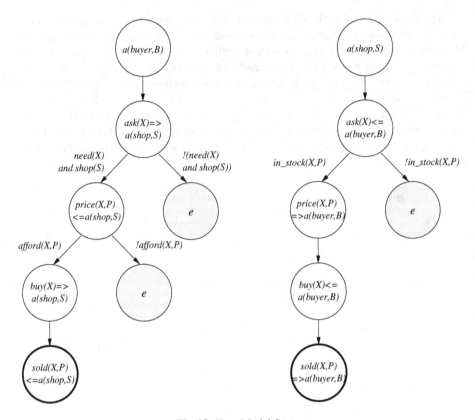

Fig. 15. Shop Model States

cannot be satisfied. For example, if the buyer cannot afford the item, the model will terminate prematurely. In many cases, we will want to avoid this possibility, and so we would amend the model with additional behaviours and recheck the state space.

Our initial model checking experiments with LCC have focused principally on the *termination* of the interaction models. This is an important consideration in the design of models, as we do not normally want to define models that cannot conclude. Non-termination can occur as a result of many different issues such as deadlocks, livelocks, infinite recursion, and message synchronisation errors. Furthermore, we may also wish to ensure that models do not simply terminate due to failure within the model, as in our shop example. The termination condition is the most straightforward to verify by model checking. Given that progress is a requirement in almost every concurrent system, the SPIN model checker automatically verifies this property by default. The termination condition states that every process eventually reaches a valid end state. This can be expressed as the following LTL formula, where end1 is the end state for the first process, and end2 is the end state for the second process, etc: $\Box(\Diamond(\text{end1} \land \text{end2} \land \text{end3} \land \cdots))$. For our shop example, we may define the property $\Box(\Diamond(sold(X, P)))$ to ensure that the item is always sold to the buyer.

One of the main pragmatic issues associated with model checking is producing a state space that is sufficiently small to be checking with the available resources. Hence, it frequently is necessary to use abstraction techniques, such as we have done for the constraints, and to make simplifying assumptions. Other researchers have also considered this problem. For example, [10] proposes a program-slicing technique to improve the efficiency of the model checking process. Model checking is also restricted to finite models, and therefore we must ensure that our models are bounded. Nonetheless, our initial experiments with this approach have proved promising [56].

The encoding which we have outlined here is designed to perform *automatic* checking of LCC models. This makes the system suitable for use by non-experts who do not need to understand the model checking process. However, our approach places restrictions on the kinds of properties of the models that we can check. In particular, we cannot automatically verify properties which are specific to the domain of the model. For example, to verify that the highest bidder always wins in an auction model. Our current research is aimed at extending the range of properties that can be checked with model checking. For this, we need to retain the constraints in the model, and we must define additional formulae over these constraints. This should result in a greater ability to predict the outcome of our LCC models.

9   Implementation: The Current OpenKnowledge System

The ideas described in Sections 3 to 8 are part of the foundation for the OpenKnowledge peer to peer knowledge sharing system, for which a prototype has been built. By building the OpenKnowledge system[14, 50], we aim to demonstrate that sharing interaction models at very low cost to consumers and suppliers is possible. The novelty of the OK system lies in (1) the interaction centric approach, where interactions are published and efficiently stored in a P2P network, (2) decoupling interactions and roles from the services that execute these roles, and (3) a distributed way of finding coordinators that coordinate IMs. Within the design of the system we try to address the (unavoidable) tasks of ontology mapping, query routing, reputation management, dynamic peer recruitment etc. This system is completely distributed using P2P technology (not discussed here).

Each peer that participates in the OK system must run a piece of code that we call the *OpenKnowledge Kernel* [13] enabling the basic functionality of finding these interaction models and the code or peers to run the services implementing the roles in the IMs. We call these services *OK components (OKCs)*. The IMs together with the OKCs and/or the peers running the OKCs are efficiently stored and retrieved in a P2P network which we call the *OK Discovery service*. Additionally, due to the fact that the tasks are formally described, the OK system offers the functionality to select a peer to coordinate a task by executing the IM, selecting peers running the desired OKCs to fulfill a role and recruiting alternative peers in case of failure. The users of the OK-system can *publish* IMs, write interfaces to services, and subscribe these interfaces to play roles in the IMs. The system helps these users by providing tools to ease re-use of existing IMs or by helping connect two services via mappings in case the output of one does not match the input of the other. A reputation management mechanism is under development to help the user in selecting IMs, OKCs, Coordinators and peers running OKCs.

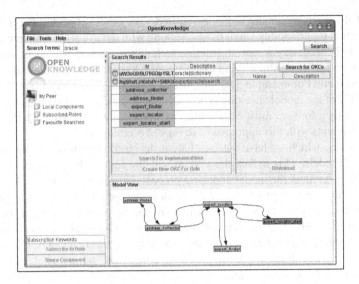

Fig. 16. Part of the Open Knowledge GUI

The first prototype provides the basic user interface shown in Figure 16) with the following functionality: (1) using the GUI, a user can wrap a piece of JAVA code into an OKC, (2) a user can share it by publishing it to the (distributed) discovery service, (3) a user search for interaction models by typing keywords (note that IMs are currently annotated by keywords), (4) a user can decide to subscribe to an OKC, meaning that it listens to function calls implementing the role for the linked IM, and finally (5) the discovery service selects a Coordinator to execute the IM when all roles are initiated by peers (the coordinator first will ask all peers with whom they want to play the interaction, after that it selects the optimal solution and if possible starts the interaction). New peers can subscribe for a running interaction, and in case of failing peers, they may be selected to take over. Figure 17 summarizes the architecture of the OK system.

To relate the possibilities of the system to sections in the system, we again use the scenario from Section 1.1.

– **Ontology matching**
   Her ontology for describing a camera purchase had to be matched to those of available services which is described in Section 3.
   The matching service implements this requirement by (1) mapping via ontologies terms from her query with the (currently term) descriptions of the interaction models and (2) mapping the capabilities of the peers with the role descriptions in the IMs.

– **Recommendations**
   Her recommendation service had to know which services might best be able to interact and enable them to do so (Sections 2 and 4).
   For this we implemented the Discovery Service which efficiently stores OKCs, IMs, peer subscriptions to OKCs and Coordinators. Together with the Mapping

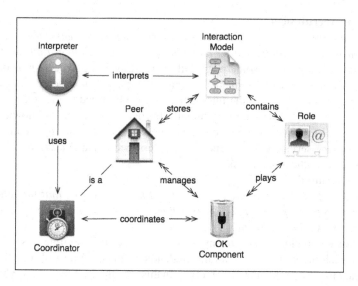

Fig. 17. Open Knowledge Architecture

Service and the Reputation strategies, the results can be ranked in order to facilitate the recommendation service. The peers that want to perform some tasks, such as buying a book or selling them, search for published interaction models for the task, and then advertise their intention of interpreting one of its roles to the discovery service for the specific task. For example, a bookseller will subscribe to perform the role of vendor in a purchase interaction, specifying that the topic is "books,novels,texts", while a peer searching a book will subscribe as customer, for a task described similarly (for example, just "book"). When all the roles are filled, the discovery service matches the peers that subscribed for the same or similar tasks (for example, "books,novels,texts" and "book").

– **Coordination**
When they interact the contextual knowledge needed to interact reliably should propagate to the appropriate services as part of the interaction (Section 5).

When the discovery service has enough subscriptions for an interaction model, it selects a coordinator to execute the IM. The coordinator sends a message to each subscribed peer containing relevant contextual information about the other peers. In this way, peers can communicate back to the coordinator their preferences about with whom they want to interact and not. Once all the roles in the IM are fulfilled by OKCs, the coordinator starts parsing the LCC in a centralised manner. When a constraint is encountered while parsing, the coordinator sends a message to the OKC fulfilling the role that must solve the constraint. This message contains the information associated to the constraint, and the interaction state. Upon receiving the constraint solving request message from the coordinator, the OKC will execute the code that solves the given constraint and return a message to the coordinator with the modified state. This process goes on repeatedly until the IM terminates.

– **Dynamic recruitment**

When the services are being coordinated then it might be necessary to reconcile their various constraints in order to avoid breaking the collaboration (Section 6).

It could happen that some peers fail or don't meet the expectations of Sarah, then either automatically or manually some other peers can be selected that (either at the beginning or during the interaction) have subscribed to the roles for which another peer needs to be found. Eventually, the discovery service acknowledges the coordinator of the running interaction with new peers. Also, during the interaction, the discovery service may be queried to return peers that subscribed to the role however with other constraints.

– **Feedback round**

Before she started, Sarah might have wanted to be reassured that the interaction conforms to the requirements of her business process (Section 7) and that her interaction was reliable (Section 8).

As stated previously, before any interaction starts, the coordinator sends a message to each individual peer with the credentials and constraints of the other peers that want to play a role in the interaction. In this way the user can check which ones suit the desired characteristics.

Summarizing, the OK-system tries to implement the methodology described in this paper. Currently we have a first version running, providing the basic functionality as described above (user interface, code wrapping, publishing of OKCs and IMs together with efficient and fully distributed storage and retrieval, coordinator selection and execution of the IMs. What is still missing and where we currently are working on, is the mapping mechanisms, reputation mechanisms and dynamic recruitment of peers.

10 Related Views of Coordination

The need for coordination is prevalent and the drive for large, complex, distributed systems maintains a pressure to improve the way in which coordination is programmed. The focus of our review below is on methods that can (without too much stretch of the imagination) be considered programming, so for example we ignore methods which have been used to specify interaction-related problems but for which no means of practical computation has been provided. By the same token we do not focus upon explorations of relevant foundational theories of interaction and coordination - for example those originating in theoretical computing science (such as [20, 38, 53, 58]), planning (such as [21]) and multi-agent systems theory (such as [52, 61]). Even with this limit there remain many related approaches so our review describes categories of system with illustrative examples selected from the many available.

10.1 Restricted Languages and Specialist Infrastructures

We have presented a generic language for coordination. By choosing a more specific language in which to express desired patterns of coordination, however, it may be possible build tools that exploit those language restrictions in order to build a limited variety of interactions with less effort. This follows the tradition of domain specific synthesis of traditional programs (such as [49] in ecological modelling; [34] in astrophysics) and in visual languages used to describe the structure of process and workflow systems.

Examples of this in an Internet setting are workflow editing and enactment systems to support scientific computing over large data sets - for instance the Taverna [40] and Kepler [2] workflow systems. These systems aim to support scientists (who have little knowledge of the intricacies of computing on computational grid architectures) with a high level visual language for designing experiments expressed as workflows, and then executing these in a manner that allows the sorts of provenance attribution and runtime monitoring that their communities of scientists demand. These systems are effective because they constrain the task and domain. Neither Kepler nor Taverna is intended as a programming language - they are languages for Grid service connection and provide graphical interfaces for this purpose. Although it is possible to invent a visual language that can be generically used for specifying logic-based programs (see [1] for an example) there is no evidence that a generic visualisation is easier for human communities of practise to understand than the mathematical representation from which it originated - hence to need to specialise by task and/or domain.

10.2   Coordination Via Finite State Models

Throughout this paper we have taken a view of computation that emphasises process rather than state. There is, however, a strong interaction between process and state so there are deep similarities between LCC and interaction models described by finite state machines. An example is the Islander system [16]. The framework for describing agent interactions in Islander relies upon a (finite) set of state identifiers representing the possible stages in the interaction. Agents operating within this framework must be allocated roles and may enter or leave states depending on the locutions (via message passing) that they have performed. In order to structure the description, states are grouped into scenes. An institution is then defined by a set of scenes and a set of connections between scenes with constraints determining whether agents may move across these connections. A scene is defined as a collection of the following sets: roles; state identifiers; an initial state identifier; final state identifiers; access state identifiers for each role; exit state identifiers for each role; and cardinality constraints on agents per role.

Systems like Islander have been used in what is essentially a client-server mode, where interacting peers must connect through a central server that enforces whatever synchronisation and sequencing the chosen interaction model requires. This is a very different way of using interaction models from that described using LCC but there is nothing to prevent state machine models from being used with the LCC style of deployment (or vice versa). A more fundamental difference is in the way models are described which, for LCC, is in the style of a declarative programming language.

10.3   Coordination Controlled by Local Constraints

Policy languages, such as those described in [27], are a means of specifying requirements imposed locally by a peer as conditions for interacting with other peers in different contexts. Such specifications are useful because they provide a way of determining some of the constraints on interaction in advance of actually interacting. Re-interpreted from our viewpoint, this sets constraints locally that could interact with the global constraints set by an interaction model, affecting the likelihood that it would succeed. In this view, policies are complementary to our approach but they offer an additional opportunity (and problem) not discussed elsewhere in this paper. Policies allow the possibility of making better guesses about the appropriateness of peers to perform roles required by interaction models but, to take advantage of this, it is necessary to have rapid and automatic ways of checking the satisfiability of local constraints with respect to (partially complete) interaction models. In [42] we describe early, encouraging results in applying a form of model checking to this problem.

10.4   Coordination Via Shared Task or Service Specifications

In this paper we have assumed that entire interaction models can be shared and this is the basis for coordination between peers as well as other features, such as matchmaking, desirable in open peer-to-peer environments. An alternative view, promoted by specification languages such as WSMO [18] and automated by systems such as IRS [39], is that only task specifications should be shared between peers. These task specifications differ from interaction models in that they do not describe the course of the interaction, only the outcome that is desired by the peer posting the task. It is then necessary for tasks posted by peers to be connected to other peers capable of perhaps performing those tasks. This is facilitated by special purpose components known as mediators, that relate task specifications to problem solver specifications posted by peers that wish to perform tasks. Programming of interactions independently of peers (our focus in this paper) is not the aim of this sort of system, in which control of interaction remains local to each peer. Instead, standardisation of task and problem solver specifications (using a specification language originating in UPML [17]) is used to make tasks more flexibly shared.

OWL-S is a domain-independent OWL ontology for describing web services such that they can be reasoned about by users and middleware. With [35], a service is specified by three facets: the service *profile*, service *model* model, and service *grounding*. The profile specifies what the service does and who provides it. The functionality is described by typing inputs and outputs of the service, using concepts in some domain-specific ontology external to OWL-S, the intention being that the ontology can be used to find those services most closely resembling the requested one. For instance, a photographer looking to buy a new 35mm SLR film camera might happily accept a digital 35mm SLR (closely related in the ontology), but would be less satisfied to be given a 35mm point-and-shoot film camera, which would be ontologically more distant. The service model describes how the service operates, by means of atomic processes and a workflow-like language to combine those processes into composite ones. The grounding describes how to map from this high-level description to the low-level implementation, that is, how to invoke the service.

OWL-S is a language so does not, itself, specify any matchmaking process or infrastructure: service discovery and matchmaker querying, while enabled by design, are not defined. Others, however, have built inference systems that use OWL-S. Perhaps the best known matchmaker system is Semantic MatchMaker [45]. Here, service discovery is achieved by inserting OWL-S descriptions into user-defined fields in the UDDI record [43]. A semantic matchmaker is responsible for interpreting this to find matches between clients and providers. More troublesome is the lack of provision in OWL-S for changing or replacing the executing process. To resolve this 'broker paradox' [44], Semantic MatchMaker introduces an 'exec' primitive to indicate that a new, brokered, process should be substituted in place of the executing one which negotiated the matchmaking. The broker paradox is not a problem in LCC, since we can dynamically alter the current interaction model, or instruct a participant to execute a new one.

10.5   Coordination as Logic Programming

LCC can be viewed as an unusual from of logic programming in which the subgoals of clauses are message passing subgoals or role changes. This, we believe, is a strength of the approach because there exists a large body of engineers trained in this form of computational logic. Other efforts, however, have produced different solutions to coordination that also draw inspitation from logic programming. The Go! language [11], for example, provides a multi-threaded environment in which agents are coordinated via a shared memory store of beliefs, desires and intentions. This is a form of agent oriented programming, using a standardised architecture, and therefore local agent beliefs (rather than interaction models as in LCC) are the anchor point for coordination. Perhaps closer to LCC is the work being done on modelling multi-agent coordination using logic programs, for example in [4] where the Event Calculus is used to specify and analyse social constraints between agents (the motivation for this being similar to that of [37]). For a logic programming view of LCC see [47].

11   Conclusion: What Is the Simplest Thing That Could Possibly Work?

In Section 3 to 7 we discussed six problems encountered when automated reasoning systems must interact in large, open, distributed systems. These problems are: ontology alignment; coalition formation; outcome prediction; maintaining shared knowledge; respecting local constraints; and relating interaction to process requirements. Our contribution is to provide a novel integrative view across all these problems by turning control of interaction into a declarative programming problem, thus bringing it within scope of the existing tools and techniques. This is not to say that any of our six problems are entirely solved in this way (they are, arguably, too general to be solved definitively) but by making models of interaction explicit, via the LCC language of Section 2.1, we are able to tackle aspects of each of the problems that are difficult to address using conventional declarative or agent oriented programming.

What is the simplest thing that could possibly work? The LCC language of Section 2 is pared down to a minimal set of concepts that we have argued are essential to describe intearctions in an executable form. Ontology matching using the interaction-specific method described in Section 3 need not involve more than each individual peer maintaining lists of matches between expressions that are useful specifically for the interactions in which it is engaged. The event based matchmaker of Section 4 relies on counting successes and failures generated from the underlying interaction mechanism so this is no more complex from a user's point of view than page ranking in the current Web. Interaction context is maintained, in Section 5, by using parameters to roles in the interaction model. Getting this right can be a sophisticated task for the designer of an interaction model but those using interaction models need not be aware of this sophistication (just like the muddy children in our example needn't have been aware that the interaction model in which they were involved was constructed in a devious way). One simple reaction to brittleness of interaction (our topic in Section 6) is to endure it as a fact of life in the same way as we tolerate brittleness in conventional Web services. If this proves too brittle, however, our declarative style of modelling allows us to increase flexibility in some respects (via constraint handling and adaptation) without requiring individual peers to become significantly more sophisticated. All of the methods described in these sections could be made more complex as needs demand but none of them require great sophistication from individual users of the peer to peer system.

Simplicity may also be achieved through familiarity, and here the issue is whether those who wish to describe (rather than only use) interactions would adopt any language other than the one with which they are already familiar. Section 7 demonstrates one way of bridging this gap by write interpreters for the community-specific languages in LCC. Writing each interpreter is complex, but done only once, while using it is then simple. Furthermore, some of the complexity of model design can be reduced by applying traditional methods of formal verification to interaction models, as we describe in section 8.

Acknowledgements

This work was supported by the UK EPSRC Advanced Knowledge Technologies Interdisciplinary Research Collaboration (GR/N15754/01) and by the EU OpenKnowledge project (FP6-027253).

References

1. Agusti, J., Puigsegur, J., Robertson, D.: A visual syntax for logic and logic programming. Journal of Visual Languages and Computing 9 (1998)
2. Altintas, I., Birnbaum, A., Baldridge, K., Sudholt, W., Miller, M., Amoreira, C., Potier, Y., Ludaescher, B.: A framework for the design and reuse of grid workflows. In: Herrero, P., Perez, M., Robles, V. (eds.) SAG 2004. LNCS, vol. 3458, pp. 120–133. Springer, Heidelberg (2005)
3. Andrews, A., Curbera, F., Dholakia, H., Goland, Y., Klein, J., Leymann, F., Liu, K., Roller, D., Smith, D., Thatte, S., Trickovic, I., Weerawarana, S.: Business process execution language for web services, version 1.1 (2003)
4. Artikis, A., Pitt, J., Sergot, M.: Animated specifications of computational societies. In: Castelfranchi, C., Lewis Johnson, W. (eds.) Proceedings of the 1st International Joint Conference on Autonomous Agents and MultiAgent Systems, Bologna, Italy, July 15–19, 2002, pp. 1053–1061. Association for Computing Machinery (2002)
5. Barker, A., Mann, R.: Integration of multiagent systems to AstroGrid. In: Proceedings of Astronomical Data Analysis Software and Systems XV, European Space Astronomy Centre, Spain (2005)

6. Barwise, J.: Scenes and other situations. Journal of Philosophy 78(7), 369–397 (1981)
7. Benerecetti, M., Giunchiglia, F., Serafini, L.: Model checking multiagent systems. Journal of Logic and Computation 8(3), 401–423 (1998)
8. Besana, P., Robertson, D., Rovatsos, M.: Exploiting interaction contexts in p2p ontology mapping. In: 2nd International Workshop on Peer to Peer Knowledge Management, San Diego, California, USA. CEUR Workshop Proceedings, pp. 1613–1673 (July 2005); ISSN 1613-0073, online CEUR-WS.org/Vol-139/2.pdf
9. Bordini, R.H., Fisher, M., Pardavila, C., Wooldridge, M.: Model checking agentspeak. In: Proceedings of the Second International Conference on Autonomous Agents and Multiagent Systems, Melbourne, Australia. ACM Press, New York (2003)
10. Bordini, R.H., Fisher, M., Visser, W., Wooldridge, M.: State-space reduction techniques in agent verification. In: Proceedings of the Third International Conference on Autonomous Agents and Multiagent Systems, pp. 896–903. ACM Press, New York (2004)
11. Clark, K.L., McCabe, F.G.: Go! for multi-threaded deliberative agents. In: Leite, J.A., Omicini, A., Sterling, L., Torroni, P. (eds.) DALT 2003. LNCS (LNAI), vol. 2990, pp. 54–75. Springer, Heidelberg (2004)
12. Clarke, E.M., Grumberg, O., Peled, D.A.: Model Checking. MIT Press, Cambridge (1999)
13. de Pinninck, A.P., Dupplaw, D., Kotoulas, S., Siebes, R.: The openknowledge kernel. In: Proceedings of the IX CESSE conference, Vienna, Austria (2007)
14. de Pinninck, A.P., Dupplaw, D., Kotoulas, S., Siebes, R., Roberson, D., van Harmelen, F.: The architecture of the open-knowledge system. Technical report, Open-knowledge consortium (2006)
15. Decker, K., Sycara, K., Williamson, M.: Middle-agents for the internet. In: Proceedings of the 15th International Joint Conference on Artificial Intelligence, Nagoya, Japan (1997)
16. Esteva, M., de la Cruz, D., Sierra, C.: Islander: an electronic institutions editor. In: Proceedings of the 1st International Joint Conference on Autonomous Agents and MultiAgent Systems, pp. 1045–1052 (2002)
17. Fensel, D., Benjamins, R., Motta, E., Wielinga, R.: A framework for knowledge system reuse. In: Proceedings of the International Joint Conference on Artificial Intelligence, Stockholm, Sweden (1999)
18. Fensel, D., Bussler, C.: The web service modellign framework. Electronic commerce: Research and applications 1, 113–137 (2002)
19. Giunchiglia, F., Shvaiko, P., Yatskevich, M.: S-match: an algorithm and an implementation of semantic match. In: Proceedings of the European Semantic Web Symposium, pp. 61–75 (2004)
20. Goldin, D., Smolka, S., Attie, P., Sonderegger, E.: Turing machines, transition systems and interaction. Information and Computation 194(2) (2004)
21. Grosz, B., Kraus, S.: Collaborative plans for complex group action. Artificial Intelligence 2 (1986)
22. Halpen, J.Y., Moses, Y.: Knowledge and common knowledge in a distributed environment. Journal of the ACM 37(3), 549–587 (1990)
23. Hassan, F., Robertson, D.: Constraint relaxation to reduce brittleness of distributed agent ppotocols. In: Proceedings of the ECAI Workshop on Coordination in Emergent Agent Societies, Valencia, Spain (2004)
24. Hassan, F., Robertson, D., Walton, C.: Addressing constraint failures in an agent interaction protocol. In: Proceedings of the 8th Pacific Rim International Workshop on Multi-Agent Systems, Kuala Lumpur, Malasia (2005)
25. Holzmann, G.J.: The SPIN Model Checker: Primer and Reference Manual. Addison Wesley, Reading (2003)
26. Jackson, D., Wing, J.: Lightweight formal methods. IEEE Computer (April 1996)

27. Kagal, L., Finin, T., Joshi, A.: A policy language for pervasive systems. In: Fourth IEEE International Workshop on Policies for Distributed Systems and Networks (2003)
28. Kalfoglou, Y., Schorlemmer, M.: Ontology mapping: the state of the art. Knowledge Engineering Review (2003)
29. Kavantzas, N., Burdett, D., Ritzinger, G., Lafon, Y.: Web services choreography description language version 1.0, 2004. W3C Working Draft (October 12, 2004)
30. Klusch, M., Sycara, K.: Brokering and matchmaking for coordination of agent societies: a survey. In: Coordination of Internet agents: models, technologies, and applications, pp. 197–224. Springer, Heidelberg (2001)
31. Lambert, D., Robertson, D.: Matchmaking and brokering multi-party interactions using historical performance data. In: Fourth International Joint Conference on Autonomous Agents and Multi-agent Systems (2005)
32. Li, G., Chen-Burger, J., Robertson, D.: Mapping a business process model to a semantic web services model. In: Proceedings of the IEEE International Conference on Web Services, San Diego (2004)
33. Li, G., Robertson, D., Chen-Burger, J.: A novel approach for enacting distributed business workflow on a peer-to-peer platform. In: Proceedings of the IEEE Conference on E-Business Engineering, Beijing (2005)
34. Lowry, M., Philpot, A., Pressburger, T., Underwood, I.: A formal approach to domain-oriented software design environments. In: Proceedings of the 9th Knowledge-Based Software Engineering Conference, Monterey, California, pp. 48–57 (1994)
35. Martin, D., Burstein, M., Hobbs, J., Lassila, O., McDermott, D., McIlraith, S., Narayanan, S., Paolucci, M., Parsia, B., Payne, T., Sirin, E., Srinivasan, N., Sycara, K.: owl-s 1.1 (2004)
36. McGinnis, J., Robertson, D.: Realising agent dialogues with distributed protocols. In: van Eijk, R.M., Huget, M.-P., Dignum, F.P.M. (eds.) AC 2004. LNCS (LNAI), vol. 3396, pp. 106–119. Springer, Heidelberg (2005)
37. McIlraith, S., Son, T.: Adapting golog for composition of semantic web services. In: Proceedings of the Eighth International Conference on Knowledge Representation and Reasoning, pp. 482–493 (2002)
38. Milner, R., Parrow, J., Walker, D.: A calculus of mobile processes, part I/II. Information and Computation 100(1), 1–77 (1992)
39. Motta, E., Domingue, J., Cabral, L., Gaspari, M.: Irs-ii: A framework and infrastructure for semantic web services. In: Proceedings of the Second International Semantic Web Conference, Florida, USA (2003)
40. Oinn, T., Addis, M., Ferris, J., Marvin, D., Senger, M., Greenwood, M., Carver, T., Glover, K., Pocock, M., Wipat, A., Li, P.: Taverna: a tool for the composition and enactment of bioinformatics workflows. Bioinformatics 20(17), 3045–3054 (2004)
41. Osman, N.: Addressing Constraint Failures in Distributed Dialogue Protocols. Ph.D thesis, School of Informatics, University of Edinburgh, M.Sc Thesis (2003)
42. Osman, N., Robertson, D., Walton, C.: Run-time model checking of interaction and deontic models for multi-agent systems. In: Proceedings of the Third European Workshop on Multi-agent Systems, Brussels, Belgium (2005)
43. Paolucci, M., Kawamura, T., Payne, T., Sycara, K.: Importing the semantic web. In: UDDI (2002)
44. Paolucci, M., Soudry, J., Srinivasan, N., Sycara, K.: A broker for OWL-S web services. In: Proceedings of the 2004 AAAI Spring Symposium on Semantic Web Services (2004)
45. Paulucci, M., Kawamura, T., Payne, T.R., Sycara, K.: Semantic matching of web services capabilities. In: Proceedings of the International Semantic Web Conference (2002)
46. Robertson, D.: A lightweight coordination calculus for agent social norms. In: Proceedings of Declarative Agent Languages and Technologies workshop at AAMAS, New York, USA (2004)

47. Robertson, D.: Multi-agent coordination as distributed logic programming. In: International Conference on Logic Programming, Sant-Malo, France (2004)
48. Robertson, D., Agusti, J.: Software Blueprints: Lightweight Uses of Logic in Conceptual Modelling. Addison Wesley/ACM Press (1999) ISBN 0201398192
49. Robertson, D., Bundy, A., Muetzelfeldt, R., Haggith, M., Uschold, M.: Eco-Logic: Logic-Based Approaches to Ecological Modelling. Logic Programming Series. MIT Press, Cambridge (1991)
50. Siebes, R., Dupplaw, D., Kotoulas, S., de Pinninck, A.P., Roberson, D., van Harmelen, F.: The functional description of the open-knowledge system. Technical report, Open-knowledge consortium (2006)
51. Smith, R.G.: The contract net protocol: high-level communication and control in a distributed problem solver. In: Distributed Artificial Intelligence, pp. 357–366. Morgan Kaufmann Publishers Inc., San Francisco (1988)
52. van der Hoek, W., Wooldridge, M.: On the logic of cooperation and propositional control. Artificial Intelligence 164(1-2) (2005)
53. van Leeuwen, J., Wiedermann, J.: A computational model of interaction in embedded systems. Technical Report UU-CS-02-2001, Dept. of Computer Science, University of Utrecht (2001)
54. Walton, C.: Model checking agent dialogues. In: Proceedings of the 2004 Workshop on Declarative Agent Languages and Technologies, New York, USA (2004)
55. Walton, C.: Model checking multi-agent web services. In: Proceedings of AAAI Spring Symposium on Semantic Web Services, California, USA (2004)
56. Walton, C.: Model checking multi-agent web services. In: Proceedings of the AAAI Spring Symposium on Semantic Web Services, Stanford, USA. AAAI, Menlo Park (2004)
57. Walton, C., Barker, A.: An agent-based e-science experiment builder. In: Proceedings of the 1st International Workshop on Semantic Intelligent Middleware for the Web and the Grid, Valencia, Spain (August 2004)
58. Wegner, P.: Why interaction is more powerful than algorithms. Communications of the ACM 40(5) (1997)
59. Wong, H., Sycara, K.: A taxonomy of middle-agents for the internet. In: Proceedings of the International Conference on Multi-agent Systems (2000)
60. Wooldridge, M., Fisher, M., Huget, M.P., Parsons, S.: Model checking multiagent systems with MABLE. In: Proceedings of the First International Conference on Autonomous Agents and Multiagent Systems, Bologna, Italy (2002)
61. Wooldridge, M., Jennings, N.R.: The cooperative problem solving process. Journal of Logic and Computation 9(4) (1999)
62. Zeng, L., Benatallah, B., Dumas, M., Kalagnanam, J., Sheng, Q.: Quality driven web services composition. In: Proceedings of the twelfth international conference on World Wide Web, pp. 411–421. ACM Press, New York (2003)
63. Zhang, Z., Zhang, C.: An improvement to matchmaking algorithms for middle agents. In: Proceedings of the first international joint conference on Autonomous agents and multiagent systems, pp. 1340–1347. ACM Press, New York (2002)

Extraction Process Specification for Materialized Ontology Views

Carlo Wouters[1], Tharam S. Dillon[2], Wenny Rahayu[1], Robert Meersman[3],
and Elizabeth Chang[2]

[1] Department of Computer Science and Computer Engineering
La Trobe University, Bundoora, Victoria 3086, Australia
{cewouter,wenny}@cs.latrobe.edu.au
[2] Digital Ecosystems and Business Intelligence Institute,
Curtin University of Technology,
Perth, Australia
tharam.dillon@cbs.curtin.edu.au,
elizabeth.chang@cbs.curtin.edu.au
[3] STARLab, Department of Computer Science
Vrije Universiteit Brussel, Brussel, 1050, Belgium
Robert.Meersman@vub.ac.be

Abstract. The success of the semantic web relies heavily on ontologies. However, using ontologies for this specific area poses a number of new problems. One of these problems, extracting a high quality ontology from a given base ontology, is currently receiving increasing attention. Areas such as versioning, distribution and maintenance of ontologies often involve this problem. Here, a formalism is presented that enables grouping ontology extraction requirements into different categories, called optimization schemes. These optimization schemes provide a way to introduce quality in the extraction process. An overview of the formalism is discussed, as well as a demonstration of several example optimization schemes. Each of these optimization schemes meets a certain requirement, and consists of rules and algorithms. Examples of how the formalism is deployed to reach a high-quality result, called a materialized ontology view, are covered. The presented methodology provides a foundation for further developments, and shows the possibility of obtaining usable ontologies in a highly automated way.

ACM Subject Descriptors ('98): H.3.5 [INFORMATION STORAGE AND RETRIEVAL]: Web-based services; I.1.2 [SYMBOLIC AND ALGEBRAIC MANIPULATION]: Algorithms; I.2.4[ARTIFICIAL INTELLIGENCE]: Semantic networks --- Representation languages.

Additional Keywords: Ontology Extraction.

1 Motivation

In recent years, the unstructured storage of data, especially on the World Wide Web, and the difficulties experienced with retrieving relevant data with the existing search

T.S. Dillon et al. (Eds.): Advances in Web Semantics I, LNCS 4891, pp. 130–175, 2008.

engines, have triggered new research aimed at ameliorating information retrieval and storage. New ways of storing information meant for the Internet were developed, such as XML [W3C 1999], HTML[a] [Fensel, Decker et al. 1998], DTD and RDF. These languages provide a tool to store the information in a structured way, but with that another problem arose; everyone was free to develop there own taxonomy of how they want to categorize their information, e.g. [Heflin, Hendler et al. 1999; Van Harmelen and Fensel 1999]. It is clear that widely accepted standards should be used as metadata to define how the actual information is split up, no matter what language or syntax is used to implement this. These widespread standards are formulated as ontologies.

The first wave of ontology applications and researchers mainly concentrated on getting an effective system up, solving the apparent issues that had been holding back knowledge acquisition from the Internet and related resources. A number of these have turned out to be beneficial, without any of them clearly standing out, and no single standard has been agreed upon [Hovy 1998]. Since then we have seen merging of some of the standards – e.g. OIL incorporating elements of OKBC, XOL and RDF [Fensel, Horrocks et al. 2000], Ontolingua using KIF [Genesereth 1991; Genesereth and Fikes 1992; Gruber 1992], DAML and OIL into DAML-OIL [Berners-Lee and Al 2001] (currently being reworked into OWL [W3C 2002c]) – and diversification of others.

Now that the first generation of ontology applications has settled in, more complicated issues and considerations have reared their heads, such as the quality of ontologies in all its facets [Colomb and Weber 1998; Hahn and Schnattinger 1998; Kaplan 2001; Guarino and Welty 2002; Holsapple and Joshi 2002; Wouters, Dillon et al. 2002b]. Improvements need to be made to the systems that are already in place, and theoretical and practical modifications have to be made to cater for versioning, maintenance and distribution of ontologies [Klein, Fensel et al. 2002]. Furthermore, a further integration of different existing systems is needed [McGuinness, Fikes et al. 200; Noy, Sintek et al. 2001].

The ontologies used also tend to grow larger, to a point where ideally the entire world is modeled in one super-ontology (through the use of upper ontologies [Lenat 1995]), providing great compatibility and consistency across all sub-domains, but practically it introduces the new problem of being too vast to be used in its entirety by any application. Considering the internet as a data repository, it seems clear that users with a very slow or costly connection to this repository might opt to get a local, modified copy of the repository to base views upon, and to query in other ways. It seems highly unlikely that someone will be able to copy the entire contents of the World Wide Web to a local repository, and even more unlikely that all this data will be actually used in whatever application the user might intend it to be used for. If a business just needs access to information on the share market, it would not benefit from all the other information that their local copy would contain. This is just one of the many reasons why a complete ontology might not be a valid structuring option for certain users. Another reason can be found in varying levels of security and confidentiality – not necessarily every user of an ontology has the same access rights, and using a smaller ontology, merely containing the appropriate parts of the base ontology might enable local copies. Efficiency of querying repositories might be another reason for having a simplified, local version of an ontology, and there are many more.

It is imperative that when an ontology view is derived, the quality of the resulting ontology is as high as possible. First of all, the intentions of the ontology engineer should obviously be satisfied, and the resulting design should be a consistent, cohesive, complete and well-formed ontology. Secondly, the result should be further pruned, i.e. is the obtained solution one of the most efficient, flexible, simple, versatile, etc. solutions (varying with the specific needs of the ontology engineer).

The resulting ontology should be usable as the base for an independent system, i.e. be an ontology in its own right. This article aims to identify the processes involved in this extraction. It is set up in such a way that the theoretical definitions and processes can also be readily transformed and applied to other new research, such as versioning and distribution of ontologies [Wouters, Dillon et al. 2002a]. While establishing the details of the sub-processes used for the extraction, great care is taken to obtain a result that adheres to the quality assurance issues raised before. Finally, a brief outline of future research will be given, eventually leading up to an integrated transformation environment for ontologies, interfacing with existing international standards.

2 Defining the Ontology

Although different standards have emerged, and none of them has achieved the status of the ultimate ontology definition, this does not pose a huge problem. The use of wrappers, which can be used to extract data – and semantic information – from a certain type of ontology definition and implementation, and translate it to another, helps with this. Another way in which wrappers can be used is in trying to transform the normal natural language, textual information resources on the web to more structured resources [Muslea 1999].

None of these approaches however, provides a consistent, reliable solution towards the future. There is a clear need to steer towards a more solid foundation in terms of theoretical definitions and international standards.

2.1 Overview of Ontology Definitions

A problem that occurs when doing any research in the field of ontologies is that no single definition for an ontology exists that can be completely utilized in Information Technology. Definitions have been modified, and diversified over the last couple of years, but the problems remain; either a proposed definition is too general to have any practical value [Gruber 1993; Fensel 2001], or the definition provides enough preciseness, but only suits a small number of models and systems in use today [Noy and Hafner 1997]. In practical applications this sometimes is overcome by introducing several 'layers' for ontologies, for example the ontological commitments in [Spyns, Meersman et al. 2002]. It is not within the scope of this article to produce a final solution to this problem, but there is still the necessity for clarifying what definitions will be used throughout this article. The initial definition will be extended as new definitions are introduced, incorporating stricter requirements for an ontology.

The first definition clarifies what cardinalities are used. Please refer to Appendix A for the multiplication table of these cardinalities.

Definition 1

$$\textit{The set of cardinalities card} \equiv \{0, 1, m\} \tag{1}$$

Definition 2

$$\vartheta \equiv \{The\ set\ of\ all\ ontologies\} \tag{2}$$

Definition 3. Let an ontology O be a six tuple (<C, A, attr, B, M_a, M_b>),consisting of a non-empty, finite set of concepts C, a finite set of attributes A, a mapping attr, a set of binary relationships B, an attribute cardinality mapping M_a, and a relationship mapping M_b, with attr the mapping of concepts onto elements of A, B the disjoint union of the binary associative relationships B_s, the binary inheritance relationships B_i, and the binary aggregate relationships B_{agg}.

$$O \in \vartheta \Leftrightarrow O \equiv <C,\ A,\ attr,\ B,\ M_a,\ M_b > \tag{3}$$

with

$C = finite \wedge C \neq \varnothing$

$A = finite$

$attr{:}C \rightarrow A$

$B \subseteq C \times C \wedge B = B_s \cup B_i \cup B_{agg}$

$M_a{:}attr \rightarrow card^2$

$M_b{:}B \rightarrow card^4$

Intuitively, ontologies conceptually represent a perceived world through concepts, attributes and relationships. Concepts may represent the different higher-level components of this world, and each component may have attributes. These attributes may be derived from the characteristics of components of the world. Relationships may also hold between these concepts.

For practical reasons, only binary relationships are considered here. Note that binary relationships occur most frequently in current modeling. Unary models are not modeled, since these are taken care of by forming subtypes. N-ary relationships are not considered as transformation of this type of relationship to binary relationships is possible. For a more comprehensive treatment of arity of relationships see [Halpin 1995].

Some useful notations are given here to enhance readability throughout the rest of the article.

Notation 1

Given an ontology O=<C, A, attr, B, M_a, M_b > and an attribute mapping (4)
t=(c,a), with c∈C, a∈A

we denote

$\pi_1(t){=}c,$

$\pi_2(t){=}a,$

$attr(c){=}\{a{\in}A|\ \exists t{\in}attr :t{=}(c,a)\}$

Notation 2

> *Given an ontology $O=<C, A, attr, B, M_a, M_b>$ and a binary relationship*   (5)
> $$b=(c_1, c_2), \text{ with } c_1, c_2 \in C$$

we denote

$\pi_1(b)=c_1$

$\pi_2(b)=c_2$

Notation 3

> *Given an ontology $O=<C, A, attr, B, M_a, M_b>$ and*   (6)
>
> *An attribute cardinality $m=(n_1, n_2)$ for an attribute mapping $t \in attr$*
>
> *with $n_1, n_2 \in card$, and t the attribute mapping for attribute a*

we denote

$m(t)=m(a)=(n_1, n_2)$

$m_{min}(t)=m_{min}(a)=n_1$

$m_{max}(t)=m_{max}(a)=n_2$

Notation 4

> *Given an ontology $O=<C, A, attr, B, M_a, M_b>$ and*   (7)
>
> *a relationship cardinality $m=(n_1, n_2, n_3, n_4)$ for a binary relationship b*
>
> *with $n_1, n_2, n_3, n_4 \in card$*

we denote

$m(b)=(n_1, n_2, n_3, n_4)$

$m_{min1}(b)=n_1$

$m_{max1}(b)=n_2$

$m_{min2}(b)=n_3$

$m_{max2}(b)=n_4$

Figure 1 gives an example of a UML diagram, and how this can be interpreted using the notations defined above, is shown in Table 1.

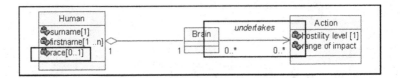

Fig. 1. Example UML Diagram

Table 1. Mapping from UML Diagram to Defined Notation

Statement in UML	Notation
Race is attribute	$Race \in A$
Race has cardinality [0..1]	$m(race) = (0,1)$ $m_{min}(race) = 0$ $m_{max}(race) = 1$
'undertakes' is relationship	$undertakes \in B$
Brain and Action are the connected concepts	$undertakes = (Brain, Action)$ $\pi_1(undertakes) = Brain$ $\pi_2(undertakes) = Action$
undertakes has cardinalities 0..* and 0..*	$m(undertakes) = (0, n, 0, n)$ $m_{min1}(undertakes) = 0$ $m_{max1}(undertakes) = 1$ $m_{min2}(undertakes) = 0$ $m_{max2}(undertakes) = 1$

Obviously, the minimum cardinality for an attribute mapping should never be more than the maximum cardinality, so an additional rule always applies:

Transformation Rule 1

$$Given\ an\ ontology\ O = <C,\ A,\ attr,\ B,\ M_a,\ M_b > \qquad (8)$$

$$\forall m \in M_a : m_{min}(t) \le m_{max}(t),\ with\ t \in attr$$

$$\forall m \in M_b : m_{min1}(t) \le m_{max1}(t),\ with\ t \in attr$$

$$\forall m \in M_b : m_{min2}(t) \le m_{max2}(t),\ with\ t \in attr$$

Following [Spyns, Meersman et al. 2002] the 'semantics' of an ontology is the range of interpretation mapping of an application environment onto an ontology. Note that the semantics of the real world problem are replaced by an ontology. Some examples of an application environment are RDBMS, software applications, documents, website, etc. Whilst it is recognized that there are important differences between an ontology, and a conceptual model, for the purpose of deriving sub-ontologies these are immaterial. Frequently, a conceptual model can be considered to be an ontology expressed in a chosen syntax. However, this syntax should not impact the definitions and theorems presented, so for the purpose of this paper this difference is irrelevant.

Throughout this paper UML [Rumbaugh, Jacobson et al. 1999] is sometimes used to graphically represent an ontology, but it is not the intention to show the suitability of UML for the modeling of ontologies. UML is merely a convenient modeling notation that for practical reasons was chosen to highlight aspects of ontologies. There should be no confusion as to the difference of the ontology and the modeling notation used to represent it, and by no means is it the intention of the authors to promote this modeling notation to a higher status. UML is used to represent object oriented models, and as our definition of an ontology (section 2) has concepts (similar to classes), attributes and relationships, it was found convenient to use this easy to read data model to illustrate aspects of semantics of ontologies. One could, however, choose any alternative notation such as semantic nets [Feng, Chang et al. 2002] to illustrate the issues. This does not distract from the methodologies in this paper.

2.2 Concepts in Detail

What is referred to as 'concept' in the definition is a very broad ontological element. It can be defined in a variety of ways, potentially providing a lot of additional information about itself, and its relationships (and topological proximity) to other elements. Through an inheritance structure concepts can be made specializations and generalizations of other concepts. The conventions used for a 'concept' are exactly the same as what is labeled a 'class' in OWL [W3C 2002c]. Instead of 'definition' of a concept, such declarations of a concept are called 'axioms'[1]. An example of an axiom defining a concept (labeled 'owl:class') is given in Figure 2, where a "Wildcard" is axiomatically defined as the result of Boolean operations on subsets of other classes.

```
<owl:Class rdf:ID="Wildcard">
    <owl:IntersectionOf>
        <owl:subClassOf rdf:resource="#Finalist">
            <owl:restriction>
                <owl:onProperty rdf:resource="#year"/>
                <owl:hasValue>
                    <owl:oneOf rdf:parseType="Collection">
                        <owl:hasValue>2002</owl:hasValue>
                        <owl:hasValue>2001</owl:hasValue>
                        <owl:hasValue>2000</owl:hasValue>
                        <owl:hasValue>1999</owl:hasValue>
                        <owl:hasValue>1998</owl:hasValue>
                    </owl:oneOf>
                </owl:hasValue>
            </owl:Restriction>
        </owl:subClassOf>
        <owl:subCLassOf rdf:resource="#Tournament_Winner"/>
    </owl:IntersectionOf>
</owl:Class>
```

Fig. 2. Example OWL statements in RDF for axioms[2] [W3C 2002b]

Note in Figure 2 that a restriction results in a concept, which is here unnamed, as it is only used as a mechanism to define another (labeled) concept.

2.3 Attributes and Relationships in Detail

Both Attributes and Relationships can be regarded as properties that belong to a concept. In fact, in [W3C 2002a; W3C 2002b] the naming convention used is indeed "Property" for both these terms. However, it is not uncommon to make a further

---

[1] Note that throughout the rest of this article the term 'axiom' is used as specified in the OWL definition W3C (2002c). OWL Web Ontology Language 1.0 Reference. *W3C Working Draft.* However, readers with a background in A.I. and mathematics should be aware of the broader meaning of this term here, otherwise it could lead to erroneous interpretation of the definitions and theorems presented.

[2] As the final syntax of OWL still has to be finalized, this example should be taken as an indication of a possible representation, rather than an exact OWL RDF file example.

distinction between the two types of properties. For example, [W3C 2002a] groups the properties in "DatatypeProperties" (i.e. attributes) and "ObjectProperties" (i.e. relationships). Because of the distinction in semantic meaning between these two types of properties, the names "Attribute" and "Relationship" were preferred, as they clearly indicate the semantic meaning, and relate closer to the naming convention as used in Object Oriented Model, and Relational Database Management Schemas.

2.4 Restrictions to Obtain Practical Ontology Definition

The formal definition as proposed above is at too high a level of abstraction to be practically usable. To make the ontology definition more practical, a couple of refinements are introduced.

The first restriction that is made, which is an integral part of the ontology definition is the fact that an attribute that is a part of an ontology should at least belong to one concept. This is given in the following extension to the definition of an ontology.

Definition 4

$$\forall O \in \vartheta : O = <C, A, attr, B, M_a, M_b> \ \Rightarrow \ \forall a \in A, \ \exists c \in C: a \in attr(c) \qquad (9)$$

This definition extension ensures that there is always a semantic meaning of the attribute in relation to the rest of the ontology. Note that some standards do not have this restriction (e.g. OWL [W3C 2002c]). Here it is important to regard the ontology as semantically interlinked, as this provides a solid lower boundary for algorithms.

Furthermore, the boundaries of a single ontology need to be established, i.e. when does one ontology cease to be one ontology, and become multiple ontologies. The following definitions aim to support the notion of a well-formed ontology.

Definition 5. Let a graph G consist of a set of vertices V, and a set of edges E, with an edge defined by two vertices in V.

$$G \equiv <V, E> \qquad (10)$$

with

$$V = finite \ \wedge \ V \neq \varnothing$$

$$E \subseteq V \times V$$

This is just a standard definition of a graph [Biggs, Lloyd et al. 1976], and is merely given to facilitate the consequent definitions and theorems. For more information on Graph Theory, a number of sources are readily available (e.g. [Von Staudt 1847; Biggs, Lloyd et al. 1976]).

Definition 6

Given a graph G=<V, E> $\qquad (11)$

we define

$$G' \ an \ island \ in \ G \ \Leftrightarrow G' \equiv <V', E'>$$

with

> $V' \neq \emptyset$
>
> $V' \subseteq V$
>
> $E' \subseteq E$
>
> $G' \neq G$
>
> $\forall v \in V', \ \forall v' \in V \backslash V' : (v, v') \notin E$

An island is another term for what is called 'proper component' of a graph [Biggs, Lloyd et al. 1976], with slight modification to suit our purposes. Next, an Ontology Graph is defined. Ontology graphs will aid in the more 'geographical' or 'topological' characteristics of an ontology, and their implications.

Definition 7

Given an ontology $O = <C, A, attr, B, M_a, M_b>$ (12)

We define

> G_O *is an Ontology Graph for $O \Leftrightarrow G_O \equiv <V_O, E_O>$*

with

> $V_O = C$
>
> $E_O = B$
>
> G_O *has no islands*

It is important to note here that a valid Ontology Graph cannot contain an island. Through an extension to the definition of a (valid) ontology, this requirement for an Ontology Graph becomes a requirement for an ontology as well.

Definition 8

$$\forall O \in \vartheta \Rightarrow \exists \ an \ Ontology \ Graph \ G_O \ for \ O$$ (13)

This definition says that an ontology is not considered valid unless there is an Ontology Graph that can be associated with it. Because an Ontology Graph cannot have an island, this means that a valid ontology has to have a corresponding graph representation without islands.

For completeness, the full definition for an ontology is given here, i.e. the addition of Definition 4 and Definition 8 to Definition 3.

Definition 9

$$Given \ a \ set \ of \ concepts \ C$$ (14)

We define

> *An ontology over C, $O \in \vartheta \Leftrightarrow O \equiv <C, A, attr, B, M_a, M_b>$*

with

$C = finite \wedge C \neq \emptyset$

$A = finite$

$Attr: C \rightarrow 2^A$

$B \subseteq C \times C \wedge B = B_r \cup B_i \cup B_{agg}$

$M_a: attr \rightarrow card^2$

$M_b: B \rightarrow card^4$

$\forall a \in A, \exists c \in C: a \in attr(c)$

$\exists\,an\ Ontology\ Graph\ G_O\ for\ O$

2.5   Ontology Complement Theorem

In order to define a complement of an ontology for a given base ontology, first the notion of subsets needs to be defined. The following two definitions indicate a subtle but important difference; not all subsets are ontologies, the subsets that are ontologies in their own right are called sub-ontologies.

Definition 10

Given an ontology $O = <C, A, attr, B, M_a, M_b>$ (15)

we define

a six tuple S is a subset of O $\Leftrightarrow$ S$\equiv$< $C_1, A_1, attr_1, B_1, M_{a1}, M_{b1}$>

with

$C_1 \subseteq C$

$A_1 \subseteq A$

$attr_1 \subseteq attr$

$B_1 \subseteq B$

$M_{a1} \subseteq M_a$

$M_{b1} \subseteq M_b$

Notation 5

Given an ontology $O = <C, A, attr, B, M_a, M_b>$ *and a subset S of O*   (16)

we denote

$S \subseteq O$

The definition of a subset is expanded to incorporate the requirements for a valid ontology. Although 'a subset that is a valid ontology' is sufficient for a definition, we will expand the definitions into the individual requirements, and combine them into one more detailed definitions for a sub-ontology.

Definition 11

> Given an ontology O (17)

> we define

> O' is a sub-ontology of $O \Leftrightarrow O' \in \vartheta \wedge O' \subseteq O$

Notation 6

> Given an ontology O and a sub-ontology O' of O (18)

> we denote

> $O' \in O$

Two very basic Ontologies, O_1 and O_2 are presented in Figure 3. The ontology to the right (O_2) is a sub-ontology of the ontology to the left (O_1), i.e. it is a valid ontology in its own right, as well as being a subset of O_1.

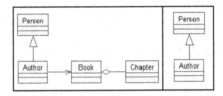

Fig. 3. Two simplified base ontologies (left: O_1, right: O_2)

Definition 12

> Given ontologies $O_1 = <C_1, A_1, attr_1, B_1, M_{a1}, M_{b1}>$, $O_2 = <C_2, A_2, attr_2, B_2,$ (19) $M_{a2}, M_{b2}>$

> with

> $O_2 \in O_1$

> We define

> $O_1 \backslash O_2 \equiv < C_1 \backslash C_2, A_1 \backslash A_2, attr_1 \backslash attr_2, B_1 \backslash B_2, M_{a1} \backslash M_{a2}, M_{b1} \backslash M_{b2} >$

In this definition the characteristics of the complement of a sub-ontology are presented. An interesting issue is raised by this definition, namely is the complement ($O_1 \backslash O_2$) always a valid ontology, or not ? The answer to this question is negative, as following example illustrates a complement that is not a valid ontology.

Looking back at Figure 3, the first thing to note is that the basic requirement (antecedent of formula (19)) is fulfilled, and that O_2 is a sub-ontology of O_1 ($O_2 \in O_1$). Secondly, in this simplified form, O_1 counts 7 elements (or 7 that matter in the example), being 4 concepts and 3 relationships. On the other hand, O_2 only has 3; 2 concepts and 1 relationship.

Looking at the complement of O_2 ($O_1\backslash O_2$), it can easily be visually verified in Figure 4 that there is an invalid relationship.

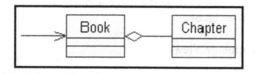

Fig. 4. The resulting graphical representation of set $o_1\backslash o$

The relationship between 'Author' and 'Book' is an element of the binary relationship set of O_2. From the definition this set B_2 is a subset of $C_2 \times C_2$, but one of the concepts connected by the relationship is an element of C_1, and from this definition cannot be an element of C_2, hence $O_1\backslash O_2$ is not an ontology.

The obvious question to answer now is when this complement would be an ontology. The following theorem provides the key to solving this problem.

Theorem

Given an ontology $O=<C, A, attr, B, M_a, M_b>$, and its corresponding ontol-   (20)
ogy graph $G_O=<V, E>$ (with $V=C$, $E=B$)

$$\forall O_1 \in \vartheta : O_1 \in O \Rightarrow O\backslash O_1 \notin \vartheta$$

Proof

1. $O_1=O$

    $\Rightarrow O\backslash O_1= O\backslash O=\varnothing$

    $\wedge \varnothing \notin \vartheta$ ($C \neq \varnothing$ for valid ontology)

    $\Rightarrow O\backslash O_1 \notin \vartheta$

2. $O_1 \neq O$

    We give a proof by contradiction. Following statement is proven wrong:

    $$\exists O_1 \in \vartheta : O_1 \in O \Rightarrow O\backslash O_1 \in \vartheta$$

    Let $O_d = O\backslash O_1$

    If $O_d \in \vartheta$ then, per definition (of ontology) there exists an ontology graph $G_d=<C\backslash C_1, B\backslash B_1>$. Let $V_d= C\backslash C_1$, $E_d= B\backslash B_1$

    We now show that G_d is an island in G_O, which is not permitted (see section 2.4).

    Indeed, let $v_d \in V_d$ and be $v \in V_1$ arbitrary

    Clearly;

    Either

    $(v_d, v) \in Ed$ - not permitted because $v \in V_1 (=C_1)$, so $v \notin V_d (=C\backslash C_1)$

    or $(v_d, v) \in E1$ - not permitted because $v_d \in V_d (=C\backslash C_1)$, so $v_d \notin V_1 (=C_1)$

    or $(v_d, v) \notin E$ - Only remaining option ($E=E_d \cup E_1$), so must be this

In other words, there is no edge that can be found that has a vertex in V_d and one in V_l. This means that the two corresponding graphs have no connection, and are thus separate islands. Thus $G_O(=G_l \cup G_d)$ itself would consist of two islands, and thus is in conflict with the definition of ontology graphs (section 2.4).

The above proof shows that it is true that the complement of an ontology and one of its sub-ontologies is never an ontology.

2.6 Materialized Ontology Views

The result of an extraction process is not just simply referred to as an extracted ontology, but rather an extracted materialized ontology view. This section will discuss the term 'materialized ontology view', and why it is used to describe the result of the extraction process.

In the extraction process, no new information should be introduced (e.g. adding a new concept). However, it is possible that existing semantics are represented in a different way (i.e. a different view is established). This is similar to the use in the database area, where views do exactly that [Chen 1976; Date 2000], and thus the notion of an ontology view can be obtained by an analogy to the database view definition, with some minor modifications.

Definition 13

> A *Materialized Ontology View* of a base ontology is a (valid) ontology   (21)
> that consists solely of projections, copies, compressions, and/or combinations of elements of the base ontology, presenting a varying and/or restricting perception of the base ontology, without introducing new semantic data.

Intuitively, the definition states that – starting from a base ontology – elements can be left out and/or combined at will, as long as the result is a valid ontology again. In the process, no new elements should be introduced (unless the new element is the combination of a number of original elements, i.e. the compression of other elements).

A *materialized* ontology view is required, as the resulting ontology should be an independent ontology, i.e. be a valid ontology, even if the base ontology is taken away. This requires a materialization of the result, as opposed to leaving it as virtual, which would result in a dependent view.

3   The Ontology Extraction Process

The main issues covered in the introduction concerning current research on ontology transformations, such as versioning, distribution, evolution, bear numerous similarities. For instance, an existing ontology (from now on referred to as "base ontology") is present and used as a starting point, and through some changes a new ontology needs to be established. It is vital to identify the different elements of this process, so that one can concentrate on the appropriate issues in the research which need to be resolved.

Naturally, there are certain requirements and constraints that a derived ontology or materialized view must abide by. One of the constraints that comes to mind as being of great importance in versioning and distribution is backward compatibility, or compatibility in general [Heflin and Hendler 2000; Klein and Fensel 2001; Kim 2002]. Some of the lower level requirements to serve purposes like compatibility are explored here. Before these can be explored it is necessary to introduce a labeling of an Ontology here.

3.1 Labeling an Ontology

In this section, a new six tuple $<C', A', attr', B', M_a', M_b'>$ is constructed by applying a labeling to the base ontology, and then only using elements with certain labels. In other words, every element of the ontology receives a certain label, and this label determines whether it will be a part of the new six tuple or not. There are three main reasons for introducing such a labeling throughout the extraction process. Firstly, this labeling facilitates user manipulation of the extraction process, as it provides a way for the user to specify key elements.

Secondly, in the optimization step this labeling is also re-applied (modification of present labeling) by every optimization scheme. This is the standard way that different components of the extraction process can communicate with each other. Every optimization scheme is able to read in a labeling, incorporate it into its algorithms, and modify the labeling according those algorithms. The next optimization scheme in line uses this modified labeling as its input. This means that the changes one optimization scheme wants to make effectively are communicated to the next optimization scheme.

Thirdly, the labeling is also used to produce a final result. As indicated previously, this is done by constructing a new six tuple through the use of a label filter.

A few necessary definitions are given next, followed by the automated way of labeling the attribute mapping that is assumed in examples throughout this paper.

Definition 14

> *the selection set S is defined as* (22)
>
> $$S \equiv \{selected, deselected, void\}$$

From the previous definition it can be noted that there are three possible labels that are used; selected, deselected and void.

Definition 15

> *A labeling of an ontology* $O=<C, A, attr, B, M_a, M_b>$ *is a mapping* σ (23)
> *such that*
>
> $$\sigma(O) = \sigma(<C, A, attr, B, M_a, M_b>) = <C', A', attr', B', M_a', M_b'>$$
>
> *with*
>
> $$C' = \sigma_C(C) = \{c \in C \mid \sigma_C(c) \neq 'deselected'\},$$
>
> $$A' = \sigma_A(A) = \{a \in A \mid \sigma_A(a) \neq 'deselected'\},$$

$$attr'=\sigma_{attr}(attr)=\{t\in attr| \sigma_{attr}(t)\neq\text{'deselected'}\},$$

$$B'=\sigma_B(B)=\{b\in B| \sigma_B(b)\neq\text{'deselected'}\},$$

$$M_a'=\sigma_{Ma}(M_a)= M_a \text{ and } M_b'=\sigma_{Mb}(M_b)= M_b$$

This labeling is crucial in the interaction between humans and algorithms, and algorithms amongst themselves. An algorithm might optimize a solution set in a certain way, and needs to pass this on to the next algorithm. Often there are practical subjective influences that cannot be translated in an algorithm, and the labeling allows a user to express his/her wishes. For instance, certain information might not be made available to the materialized view, and thus by using this labeling system, this requirement can readily be communicated to the system. In this case, the unavailable elements would be deselected, so that for an undesirable ontological element e, $\sigma(e)$=deselected.

Some further definitions are needed to enable quick reference to a certain set of ontological elements. As the definitions for all the sets in the ontology six tuple are very similar, one generic definition is given (rather than six slightly varying ones):

Definition 16

$$\text{Given an ontology } O=<C, A, attr, B, M_a, M_b > \tag{24}$$

$$\sigma_X:X\rightarrow S$$

with

$$X\in\{C, A, attr, B, M_a, M_b\}$$

Next we demonstrate how the previous definitions are applied in the ontology extraction process. This is done by giving a small example. The labeling will reoccur in many examples throughout this paper, as it is an essential aid in the extraction process, and serves multiple purposes (as discussed previously).

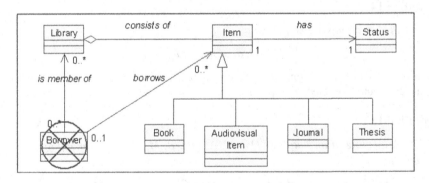

Fig. 5. UML Representation of University Example (Simplified)

When a university decides to develop a search system for its students, they use a certain library ontology as a basis, i.e. the library ontology is the base ontology for their materialized ontology view. The data they use comes from their own library which also uses the library ontology. However, the library notifies them that certain types of data will not be available, as that would infringe the privacy of the library members. Specifically, a concept named "Borrower" is off limits for the university. Although the university still wants the optimization algorithms to decide what a good ontology is to use, they now have to make this requirement clear to the algorithms, and the labeling does exactly that. By applying the labeling $\sigma_C("Borrower")=deselected$ the algorithms have the appropriate information to further process the transformation.

3.1.1 Automation of attr Labeling

Although not necessary, it is convenient to use an automatic labeling of the *attr* mapping. Here we will examine all the possible combinations of the related concept-attribute pair, and give a desired labeling for the attribute relationship between them. Afterwards these results are summarized in a couple of rules.

given an ontology $O=<C, A, attr, B, M_a, M_b>$
$\forall t \in attr$:

1) $\sigma_C(\pi_1(t))=$ *selected* $\wedge$
    a. $\sigma_A(\pi_2(t))=$ *selected*
        $\Rightarrow \sigma_{attr}(t)=$ *selected*
    b. $\sigma_A(\pi_2(t))=$ *deselected*
        $\Rightarrow \sigma_{attr}(t)=$ *deselected*
    c. $\sigma_A(\pi_2(t))=$ *void*
        $\Rightarrow \sigma_{attr}(t)=$ *void*

2) $\sigma_C(\pi_1(t))=$ *deselected* $\wedge$
    a. $\sigma_A(\pi_2(t))=$ *selected*
        $\Rightarrow \sigma_{attr}(t)=$ *deselected*
    b. $\sigma_A(\pi_2(t))=$ *deselected*
        $\Rightarrow \sigma_{attr}(t)=$ *deselected*
    c. $\sigma_A(\pi_2(t))=$ *void*
        $\Rightarrow \sigma_{attr}(t)=$ *deselected*

3) $\sigma_C(\pi_1(t))=$ *void* $\wedge$
    a. $\sigma_A(\pi_2(t))=$ *selected*
        $\Rightarrow \sigma_{attr}(t)=$ *void*
    b. $\sigma_A(\pi_2(t))=$ *deselected*
        $\Rightarrow \sigma_{attr}(t)=$ *deselected*
    c. $\sigma_A(\pi_2(t))=$ *void*
        $\Rightarrow \sigma_{attr}(t)=$ *void*

The previous list gives the desired automated mapping result for each individual case, and now we give a more comprehensive set of rules that have the same effect.

Attribute Mapping Automation Rule 1

$$given \ an \ ontology \ O=<C, A, attr, B, M_a, M_b> \hspace{2cm} (25)$$

$\forall t \in attr$: $\sigma_C(\pi_1(t))=selected \wedge \sigma_A(\pi_2(t))=selected \Leftrightarrow \sigma_{attr}(t)= selected$

Attribute Mapping Automation Rule 2

$$given\ an\ ontology\ O=<C,\ A,\ attr,\ B,\ M_a,\ M_b> \tag{26}$$

$$\forall t \in attr:\ \sigma_C(\pi_1(t))=\ deselected \lor \sigma_A(\pi_2(t))=\ deselected \Leftrightarrow \sigma_{attr}(t)=\ deselected$$

Attribute Mapping Automation Rule 3

$$given\ an\ ontology\ O=<C,\ A,\ attr,\ B,\ M_a,\ M_b> \tag{27}$$

$$\forall t \in attr:\ (\sigma_C(\pi_1(t))=\ void \land \sigma_A(\pi_2(t)) \neq deselected) \lor (\sigma_C(\pi_1(t)) \neq deselec\text{-}$$
$$ted \land \sigma_A(\pi_2(t))=void) \Leftrightarrow \sigma_{attr}(t)=\ void$$

Note that the last rule follows directly from the two previous rules. As there are only three mapping possibilities, if the first two don't apply, it has to be the third mapping. Thus, with only the first two rules a full mapping can be established (if we assume a default 'void' labeling), but the third rule was added for reasons of completeness.

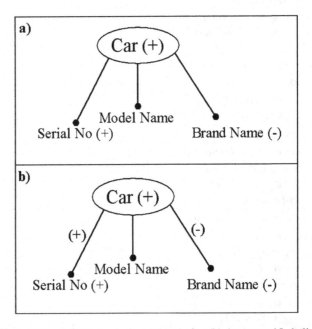

Fig. 6. Attribute Mapping before (a) and after (b) Automated Labeling

Some examples of these labeling rules are demonstrated in Figure 6. The concept "Car" is selected, and it has three attributes with different labels. Note that a missing indication means that the element has a "void" label, while a "+" means selected, and a "-" stands for deselected. The specific notation used is irrelevant for now, but will be discussed in another example (see Figure 10). For "Serial No" the first rule applies, as both concept and attribute are selected. In (b) it can be seen that the connection between these two also receives the "selected" label. The connection between "Brand

Name" and "Car" receives a "deselected" label, as one of the elements (i.e. "Brand Name") is deselected. Finally, nothing is changed to the connection between "Car" and "Model Name" (i.e. it keeps its "void" label), as the two first rules do not apply here. Alternatively, it can be said that the third rule applies.

3.2 Optimization Criteria Related to Quality of an Ontology

There are many possible optimization schemes that seek to optimize one or more criteria that are related to the quality of the ontology. However at the outset it is necessary to make precise the notion of the quality of an ontology and the related criteria that might be used in any optimization scheme to achieve this quality. Generally one can note that quality of an ontology is multifaceted and the emphasis that one might give to a particular facet will be subjective.

We have identified the following facets related to the quality of the materialized ontology view. Namely:

(i)   Consistency of the Ontology with the Requirements
(ii) Well Formedness of the Ontology
(iii) Semantic Completeness of the Ontology
(iv) Overall Simplicity of the Ontology

Each of these facets will separately be used as an optimization criterion and lead to an optimization scheme that optimizes with respect to this criterion and these are discussed in turn below, in the remainder of this section. The list presented is not a full list, as it is possible to construct new optimization schemes according to specific needs that might not be catered for through a combination of any of the existing optimization schemes.

The goal is to arrive at an ontology that is optimized for a certain application (or group of applications). To achieve this, first a number of optimization schemes are presented, i.e. the separate components for specifying what is considered an optimized ontology in a particular case. These components are then applied on the base ontology in the specified order. This means that they work as a chain, one scheme taking the output from a previous scheme as its input. Obviously these separate components have to be able to communicate with each other, and the labeling introduced in section 3.1 is used for this.

Because of the labeling (providing a standard language for communication) and the independence of the optimization schemes (as each optimization scheme can function on its own) it is easy to integrate new optimization schemes. The optimizations schemes presented here are intended to be used on the ontology standards as defined previously. Note that other standards (such as DAML-OIL, or OWL) can reuse many of the optimization schemes presented here, as well as future ones developed for various standards (although sometimes with small modifications). Next we discuss the optimization schemes with respect to the above four criteria in turn.

3.2.1 Requirement Consistency Optimization Scheme (RCOS)
This optimization scheme ensures that the requirements as expressed by a user or optimization scheme (through the applied labeling) are consistent, i.e. there are no contradicting

statements in the labeling. If contradicting statements are present, there is no way for an algorithm – or human user for that matter – to determine what the intention is.

Here, a number of rules will be specified to ensure that a solution set is consistent in its requirements.

RC Rule 1

Given an ontology $O=<C, A, attr, B, M_a, M_b>$ with a labeling σ applied to O   (28)

$$\forall b \in B: \sigma_C(\pi_1(b))=deselected \vee \sigma_C(\pi_2(b))= deselected \Rightarrow \sigma_B(b) \neq selected$$

A binary relationship provides additional meaning to how two concepts relate to each other. Having $\sigma_B(b)= selected$ indicates that this meaning has to be used in the 'target' ontology. However, if both concepts that are related to each other by b are deselected, this is saying that no information about these concepts should be present in the 'target' ontology. RC rule 1 sets out how these two statements are utilized in development of the optimized ontology.

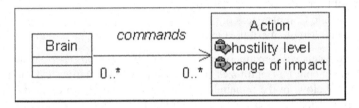

Fig. 7. UML Representation of an Ontology (Small Part)

A small example of the rule in practice is given through Figure 7. Applying two sets of labels – given in Table 2 – produces a valid and invalid combination according to this rule.

Table 2. Valid and Invalid Labeling for RCOS Rule 1

Valid Labeling (void labels not shown)	Invalid Labeling (void labels not shown)
$\sigma_C(Brain)=deselected$ $\sigma_B(commands)=deselected$	$\sigma_C(Brain)=deselected$ $\sigma_B(commands)=selected$

RC Rule 2

Given an ontology $O=<C, A, attr, B, M_a, M_b>$ with a labeling σ applied to O   (29)

$$\forall t \in attr: \sigma_C(\pi_1(t))=deselected \vee \sigma_A(\pi_2(t))= deselected \Rightarrow \sigma_{attr}(t) \neq selected$$

This transformation rule is the equivalent to RC Rule 1, applied to the special relationship between concepts and their attributes, which is represented in the mapping *attr*.

RC Rule 3

Given an ontology $O=<C, A, attr, B, M_a, M_b>$ with a labeling σ applied to O   (30)

$\forall t \in attr:\ \sigma_A(\pi_2(t))=$ *selected* $\Rightarrow \sigma_{attr}(t) \neq$ *deselected*

Here the more specific characteristic of the attribute-concept relationship is considered. No contradicting statements between the attribute and the connection are allowed, and this transformation rule together with the previous one ensures this.

Before the next rule can be given, first the notion of a path needs to be introduced and defined. Paths are very important in the specification of ontology views, as they introduce new relationships that are semantically linked to the original relationships.

Definition 17

$\qquad$ *Given an ontology $O=<C, A, attr, B, M_a, M_b>$*   (31)

we define

a path p of O $\Leftrightarrow p \equiv b_1, b_2, ..., b_n \in B^+$, *with $n \in \mathbb{N}_0$*
such that

$\pi_1(b_i)=\pi_2(b_{i-1})$, *with $i \in [2, n]$*

Definition 18

$\qquad$ *Given an ontology O*   (32)

We define

$\qquad$ *Path(O)$\equiv$\{set of all possible paths in O\}*

Notation 7

$\qquad$ *Given an ontology $O=<C, A, attr, B, M_a, M_b>$*   (33)

$\qquad$ *And a path $p \in Path(O)$, with $p= b_1, b_2, ..., b_n \in B^+$ ($n \in \mathbb{N}_0$)*

we denote

$\pi_1(p)=\pi_1(b_1)$

$\pi_2(p)=\pi_2(b_n)$

Definition 19

$\qquad$ *Given an ontology $O=<C, A, attr, B, M_a, M_b>$ and*

$\qquad$ *a labeling σ applied to O*

we define

a proper path $p \Leftrightarrow p \equiv b_1, b_2, ..., b_n \in B^+$ ($n \in \mathbb{N}_0$)   (34)

with

$p \in Path(O)$

$\sigma_c(\pi_2(p)) \neq$ '*deselected*'

$\sigma_B(b_j) \neq$ '*deselected*', *with $j \in [1,n]$*

Definition 20

Given an ontology O

(35)

We define

$PPath(O) \equiv \{set\ of\ all\ possible\ proper\ paths\ in\ O\}$

The additional definition for a more restricted set of paths $PPath(O)$ is given as these are the types of paths that are considered mostly for the concatenation and replacement cases, which will be discussed further on.

From the previous definitions it follows that the $PPath(O)$ is a subset of $Path(O)$ ($PPath(O) \subset Path(O)$).

To give a couple of examples of valid and invalid types of paths, a labeled ontology graph is presented in Figure 8.

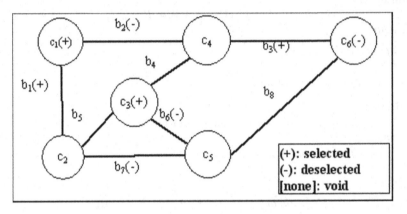

Fig. 8. Example Labeled Ontology Graph

Table 3 gives a number of paths and indicates why certain paths are invalid. Note that although not present in this table, it is possible for a valid proper path to connect a vertex (i.e. concept) that is deselected, as long as it is not the last vertex in the chain.

Table 3. Example Valid and Invalid Paths

Path	Validity	Description
$b_7b_6b_5b_1b_3b_8$	Invalid path	$\pi(b_1) \neq \pi(b_3) \rightarrow$ No concept that is connected by b_1 is connected by b_3.
$B_7b_6b_5b_1b_2b_4$	Valid path	
$B_7b_6b_5b_1b_2b_4$	Invalid Proper Path	$\sigma_B(b_7)=\sigma_B(b_6)=\sigma_B(b_2)=$deselected $\rightarrow$ Some relationships in the path are labeled 'deselected'.
$b_5b_4b_3$	Invalid Proper Path	$\pi_2(b_3)=c_6=$deselected $\rightarrow$ target concept is deselected.
$b_3b_4b_5b_1$	Valid Proper Path	

Now that the definitions for a path and a proper path have been given, the final RCOS rule can be presented. This rule specifies that if an attribute is selected, but the concept is deselected, i.e. there will be a need for distributing the attribute, there

should be a proper path to a concept that is not deselected (so the attribute can potentially be placed there).

RC Rule 4

Given an ontology $O=<C, A, attr, B, M_a, M_b>$ with a labeling σ applied to O   (36)

$$\forall t \in attr, \exists p \in PPath(O):$$

$$\sigma_A(\pi_2(t))=\text{'selected'} \land \sigma_C(\pi_1(t))=\text{'deselected'}$$

$$\Rightarrow \sigma_C(\pi_2(p)) \neq \text{'deselected'} \land \pi_1(p)= \pi_1(t)$$

An example of the combination described in the antecedent is presented in Figure 9,
where the concept "Writer" is deselected, while one of its attributes ("genre") is selected. To adhere to the rule, a proper path p needs to exist that has its final concept
not deselected, and that starts from the concept "Writer".

In this particular example there are two possible proper paths that can be used, and
they are given below. Which one of these paths will eventually be utilized (or may be

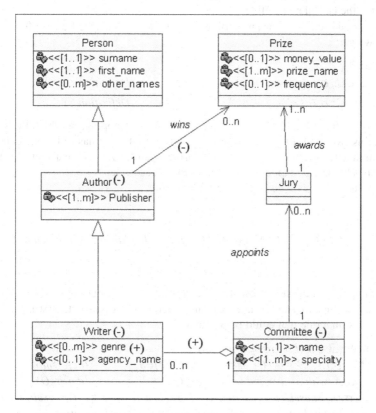

Fig. 9. Example Labeled UML Representation for RCOS Rule 4

even another possible solution might arise) is not important at this stage, merely that is possible to find at least one solution to the problem.

The first proper path that can be identified starts with the inheritance relationship between "Writer" and "Author". Because "Author" is deselected, this is not a valid end result for a path, but if the next inheritance relationship is included in the path (from "Author" to "Person"), a valid proper path is obtained, as none of the relationships is deselected, nor is the target concept ("Person"). Another possibility is the path that starts at "Writer", then goes to "Committee", and then crosses to "Jury". There are more possibilities (e.g. by extending the two paths already obtained), but as it is known for certain now that there is at least one proper path, the rule is satisfied.

3.2.2 Well Formedness Optimization Scheme (WFOS)

It might be clear what the intentions are of a certain labeling, but there possibly are statements that inevitably lead to a solution set that is not a valid ontology. The WFOS contains the proper rules to prevent this from happening.

Effectively, all the separate criteria as given in section 2 are interpreted here. It would suffice to say that per definition the set consisting of the selected elements of O (according to labeling σ) has to be a valid ontology, but here we want to establish the rules that will not only ensure this, but also are written in relation to the original ontology, i.e. the complete labeling.

A first, basic requisite for an ontology was $C \neq \emptyset$. In the specific case for the solution set we get:

$$\sigma_C(C) \neq \emptyset$$
$$\Rightarrow \quad \exists c \in C: \sigma_C(c) = 'selected' \lor \sigma_C(c) = 'void' \qquad (Definition\ \sigma_C(C))$$
$$\Rightarrow \quad \exists c \in C: \sigma_C(c) \neq 'deselected' \qquad (Definition\ S)$$

Intuitively, as there are only three possibilities for the mapping, of which the solution set contains two, and following from the criterion that there has to be one concept in the solution set for it to be a valid, well formed ontology, it is clear that at least one concept has to have a mapping that is other than 'deselected'.

This result is left as such and put into a first rule:

WF Rule 1

Given an ontology $O = <C, A, attr, B, M_a, M_b>$ with a labeling σ applied to O   (37)

$$\exists c \in C: \sigma_C(c) \neq 'deselected'$$

Another requirement that is present in the definition for an ontology is $\forall a \in A, \exists c \in C: a \in attr(c)$, stating that every attribute must belong to a concept. Starting from the adapted version a rule that could be postulated is:

$$\forall a \in \sigma_A(A), \exists c \in \sigma_C(C): a \in \sigma_{attr}(attr)(c)$$
$$\Rightarrow \quad \forall a \in \sigma_A(A), \exists c \in \sigma_C(C), \exists t \in \sigma_{attr}(attr): t(c) = a \qquad (def.\ Ontology)$$
$$\Rightarrow \quad \forall a \in A, with\ a = t(c)\ and\ t \in attr, c \in C: \qquad (def\ \sigma(O))$$
$$\neg(\sigma_A(a) = deselected) \Rightarrow \neg(\sigma_C(C) = deselected) \land \neg(\sigma_{attr}(t) = deselected)$$
$$\Rightarrow \quad \sigma_C(c) = deselected \lor \sigma_{attr}(t) = deselected \Rightarrow \sigma_A(a) = deselected$$
$$(*)$$

However this rule is not in satisfactory form as an additional requirement must also be considered. This additional requirement will then be combined with the above unsatisfactory rule to develop a basis for the additional rules.

The other requirement for the ontology that will be used next is $Attr:C \rightarrow 2^A$. Adapted to the well formedness of a solutions set, this gives;

$\sigma_{attr}(attr)$: $\sigma_C(C) \rightarrow 2 \sigma^{C(A)}$

$\Rightarrow$ *Given an attribute mapping t=(c, a) with $c \in C$, $a \in A$*

     $\neg(\sigma_{attr}(t)=deselected) \Rightarrow \neg(\sigma_C(c)=deselected) \wedge \neg(\sigma_A(a)=deselected)$

$\Rightarrow$ $\sigma_C(c)=deselected \vee \sigma_A(a)=deselected \Rightarrow \sigma_{attr}(t)=deselected$

     (**)

Combining the above two results we obtain:

(*) $\wedge$ (**)

$\Rightarrow$ *1)* $\sigma_C(c)=deselected \Rightarrow \sigma_{attr}(t)=deselected$

     2) $\sigma_A(a)=deselected \Rightarrow \sigma_{attr}(t)=deselected$

     3) $\sigma_C(c)=deselected \Rightarrow \sigma_A(a)=deselected$

     4) $\sigma_{attr}(t)=deselected \Rightarrow \sigma_A(a)=deselected$

1) $\wedge$ 3) $\Rightarrow$ $\sigma_C(c)=deselected \Rightarrow \sigma_A(a)=deselected \wedge \sigma_{attr}(t)=deselected$

2) $\wedge$ 4) $\Rightarrow$ $\sigma_A(a)=deselected \Leftrightarrow \sigma_{attr}(t)=deselected$

These last statements are taken as the rules for the well formedness optimization scheme, thus:

WF Rule 2

Given an ontology O=<C, A, attr, B, Ma, Mb> with a labeling σ applied to O   (38)

$\forall t \in attr: \sigma_C(\pi_1(t))=deselected \Rightarrow \sigma_A(\pi_2(t))=deselected \wedge \sigma_{attr}(t)=deselected$

WF Rule 3

Given an ontology O=<C, A, attr, B, M_a, M_b> with a labeling σ applied to O   (39)

$\forall t \in attr: \sigma_A(\pi_2(t))='deselected' \Leftrightarrow \sigma_{attr}(t)='deselected'$

In Figure 10 an example of these two rules is shown. As the attribute mapping is not explicitly catered for in UML (it is not a link, but a container, i.e. attributes are contained inside a class), a different notation (semantic net notation [Feng, Chang et al. 2002]) for the example was used. The concept "Person" is deselected, and so is according to the antecedent of WF Rule 2. The attribute "surname" is selected (and its connection to "Person" has no indication, so it's the default label, i.e. "void"), and that is invalid for WF Rule 2. The attribute "first_name" and its attribute mapping are both deselected, giving a valid case for WF Rule 2. As "Document" has the default – void – label, the antecedent is not met, so every case including that concept is valid according to WF Rule 2.

However, WF Rule 3 says both attributes and attribute mappings for the concept "Document" result in invalid combinations, as both the attribute as its attribute

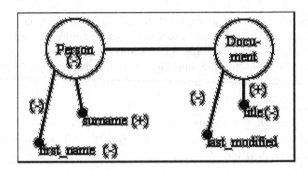

Fig. 10. WFOS Valid and Invalid Cases in Semantic Net Notation [Feng, Chang et al. 2002]

mapping should be deselected (if one of them is deselected). Looking at the attributes for concept "Person", both of them are valid. The attribute "surname" does not trigger WF Rule 3, and "first_name" does, but is a valid combination.

The next requirement for an ontology as given in the definition is $B \subseteq C \times C$. Put in the proper context this can be stated as:

$\sigma_B(B) \subseteq \sigma_C(C) \times \sigma_C(C)$
$\Rightarrow$ *given a binary relation $b=(c_1, c_2)$ with $c_1, c_2 \in C$*

     $\sigma_B(b) \neq$ *'deselected'* $\Rightarrow \sigma_C(c_1) \neq$ *'deselected'* $\wedge \sigma_C(c_2) \neq$ *'deselected'*
$\Rightarrow$ $\sigma_C(c_1)=$ *'deselected'* $\vee \sigma_C(c_2)=$ *'deselected'* $\Rightarrow \sigma_B(b)=$ *'deselected'*
$\Rightarrow$ *(1) $\sigma_C(c_1)=$ 'deselected' $\Rightarrow \sigma_B(b)=$ 'deselected'*
     (2) $\sigma_C(c_2)=$ 'deselected' $\Rightarrow \sigma_B(b)=$ 'deselected'

(1) and (2) are now used to define the next rule which is as follows:

WF Rule 4

Given an ontology $O=<C, A, attr, B, M_a, M_b>$ with a labeling σ applied to O

     1) $\forall b \in B: \sigma_C(\pi_1(b))=$ 'deselected' $\Rightarrow \sigma_B(b)=$ 'deselected'   (40)

     2) $\forall b \in B: \sigma_C(\pi_2(b))=$ deselected' $\Rightarrow \sigma_B(b)=$ 'deselected'

Additional rules could be provided here for the n-ary relationship, but through transformations of the design n-ary relationships can be written as binary relationships, so the same rules would apply to these relationships.

To be complete, the ontology graph requirement for an ontology is repeated here, although it is not developed into another rule, but rather kept in its original form.

WF Rule 5

       Given an ontology O with a labeling σ applied to O   (41)

       $\exists$ an ontology graph $G\sigma_{(O)}$ for $\sigma(O)$

An example of an invalid ontology is given in Figure 11. Although it may not be clear immediately (consider the difficulty if not impossibility of visually recognizing

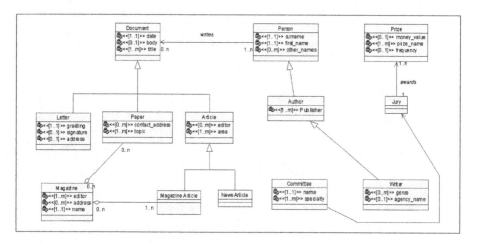

Fig. 11. UML Representation of Invalid Ontology

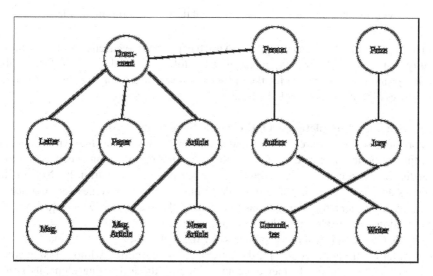

Fig. 12. Ontology Graph Representation of the Ontology (Invalidity Unclear)

a similar situation in real – complex – ontologies), even not when looking at the corresponding ontology graph (in Figure 12), this ontology is invalid as its ontology graph contains an island (or proper component in graph theory terms). This becomes clear when the vertices from the ontology graph are rearranged (no information changed, just the visual coordinates).

The fifth WFOS Rule clearly requires a valid ontology graph, and a valid ontology graph according to its definition cannot contain any islands. An alternative representation of the ontology graph is shown in Figure 13. They boxed gray area is a clear separate island, and so the resulting ontology graph (that is obtained from the ontology) is invalid, and so the ontology is invalid as well (not well formed, i.e. not adhering to WF Rule 5).

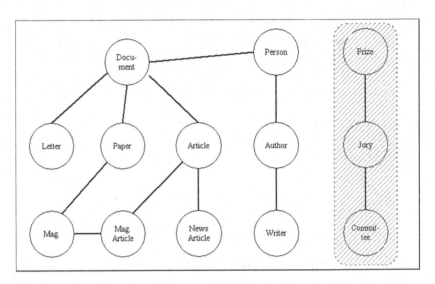

Fig. 13. Ontology Graph Representation of the Ontology (Invalidity Clear)

This optimization scheme, together with the previous one, ensures the minimal quality that is necessary for any materialized ontology view (that it is in fact a valid ontology, and that the criteria for the optimization are clear). Any meaningful extraction process should always include these two optimization schemes.

3.2.3 Semantic Completeness Optimization Scheme (SCOS)

The previous schemes were obvious, and are the minimal schemes that will be applied (except maybe in some rare cases), but the following optimization schemes, starting with the SCOS are not always wanted, or even possible concurrently. Not all the schemes are given here, and it is possible to introduce new schemes for new needs, however, the schemes given demonstrate clearly the manner in which quality – or optimization – is introduced in the extraction process.

The SCOS considers the completeness of the concepts, i.e. if one concept is defined in terms of another concept, the latter cannot be omitted without losing some semantic meaning of the former concept. Including these defining elements is not always required, but in those cases this optimization scheme should not be selected, as the ordered list of optimization schemes can be freely composed by an ontology engineer.

Firstly, some additional definitions need to be made to specify what is meant by 'defining elements'. In the definition for an ontology the set of binary relationships was further split up in different types of relationships. For some of these sets the direction of the relationship is important, so now a more detailed function π is introduced to serve the purpose of distinguishing between the linked concepts.

Additional information on the notation of some general concepts such as sub-super classes, whole-part aggregation, etc. is given next to facilitate further discussion.

Notation 8

Given an ontology $O=<C, A, attr, B, M_a, M_b>$ with a labeling σ applied to O

$$B = B_i \cup B_{agg} \cup B_r, \ \forall c_1, c_2 \in C \tag{42}$$
$$\text{We denote}$$
$$c_1 \text{ is a subconcept of } c_2$$
$$\Leftrightarrow \in b \in B_i: \ \pi_1(b)=c_1 \wedge \pi_2(b)=c_2$$

Notation 9

Given an ontology $O=<C, A, attr, B, M_a, M_b>$ with a labeling σ applied to O

$$B = B_i \cup B_{agg} \cup B_r, \ \forall c_1, c_2 \in C \tag{43}$$
$$\text{We denote}$$
$$c_1 \text{ is a superconcept of } c_2$$
$$\Leftrightarrow \in b \in B_i: \ \pi_1(b)=c_2 \wedge \pi_2(b)=c_1$$

In the example shown in Figure 14 (a) "Tree" is the sub-concept ($\pi_1(b_1)$), while "Plant" is the super-concept ($\pi_2(b_1)$).

Notation 10

Given an ontology $O=<C, A, attr, B, M_a, M_b>$ with a labeling σ applied to O

$$B = B_i \cup B_{agg} \cup B_r, \ \forall c_1, c_2 \in C \tag{44}$$
$$\text{We denote}$$
$$c_1 \text{ is a part of } c_2$$
$$\Leftrightarrow \in b \in B_{agg}: \ \pi_1(b)=c_1 \wedge \pi_2(b)=c_2$$

Notation 11

Given an ontology $O=<C, A, attr, B, M_a, M_b>$ with a labeling σ applied to O

$$B = B_i \cup B_{agg} \cup B_r, \ \forall c_1, c_2 \in C \tag{45}$$
$$\text{We denote}$$
$$c_1 \text{ contains } c_2$$
$$\Leftrightarrow \in b \in B_{agg}: \ \pi_1(b)=c_2 \wedge \pi_2(b)=c_1$$

This aggregation relationship is shown in UML notation in Figure 14(b). The concept "Computer" is the whole-concept ($\pi_2(b_2)$), while "CPU" is the part-concept ($\pi_1(b_2)$)

Notation 12

Given an ontology $O=<C, A, attr, B, M_a, M_b>$ with a labeling σ applied to O

$$B = B_i \cup B_{agg} \cup B_r, \ \forall c_1 \in C, \ \forall a_1 \in A \tag{46}$$
$$\text{We denote}$$
$$a_1 \text{ is a defining attribute for } c_1$$
$$\Leftrightarrow \exists t \in attr: t=(c_1, a_1) \wedge \exists m \in card : \neg(m_{min}(t)=0)$$

Figure 14(c) shows an example of a defining attribute. A "Person" has to have at least (here exactly) one "surname".

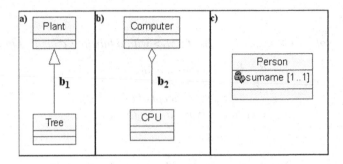

Fig. 14. Example Semantic Essential Elements

Because ontologies allow for such a diverse way of defining new elements (e.g. a new class that is the union of two existing concepts), a simplified approach is taken here. Although we will show only a limited number of ways for developing rules, extending this basic set is straight forward. As mentioned before, optimization schemes can also be constructed for various standards.

The first defining way considered here is the super-concept of another concept.

SC Rule 1

Given an ontology $O=<C, A, attr, B, M_a, M_b>$ with a labeling σ applied to O (47)

$$\forall b \in B: \sigma_C(\pi_1(b))=selected \Rightarrow \sigma_C(\pi_2(b))=selected \wedge \sigma_B(b)=selected$$

This rules states that whenever a concept is retained for a view, the superconcepts of it, and the actual inheritance relationship stating this are required as well to be semantically complete.

The following rule does the same for the part relationship, i.e. when a concept is required (mapped to 'selected') then part concepts of this concepts are retained as well.

SC Rule 2

Given an ontology $O=<C, A, attr, B, M_a, M_b>$ with a labeling σ applied to O,
and a binary aggregation relationship $b \in B$ (48)

$$\sigma_C(\pi_2(b))=selected \Rightarrow \sigma_C(\pi_1(b))=selected \wedge \sigma_B(b)=selected$$

Finally, also a rule is given here for the 'not null' attributes of a concept, as they provide information that according to the ontology specification should always be there (an empty field is not possible). Although 'defining' might be too strong a word for these attributes, they are nonetheless required for there to be a semantically complete ontology view.

SC Rule 3

Given an ontology $O=<C, A, attr, B, M_a, M_b>$ with a labeling σ applied to O,
an attribute mapping t, and a cardinality m for t (49)

$$\sigma_C(\pi_1(t))=selected \wedge m_{min} \neq 0 \Rightarrow \sigma_A(\pi_2(t))=selected \wedge \sigma_{attr}(t)=selected$$

3.2.4 Total Simplicity Optimization Scheme (TSOS)

The TSOS is an applied and modified version of Kruskal's Algorithm [Kruskal 1956] for minimal spanning trees. Applying this optimization scheme will result in the smallest possible solution that is still a valid ontology (starting from a certain solution set). This optimization scheme is included as an example of how certain requirements (smallest possible solution) are translated in an optimization scheme. Furthermore, it clearly shows as well that not all optimization schemes are desirable in every situation. Sometimes the most versatile solution is sought, not the smallest, and thus the TSOS would not be applied. The choice of what is considered important or high quality is represented by the selection of optimization schemes, and the order in which they are applied.

To get to a solution for the TSOS, a new labeling is introduced. This labeling only applies to the element previously labeled 'void'. When the methods of achieving this optimization scheme are discussed, the reasoning behind the additional labeling will be clarified. First, definitions for different sets are given to aid in the readability of the following rules and definitions.

Definition 21

Given an ontology $O=<C, A, attr, B, M_a, M_b>$ with a labeling σ applied to O

$$B' \equiv \{b \in B \mid \sigma_B(b) = void\}$$

$$C' \equiv \{c \in C \mid \sigma_C(c) = void\} \tag{50}$$

$$A' \equiv \{a \in A \mid \sigma_A(a) = void\}$$

Definition 22

Given an ontology $O=<C, A, attr, B, M_a, M_b>$ with a labeling σ applied to O

$$B'' \equiv \{b \in B \mid \sigma_B(b) = selected\}$$

$$C'' \equiv \{c \in C \mid \sigma_C(c) = selected\} \tag{51}$$

$$A'' \equiv \{a \in A \mid \sigma_A(a) = selected\}$$

For the new labeling that is needed, the labeling σ could be extended, but as not all targets are necessary, and for clarity in notation, a new labeling δ is introduced here.

Definition 23

$$T \equiv \{accepted, rejected\} \tag{52}$$

The set of possible labels contains merely two elements, but these are on top of the δ labeling.

Definition 24

$$\delta_C : C' \to T$$

$$\delta_B : B' \to T \tag{53}$$

Definition 25

$$B_a' \equiv \{b \in B' \mid \delta_{B'}(b) = accepted\}$$
$$B_r' \equiv \{b \in B' \mid \delta_{B'}(b) = rejected\}$$
$$C_a' \equiv \{c \in C' \mid \delta_{C'}(c) = accepted\}$$
$$C_r' \equiv \{c \in C' \mid \delta_{C'}(c) = rejected\}$$

(54)

Now that the differently labeled elements have been grouped in sets, the rules that follow become more readable. These rules represent what is understood by a total simplified solution. As mentioned before, Kruskal's Algorithm for minimal spanning trees is taken as a starting point, but as no weights are present, labels are used to replace them. Some of these labels – 'selected' and 'deselected' – have to be treated as fixed by the optimization scheme used, and so the best solution is sought by labeling the 'void' elements so that all of the following rules are adhered to.

TS Rule 1

Given an ontology $O = <C, A, attr, B, M_a, M_b>$ with a labeling δ applied to O,
$$<C_a' \cup C'', B_a' \cup B''> \text{ is a valid ontology graph}$$

(55)

This rule can be considered a rule that gives the lower bound, as it limits the number of elements that can be rejected. The elements accepted by the algorithm, together with the earlier labeled 'selected' elements still have to form a valid ontology graph.

An example of a seemingly valid ontology, but which is in reality invalid (because it has an invalid ontology graph), was shown in Figure 11, Figure 12 and Figure 13.

The next rule, on the other hand, determines the upper bound, stating that elements that have been labeled 'accepted' should always be necessary to have a valid ontology graph. If not, they should have been labeled 'rejected' instead, as the resulting valid ontology graph (after they are rejected) represents a smaller – i.e. more simplified – solution.

TS Rule 2

Given an ontology $O = <C, A, attr, B, M_a, M_b>$ with a labeling δ applied to O,
And an ontology graph $G_O = <V, E>$ for O
$$\forall V' \subseteq C_a, E' \subseteq B_a, O_l \in \vartheta: <V \backslash V', E \backslash E'> \text{ is not a valid ontology graph for } O_l$$

(56)

with
$$V' \neq \emptyset$$

At first, this rule seems a bit complicated, but intuitively, it states that the obtained two tuple ($<V \backslash V', E \backslash E'>$) is not a valid ontology graph for any ontology (O_l). In other words, the two tuple does not conform to all the requirements set out in the definition for an ontology graph. It is straight forward to convert elements of $V \backslash V'$ to concepts, and elements of $E \backslash E'$ to relationships. This lack of validity could be due to; i) the two tuple is not a valid graph, or ii) the two tuple is a valid graph, but contains islands (proper components).

The second case speaks for itself, but the first case needs clarification. From the elements we have in the sets an invalid graph can be obtained if at least one edge connects at least one vertex that is not an element of $V \backslash V'$ ($\exists e \in E \backslash E' : e = (v_1, v_2) \Rightarrow v_1 \vee v_2 \notin V \backslash V'$).

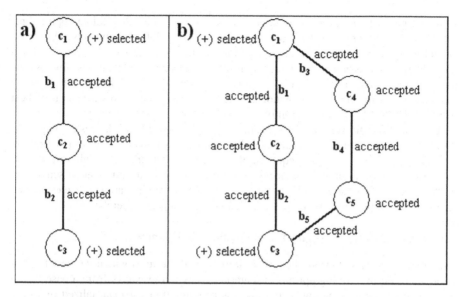

Fig. 15. Two Valid Graphs with Accepted and Selected Elements Exclusively

An example of applying this rule is demonstrated with the aid of Figure 15, show-ing two graphs with the same selected elements, but graph b) introduces a number of additional accepted elements. It is clear that graph a) is simpler than graph b), and so graph b) is not optimal and therefore not a solution (assuming they both aim to pro-vide a solution for the same base ontology in the same extraction process). The previ-ous rule can be applied, and so any combination of accepted elements can be sought, then taken away from the graph, and if the result is a possible valid ontology graph, the starting graph does not adhere to the rule. For graph a), there are only three ac-cepted (so 6 possible combinations, i.e. $\{(b_1), (b_2), (c_2), (b_1c_2),(b_1b_2),(b_2c_2)\}$, as the order is irrelevant). For graph b) however, a lot more combinations are possible, and an example can be found of a combination of accepted elements that – when they are left out – leaves a valid ontology graph. An example is the set of elements $\{b_1,b_2,c_2\}$, or $\{b_3,b_4,b_5,c_4,c_5\}$. Both these sets can be left out and the result would still be a valid ontology graph (for a certain ontology).

The last rule is a very short and strict rule, ensuring the simplest solution. It elimi-nates the possibility of having an attribute still labeled as a 'void'. The 'void' label would indicate that it is still undecided what will happen to the attribute, but as the sim-plest solution is sought, all the attributes that were undecided should be deselected now, as they are not necessary, and thus not part of the simple solution that is the goal.

TS Rule 3

Given an ontology $O=<C, A, attr, B, M_a, M_b>$ with a labeling σ applied to O,

$$\forall a \in A: \neg\; \sigma_A(a)=void \tag{57}$$

Note that potentially more than one solution is possible for a given solution set. Looking back at the non optimal ontology graph presented in Figure 15 (graph b), two different sets were given, both of which could be left out to result in a still valid and optimal result. The two outcomes of these operations are both possible solution sets. Through a weighting system a more efficient connection can be preferred (see Appendix A for a more detailed discussion of possible weights of relationships), but then it is still possible that multiple potential results have the same total sum of the weights, and thus can be considered equally optimized solutions. A practical example of this selection of best paths can be found in [Wouters, Dillon et al. (to appear December 2003)].

TSOS is different from previous optimization schemes in that it leaves a completely decided solution set, i.e. one without 'void' elements. It is only useful to put such an optimization scheme once in a sequence of optimization schemes, and only towards the end, as it would not leave an undecided element for the next optimization scheme to work with. A practical example of such a sequence of optimization schemes (called priority list) can be found in [Wouters, Dillon et al. (to appear December 2003)].

3.3 Dynamic Rules Necessary for Algorithmic Development

The previous sections set out the key definitions and the rules which define what has to be achieved in deriving optimized materialized ontology views from a base ontology. To facilitate automation of the process of extraction of materialized ontology views we need to develop algorithms which carry out the implementation of the rules. In other words, they set the 'how' rather than the 'what' of materialized ontology view extraction. As a prelude to actually working out the algorithms, we will specify a set of rules that provide the basis for the algorithms. The rules given here have a more dynamic character than the ones given in the previous section. The dynamic transformation of what replacements, modifications and even extensions can be made to achieve the static requirements is the main focus of this section.

An example is the attribute riddance rule given before. It just states that no more 'void' labels are allowed for the attributes. How we should go about achieving this is demonstrated by algorithms, such as in this case; "replace all the 'void' labels with 'deselected' labels for the TSOS".

First of all, three general rules are given in their most generic form. Further on, these rules will be utilized in algorithms.

3.3.1 Mapping Modification
This generic definition shows how the mapping in general can be modified, and thus has no impact on the actual ontology. The ontology remains exactly as it was before, i.e. there is no transformation from an ontology O to an ontology O', but only from a mapping σ to a mapping σ'. As in the case of the other rules, this rule is also defined as a transformational function with certain input parameters.

Mapping Modification Definition

Given
an ontology $O=<C, A, attr, B, M_a, M_b>$ with a labeling σ applied to O,
$X \in \{C, A, attr, B\}$, (58)
$s_1, s_2 \in S$,
$x \in X$, and
$\sigma_X(x)=s_1$

we define
$$\xi_X\sigma(x, s_2) = \sigma' =(\sigma\backslash\{\sigma_X(x)\})\cup\sigma_X'(x)$$
with

$$\sigma_X'(x)=s_2$$

Intuitively, this definition specifies that an existing mapping can be modified, resulting in a new mapping. The modification only needs an element and a label, and the resulting mapping is the same as the original, except that the label for the specified element has been changed to the new label.

3.3.2 Attribute Distribution
There already has been a mention of the necessity for redistributing attributes to obtain an optimized result (section 3.2.1). The definition of how this transformation is done is given here.

Attribute Distribution Definition

Given
  an ontology $O=<C, A, attr, B, M_a, M_b>$,
  $t\in attr$,
  $b\in B$
with
  $$\pi_1(t)=\pi_1(b)^{(1)} \vee \pi_1(t)=\pi_2(b)^{(2)}$$
we define
  $$\xi_A^O(b, t)\equiv O'\equiv<C', A', attr', B', M_a', M_b'>$$
with
  $A'=A$
  $C'=C$
  $B'=B$ (59)
  $M_b'=M_b$
  (1) $\underline{\pi_1(t)=\pi_1(b)}$
      $attr'=(attr\backslash\{t\})\cup\{t'\}$
        with $t'=(\pi_2(b), \pi_2(t))$
      $M_a'=(M_a\backslash\{m(\pi_2(t))\})\cup\{m'(\pi_2(t))\}$
        with $m'(\pi_2(t))=(m_{min}(\pi_2(t))*m_{min1}(b),m_{max}(\pi_2(t))*m_{max1}(b))$
  (2) $\underline{\pi_1(t)=\pi2(b)}$
      $attr'=(attr\backslash\{t\})\cup\{t'\}$
        with $t'=(\pi_1(b), \pi_2(t))$
      $M_a'=(M_a\backslash\{m(\pi_2(t))\})\cup\{m'(\pi_2(t))\}$
        with $m'(\pi_2(t))=(m_{min}(\pi_2(t))*m_{min2}(b),m_{max}(\pi_2(t))*m_{max2}(b))$

This definition specifies how attributes can 'travel' across a binary relationship, and what the consequences are for the cardinality of the concerned attribute.

3.3.3 Relationship Concatenation
Often it is required to drop a certain concept from a solution, but still the semantic link it provides to relationships that bridge two other concepts might be relevant. The notion of a path, which was introduced previously, is essential to the concatenation rule.

Relationship Concatenation Definition

Given
  an ontology $O=<C, A, attr, B, M_a, M_b>$,

  a path $p= b_1, b_2, ..., b_n \in B^+ (n \in N_0)$
we define
  $\xi_B^O(p) \equiv O' \equiv <C', A', attr', B', M_a', M_b'>$
with

  $A' = \{a \in A | a = \pi_2(t) \wedge \pi_1(t) \neq \pi_1(b_i), \text{ with } t \in attr, i \in [2,n]\}$

  $attr' = \{t \in attr | \pi_1(t) \neq \pi_1(b_i), \text{ with } i \in [2,n]\}$

  $C' = \{c \in C | c \neq \pi_1(b_i), \text{ with } i \in [2,n]\}$ (60)

  $B' = \{b \in B | b \neq b_i, \text{ with } i \in [2,n]\}$
  $M_a' = \{m \in M_a | m(a) = (n_1, n_2), \text{ with } a \in A', n_1, n_2 \in card\}$
  $M_b' = \{m \in M_b | m(b) = (n_1, n_2, n_3, n_4), \text{ with } b \in B', n_1, n_2, n_3, n_4 \in card\} \cup \{m(b')\}$
    with

      $b' = (\pi_1(p), \pi_2(p)) \wedge$
      $m(b') = (m_{min1}(b_1)*m_{min1}(b_2)*...*m_{min1}(b_n),$
          $m_{max1}(b_1)*m_{max1}(b_2)*...*m_{max1}(b_n),$
          $m_{min2}(b_1)*m_{min2}(b_2)*...*m_{min2}(b_n),$
          $m_{max2}(b_1)*m_{max2}(b_2)*...*m_{max2}(b_n))$

Having a valid path as an input, the resulting ontology replaces the linking concepts and relationships by a new one, calculates the cardinality for the new relationship, and discards all the attributes of the replaced concepts. The multiplication table for the cardinality set has to be specified, otherwise results cannot be calculated (see Appendix A).

4 Development of the Algorithms

The previous sections have shown what the requirements for different optimization schemes are, and what transformations we are allowed to use to modify a labeled ontology so that it complies with the requirements. This section links those two together, and for the relevant optimization schemes algorithms are given that show step by step how an application – with minimal human interaction required – can arrive at an adequate result for an optimization scheme.

4.1 Requirement Consistency Optimization Scheme

The rules set out for the RCOS ensure there is no contradiction between statements in the input user requirements. In other words, this optimization scheme checks for combinations of labels that lead to a situation that is dubious or ambiguous for the system, and that cannot be resolved without more clarification. This clarification needs to come from the user. Note that this clarification process can be automated as well, by having a rule system that can be applied in cases of inconsistency.

An example of such an automation can be that in case of inconsistency, elements labeled as 'selected' always take preference over elements labeled as 'deselected'. In case of RC rule 1 we could have a relationship b with a 'selected' labeling, but with $\pi_l(b)$ labeled as 'deselected'. In other words, the end result should definitely have the relationship, but it should not contain one of the concepts that is being related to another by the relationship. Clearly there is no solution that satisfies both of these, so the requirements are inconsistent. Without any further information an automated system would not be able to resolve this inconsistency, but if we take into account the preference rule that was added, the inconsistency can be easily resolved.

Note that this optimization scheme always is the first to be applied, and all the other optimization schemes rely on the fact that there are no inconsistencies in the requirements anymore.

The first three rules of the RCOS require algorithms that are very similar to algorithms for rules of other optimization schemes. For this reason they are not given here, but by slightly modifying other algorithms the necessary algorithms are readily obtainable.

The fourth rule needs a more complex and unique algorithm to enforce it. It uses the notion of a proper path discussed earlier.

4.1.1 RC Rule 4

1. Loop through all $c \in C$
    1.1. if ($\sigma_C(c)$== 'deselected')
        1.1.1.   loop through all attributes $a \in A$ that have an attributemapping to c
            1.1.1.1. if ($\sigma_A(a)$== 'selected')
                1.1.1.1.1.   if FIND_PROPER_PATH(c) returned false
                              // no proper path was found
                    1.1.1.1.1.1. NO SOLUTIONS
                    1.1.1.1.1.2. exit algorithm

Although seemingly very brief here, this is only because the sub algorithm FIND_PROPER_PATH was used. The particular algorithm for finding such a proper path is not given here, but an indication of how it operates can be found in the example given in Figure 9.

4.2 Well Formedness Optimization Scheme Algorithm

The algorithm shows how an ontology and mapping are checked and modified to comply with the WFOS. Every rule is sequentially visited to get to the final result, and the rules are identified in the algorithm as well, for improved understandability of the algorithm. Some input ontologies and mappings have no solutions that adhere to all requirements given by the WFOS. In the algorithm these cases are indicated as we come across them. To resolve these cases, a similar solution to the one discussed in the previous section can be used sometimes, but are not considered here.

In the algorithm it will indicated if certain cases are not considered because they are not present thanks to the prerequisite of the RCOS – remember that RCOS is a prerequisite to most optimization schemes, including WFOS.

4.2.1 WF Rule 1

1. if $(\sigma_C == \varnothing)$
    1.1. NO SOLUTIONS (all the concepts are deselected, so the end result will be empty)

4.2.2 WF Rule 2

1. While $(\exists t \in$ attr: $\sigma_C(\pi_1(t)) ==$ 'deselected' $\wedge$ $(\sigma_A(\pi_2(t)) \neq$ 'deselected' $\vee$
    $\sigma_{attr}(t) \neq$ 'deselected'))
    1.1. if there is no proper path p from $\pi_1(t)$ to another concept c_1 that is not 'dese-
          lected' $(\sigma_C(c_1) \neq$ 'deselected')
          1.1.1. if $(\sigma_A(a) ==$ 'void')   // never 'selected' because RC4
                  1.1.1.1. $\sigma \leftarrow \xi_A^\sigma(a,$ 'deselected')
    1.2. else (there is one (p) or more ($\{p_1, p_2, ..., p_n\}$) proper paths)
          1.2.1. if multiple paths
                  1.2.1.1. p $\leftarrow$ BEST_PATH($\{p_1, p_2, ..., p_n\}$)
          1.2.2. loop i from 1 to n   // with $p = b_1 b_2 ... b_n$
                  1.2.2.1. O $\leftarrow \xi_A^O(b_i, t)$
                  1.2.2.2. t $\leftarrow$ t'   // t' is constructed in Attribute Distribution Rule

In step 1.2.1.1 another algorithm BEST_PATH is called. This algorithm determines the 'best' path that is available. What is considered a good or strong path is an extensive topic, and the algorithm used here is given in Appendix A. Note that this is only one possible solution, and it is not the intention of the authors to claim this is the best or only solution. In reality it will depend on what preferences are emphasized, and a good solution for that particular case can be found, but no general 'best' solution. The algorithm for BEST_PATH supplied in the appendices suited our needs, and thus is provided merely as an example of one possibility amongst many.

Note that the algorithms that are given here are not meant to be the most efficient ones. The algorithms here serve the purpose of demonstrating how they resolve encountered problems, not how they can (or should) be implemented most efficiently.

4.2.3 WF Rule 3

2. Loop through all $t \in$ attr
    2.1. if $(\sigma_A(\pi_2(t)) ==$ 'deselected' $\wedge \sigma_{attr}(t) ==$ 'void')
                                         // never $\sigma_{attr}(t)=$'selected' because RC2
          2.1.1. $\sigma \leftarrow \xi_{attr}^\sigma(t,$ 'deselected')
    2.2. else
          2.2.1. if $(\sigma_{attr}(t) ==$ 'deselected' $\wedge \sigma_A(\pi_2(t)) ==$ 'void')
                                         // never $\sigma_A(\pi_2(t))=$'selected' because RC3
                  2.2.1.1. $\sigma \leftarrow \xi_A^\sigma(\pi_2(t),$ 'deselected')

As previously mentioned, not all possible cases are considered here, as the RCOS is a requirement, rendering some cases from not occuring here. The appropriate rules have been stated.

4.2.4 WF Rule 4

1. loop through all $b \in B$

    1.1. if $(((\sigma_C(\pi_1(b)) ==$ 'deselected' $\vee \; \sigma_C(\pi_2(b)) ==$ 'deselected') $\wedge \; \sigma_A(\pi_2(t)) ==$ 'void')

        1.1.1. $\sigma \leftarrow \xi_B{}^\sigma(b,$ 'deselected')

4.2.5 WF Rule 5

The main requirements for this rule (WF rule 5) are always correct, as they are just an assignment of certain elements to a new set. However, the third, "no islands in the graph" requirement is important here.

1. if $(ISLANDS\_EXIST(\sigma(O)) ==$ true)

    1.1. NO SOLUTIONS (a 'deselect' labeling prevents an ontology graph from being constructed)

The ISLANDS_EXIST algorithm uses Kruskal's algorithm to try to get to a minimal spanning tree, and if such a minimal spanning tree cannot be accomplished, there is an island. The algorithm attempts to find a minimal spanning tree, and such a tree has an interesting characteristic that is used here; if successful, the number of vertices is always equal to the number of edges plus one. As the resulting graph is acyclic (a minimal spanning tree cannot be cyclic), we know that there are never going to be more edges than the amount of vertices minus one. The third possibility is that there are less edges than the number of vertices minus one, indicating there must be at least two islands (if there's only one island, it is the entire ontology, logically). A simple example of these three cases is shown in Figure 16.

For a more thorough explanation of these characteristics of graphs, please refer to [Biggs, Lloyd et al. 1976], as these characteristics are taken here in a matter-of-fact manner, without explanations or proofs.

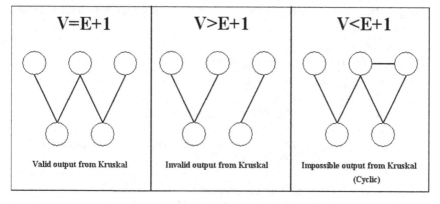

Fig. 16. The Three Cases for an Ontology Graph

4.3 Semantic Completeness Optimization Scheme Algorithm

As in the previous algorithm every rule taken from the requirements is visited and receives its own algorithm. In all the cases it is possible that through a 'deselected' label of a certain element no semantic completeness can be obtained. A mention of this particular case is made in the algorithms. It does not lead to an exit from the loop, but an indication that the ideal solution was not possible. In other words, if the ideal solution is not possible, the algorithm will inform the system of this, but it will still continue, so that the final result is as close to the ideal solution as possible (given the input labeling). The same type of preference rules as the one shown in section 4.1 can be used here, but to focus on that here would distract from the main focus of this paper.

4.3.1 SC Rule 1
1. $B_{temp} \leftarrow B_i$ // B_i is set of all binary inheritance relationships
2. while $(B_{temp} \neq \varnothing)$
   2.1. $b_{temp} \leftarrow B_{temp}[0]$
   2.2. $b_{sub} \leftarrow$ FIND_SUB$(B_{temp}, \pi_1(b_{temp}))$
   2.3. while $(b_{sub} \neq \varnothing)$
      2.3.1. $b_{temp} \leftarrow b_{sub}$
      2.3.2. $b_{sub} \leftarrow$ FIND_SUB$(B_{temp}, \pi_1(b_{temp}))$
   2.4. if $(\sigma_C(\pi_1(b_{temp})) ==$ 'selected')
      2.4.1.   if $(\sigma_B(b_{temp}) ==$ 'deselected' $\vee \sigma_C(\pi_2(b_{temp})) ==$ 'deselected')
         2.4.1.1. NO SC POSSIBLE
      2.4.2.   else
         2.4.2.1. $\sigma \leftarrow \xi_B{}^\sigma (b_{temp},$ 'selected')
         2.4.2.2. $\sigma \leftarrow \xi_C{}^\sigma (\pi_2(b_{temp}),$ 'selected')
   2.5. $B_{temp} \leftarrow B_{temp} \setminus \{b_{temp}\}$

Note that the additional algorithm FIND_SUB is a simple loop that goes through the first input parameter and tries to find a relationship that has the second input parameter as the superconcept of another concept. This (inheritance) relationship is then returned, or null if none was found. This algorithm is omitted from this article.

4.3.2 SC Rule 2
1. $B_{temp} \leftarrow B_{agg}$ // B_{agg} is the set of all the binary aggregation relationships
2. while $(B_{temp} \neq \varnothing)$
   2.1. $b_{temp} \leftarrow B_{temp}[0]$
   2.2. $b_{whole} \leftarrow$ FIND_WHOLE$(B_{temp}, \pi_2(b_{temp}))$
   2.3. while $(b_{whole} \neq \varnothing)$
      2.3.1. $b_{temp} \leftarrow b_{whole}$
      2.3.2. $b_{whole} \leftarrow$ FIND_WHOLE$(B_{temp}, \pi_2(b_{temp}))$
   2.4. if $(\sigma_C(\pi_2(b_{temp})) ==$ 'selected')
      2.4.1.   if $(\sigma_B(b_{temp}) ==$ 'deselected' $\vee \sigma_C(\pi_1(b_{temp})) ==$ 'deselected')
         2.4.1.1. NO SC POSSIBLE
      2.4.2.   else
         2.4.2.1. $\sigma \leftarrow \xi_B{}^\sigma (b_{temp},$ 'selected')
         2.4.2.2. $\sigma \leftarrow \xi_C{}^\sigma (\pi_1(b_{temp}),$ 'selected')
   2.5. $B_{temp} \leftarrow B_{temp} \setminus \{b_{temp}\}$

Very similar to FIND_SUB, the FIND_WHOLE algorithm finds the 'whole' concept of a 'part' input parameter, and because of the straight forward nature of the algorithm, is omitted here.

4.3.3 SC Rule 3

1. Loop through concepts $c \in C$ that are selected ($\sigma_C(c) ==$ 'selected')
    1.1. Loop through all $t \in$ attr with $\pi_1(t) = c$

      1.1.1.   if $(m_{min}(\pi_2(t)) \neq 0)$
          1.1.1.1. if $(\sigma_A(\pi_2(t)) ==$ 'deselected' $\vee$ $\sigma_{attr}(t) ==$ 'deselected')
              1.1.1.1.1.   NO SC POSSIBLE
          1.1.1.2. else
              1.1.1.2.1. $\sigma \leftarrow \xi_A^{\sigma}(\pi_2(t),$ 'selected')
              1.1.1.2.2. $\sigma \leftarrow \xi_{attr}^{\sigma}(t,$ 'selected')

4.4 Total Simplicity Optimization Scheme Algorithm

The Total Simplicity Optimization Scheme is different from the other optimization schemes in a number of ways, as already mentioned in section 3.2.4. Another difference with the other optimization schemes that are presented in this paper, is that the rules to adhere to are very clear as to what is optimal, but do not let themselves translate into algorithms so straight forwardly as other optimization scheme rules. Therefore, another approach is taken here, rather than providing transparent algorithms per rule.

The algorithms were first based on Kruskal's algorithm, but a more custom solution to the problem was needed, as Kruskal does not incorporate fixed elements, such as our selected and deselected elements. Inspired by the firs attempt, ontology graphs are still taken as the model to work on, instead of the actual ontology. Note that an abstraction between the ontologies and ontology graphs is made, i.e. first the ontology graph is constructed from the ontology, but from there on changes to the ontology graph do not automatically incur similar changes to the actual ontology. For instance, deleting a vertex from the ontology graph will not result in the corresponding concept being deleted from the ontology, but instead indicates a modification of the labeling of the concept to 'deselected'. To clarify these differences between operations on the ontology graph and changes that occur to the actual ontology, in the algorithm the different implications will both be described (impact on the ontology always in brackets).

To simplify the algorithm, certain combinations of elements and their labels, are explained separately, instead of in the algorithm. Also, redistribution of the attributes is not covered here, but more information on this can be found in Appendix A. After the algorithm has finished with the other elements (relationships and concepts), this redistribution of attributes takes place, according to the reallocation via the best path, as used previously.

TS algorithm
1. Construct ontology graph
2. Leave out all deselected elements
3. While **combination1** is present
    3.1. Make the edge selected (same for corresponding relationship in the ontology)

4.   While ***combination2*** is present
   4.1. Make void vertex selected (same for corresponding concept in the ontology)
5.   While ***combination3*** is present
   5.1. Replace ***combination3*** with a new vertex with a selected label (no change to ontology
   5.2. If there is another void edge between the vertices
      5.2.1.   Remove edge (corresponding relationship becomes deselected)
   5.3. Else   // edge is selected
      5.3.1.   Remove edge (no change to the ontology)
6.   While there are void vertices left, choose one (called v)
   6.1. if connectivity ==1
      6.1.1.   if connecting edge has void label
         6.1.1.1. Remove edge and vertex (deselect corresponding relationship and concept)
      6.1.2.   else   // this means edge is selected (deselected ones are already removed)
         6.1.2.1. Make vertex selected (same for corresponding concept)
         6.1.2.2. if the edge given ***combination3***
            6.1.2.2.1.   Replace ***combination3*** with a new selected vertex (no changes to the ontology)
   6.2. Else   // connectivity > 1
      6.2.1.   n ← connectivity
      6.2.2.   call edges $e_1, \ldots, e_n$
      6.2.3.   call connected vertices $v_1, \ldots, v_n$
      6.2.4.   delete v and $e_1, \ldots, e_n$ (make corresponding concept and relationships deselected in the ontology)
      6.2.5.   create C_2^n new void edges, i.e. one between every possible combination of $v_1, \ldots, v_n$ (create corresponding relationships in ontology, with a void label)
      6.2.6.   if there are multiple edges between any vertices of $v_1, \ldots, v_n$
         6.2.6.1. delete edge(s) with higher weight(s) (the corresponding relationships becomes deselected in the ontology)
7.   If any changes to the ontology graph were made, repeat step 3 -6

First, the 4 specific combinations that are used throughout the algorithm will be given here.

     - ***combination1***: A void edge connecting two selected vertices
     - ***combination2***: A selected edge with at least one void vertex
     - ***combination3***: A selected edge connecting two selected vertices

Besides the four combinations, the algorithm uses the connectivity of vertices, which is defined as the number of edges that are connected to a vertex. As a valid ontology does not contain any islands, this means that the vertices of the corresponding ontology graph always have at least one edge, and in the algorithm care is taken that when the last edge of a vertex is deleted, the vertex is removed as well (and appropriate actions are taken for their impact on the ontology). A final comment about the algorithm concerns step 6.2.6.1 where the edges with a higher weight are removed (only the minimum weight is retained). What the weights are, and how they are calculated for newly constructed edges, relies on the cardinalities of the relationships. More information about this can be found in Appendix A. As discussed previously, there are many ways in which the weight of a relationship can be calculated, leading to very different results. Here only one method is shown, merely as an example of a possibility.

5 Conclusion

Transforming ontologies is becoming an important factor in the success of the semantic web. Ontologies tend to grow larger, and this results in an extremely tedious process – maybe to the point of being impossible – if an ontology needs to be modified (e.g. new version, sub-ontology, distribution, materialized ontology view). The formalisms presented in this paper provide a way in which automated extraction of materialized ontology views becomes possible, thus preventing problems associated with the manual extraction from occurring. After some general definitions, and introducing some new concepts, optimization schemes were presented as a means of arriving at high-quality or optimized results. As the notion of quality is multifaceted, various optimization schemes were introduced, and the theory of how these different optimization schemes can be used as building blocks by ontology engineers to arrive at a solution that they consider optimized, was covered. How the rules of optimization schemes could be implemented was demonstrated by first providing some dynamic rules that could be used by a system, in the form of transformation functions, and then utilizing these transformation functions in algorithms that ensure the rules of every optimization scheme are adhered to.

At the moment, only a limited number of optimization schemes has been developed, and in future work this will be addressed, firstly by the development of more optimization schemes, but also by setting a standard of communication and working of such an optimization scheme, so anyone can use these guidelines to develop optimization schemes that can be dynamically loaded into an Extraction Framework. This will lead to a plugin architecture where optimization schemes for various standards can be used in a single framework, and where contributions to the library of optimization schemes can be made by everyone.

References

Berners-Lee, T., Al: Reference Description of the DAML+OIL Ontology Markup Language (2001)

Biggs, N.L., Lloyd, E.K., et al.: Graph Theory 1736-1936. Clarendon Press, Oxford (1976)

Chen, P.P.: The Entity-Relationship Model: Toward a Unified View of Data. ACM Transaction on Database Systems 1(1), 9–36 (1976)

Colomb, R.M., Weber, R.: Completeness and Quality of an Ontology for an Information System. In: Proceedings of International Conference on Formal Ontology In Information Systems, Trento, Italy (1998)

Date, C.J.: An Introduction to Database Systems. Addison Wesley, Reading (2000)

Feng, L., Chang, E., et al.: A Semantic Network Based Design Methodology for XML Documents. ACM Transactions on Information Systems 20(3) (2002)

Fensel, D.: The Semantic Web, Tutorial Notes. In: 9th IFIP 2.6 Working Conference on Database Semantics (2001)

Fensel, D., Decker, S., et al.: Ontobroker: Or How to Enable Intelligent Access to the WWW. In: Proceedings 11th Knowledge Acquisition for Knowledge-Based Systems Workshop, Banff, Canada (1998)

Fensel, D., Horrocks, I., et al.: OIL in a Nutshell. In: Proceedings 12th International Conference on Knowledge Engineering and Knowledge Management Methods, Juan-Les-Pins, France (2000)

Genesereth, M.R.: Knowledge Interchange Format. In: Proceedings of the Second International Conference on Principles of Knowledge Representation and Reasoning. Morgan Kaufmann Publishers, San Francisco (1991)

Genesereth, M.R., Fikes, R.: Knowledge Interchange Format, version 3.0, reference manual. Computer Science Department, Stanford University, Stanford (1992)

Gruber, T.R.: Ontolingua: A Mechanism to Support Portable Ontologies. Knowledge Systems Laboratory, Stanford University, Stanford (1992)

Gruber, T.R.: Toward principles for the design of ontologies used for knowledge sharing. In: Guarino, N., Poli, R. (eds.) Formal Ontology in Conceptual Analysis and Knowledge Representation. Kluwer Academic Publishers, Dordrecht (1993)

Guarino, N., Welty, C.: Evaluating Ontological Decisions with OntoClean. Communications of the ACM 45(2), 61–65 (2002)

Hahn, U., Schnattinger, K.: Towards Text Knowledge Engineering. In: Proceedings of the 15th National Conference on Artificial Intelligence, Madison, Wisconsin (1998)

Halpin, T.: Conceptual Schema and Relational Database Design. Prentice Hall, Englewood Cliffs (1995)

Heflin, J., Hendler, J.: Dynamic Ontologies on the Web. In: Proceedings of American Association for Artificial Intelligence Conference, Menlo Park, California (2000)

Heflin, J., Hendler, J., et al.: SHOE: A Knowledge Representation Language for Internet Applications, Dept. of Computer Science, University of Maryland (1999)

Holsapple, C.W., Joshi, K.D.: A Collaborative Approach to Ontology Design. Communications of the ACM 45(2), 42–47 (2002)

Hovy, E.H.: Combining and Standardizing Large-Scale, Practical Ontologies for Machine Translation and Other Uses. In: Proceedings of the First International Conference on Language Resources and Evaluation, Granada, Spain (1998)

Kaplan, A.N.: Towards a Consistent Logical Framework for Ontological Analysis. In: Proceedings of the International Conference on Formal Ontology in Information Systems (2001)

Kim, H.: Predicting How Ontologies for the Semantic Web will Evolve. Communications of the ACM 45(2), 48–54 (2002)

Klein, M., Fensel, D.: Ontology versioning for the Semantic Web. In: Proceedings of the International Semantic Web Working Symposium, California, USA (2001)

Klein, M., Fensel, D., et al.: Ontology versioning and Change Detection on the Web. In: Proceedings of the 13th International Conference on Knowledge Engineering and Knowledge Management, Sigüenza, Spain. Springer, Heidelberg (2002)

Kruskal, J.B.J.: On the shortest spanning subtree of a graph and the traveling salesman problem. Proc. American Mathematics Society (7), 48–50 (1956)

Lenat, D.B.: Cyc: A Large-Scale Investment in Knowledge Infrastructure. Communications of the ACM 38(11) (1995)

McGuinness, D.L., Fikes, R., et al.: An environment for merging and testing large ontologies. In: Proceedings of the Seventh International Conference on Principles of Knowledge Representation and Reasoning. Morgan Kaufmann, San Francisco (2000)

Muslea, I.: Extraction Patterns for Information Extraction Tasks: A Survey. In: AAAI 1999 Workshop on Machine Learning for Information Extraction (1999)

Noy, N.F., Hafner, C.D.: The State of the Art in Ontology Design. AI-Magazine (Fall), 53–74 (1997)

Noy, N.F., Sintek, M., et al.: Creating Semantic Web Contents with Protégé-2000. IEEE Intelligent Systems 16(2), 60–71 (2001)

Rumbaugh, J., Jacobson, I., et al.: Unified Modeling Language Reference Manual. Addison-Wesley, Reading (1999)

Spyns, P., Meersman, R., et al.: Data modelling versus Ontology engineering. SIGMOD (special issue), 14–19 (2002)

Van Harmelen, F., Fensel, D.: Practical Knowledge Representation for the Web. In: Proceedings of International Joint Conferences on Artificial Intelligence (1999)

Von Staudt, G.K.C.: Geometrie der Lage. Nurnberg (1847)

W3C. Extensible Markup Language (XML) 1.0. W3C Recommendation (1999)

W3C. Feature Synopsis for OWL Lite and OWL. W3C Working Draft (2002a)

W3C. OWL Web Ontology Language 1.0 Abstract Syntax. W3C Working Draft (2002b)

W3C. OWL Web Ontology Language 1.0 Reference. W3C Working Draft (2002c)

Wouters, C., Dillon, T., et al.: A Practical Walkthrough of the Ontology Derivation Rules. In: Hameurlain, A., Cicchetti, R., Traunmüller, R. (eds.) DEXA 2002, vol. 2453. Springer, Heidelberg (2002a)

Wouters, C., Dillon, T., et al.: Transformational Processes for Sub-Ontologies Extraction (submitted for publication, 2002b)

Wouters, C., Dillon, T., et al.: A Practical Approach to the Derivation of Materialized Ontology View. In: Taniar, D., Rahayu, W. (eds.) Web Information Systems. Idea Group Publishing (to appear, December 2003)

Appendix A

A.1 Multiplication Table

The cardinality set used throughout this article is very limited (only consisting of three elements). This set can easily be extended, but in order for the optimization schemes to be able to use a new set, it needs to define a multiplication table. This is done in a lookup matrix, so that the solution for every possible multiplication can be found. The multiplication table for the cardinality set used here is given in Table 4.

Table 4. Multiplication Table for Cardinality Set

*	0	1	n
0	0	0	0
1	0	1	n
n	0	n	n

A.2 Best Path Algorithm

The Best Path Algorithm starts with a concept, and looks for the best possible path. In this context, 'best' means the lowest multiplication result of all the cardinalities (using the multiplication table). It is possible that more than one paths are returned as 'best' paths (i.e. multiple paths with the lowest weight). In this case, branches for each possibility are created, and treated as independent extraction processes. At the end of the extraction process, they are brought back together (and compared). In other words, it is possible that an extraction process produces several results, all equivalents in terms of quality.

Best_Path(concept c_1)
- **weight** ← infinity
- $\mathbf{P_{remaining}}$ ← all relationships connected to c_1
- While there are paths left in $\mathbf{P_{remaining}}$
  - o Take first path in list (call it $p_i = b_{i0},\ldots,b_{ij}$)
  - o If $\sigma(b_{ij}) \neq$ deselected
    - ▪ If $\pi_2(p_i) \neq$ deselected   // a solution found
      - • If $P_{sol} = \varnothing$
        - o **Weight** ← *calc_weight*(p_i)
        - o $\mathbf{P_{sol}}$ ← { p_i }
      - • Else
        - o If (**weight** = *calc_weight*(p_i))
          - ▪ $\mathbf{P_{sol}}$ ← $\mathbf{P_{sol}} \cup$ { p_i }
        - o If (**weight** > *calc_weight*(p_i))
          - ▪ $\mathbf{P_{sol}}$ ← { p_i }
          - ▪ **Weight** ← *calc_weight*(p_i)
      - • Loop through $\mathbf{P_{remaining}}$ (call it p_n)
        - o If (*calc_weight*(p_n) > **weight**)
          - ▪ $\mathbf{P_{remaining}}$ ← $\mathbf{P_{remaining}} \setminus$ { p_n }
    - ▪ Else   // $\pi_2(p_i)$ = deselected
      - • $\mathbf{R_{temp}}$ ← all relationships connected to $\pi_2(p_i)$
      - • $\mathbf{R_{temp}}$ ← $\mathbf{R_{temp}} \setminus$ { p_i }
      - • Loop through $\mathbf{R_{temp}}$ (call it bx)
        - o $\mathbf{P_{remaining}}$ ← $\mathbf{P_{remaining}} \cup$ { $b_{i0},\ldots,b_{ij},b_x$ }
  - o $\mathbf{P_{remaining}}$ ← $\mathbf{P_{remaining}} \setminus$ { p_i }
- Return $\mathbf{P_{sol}}$

As discussed previously, this is merely an example of an algorithm that was used throughout this article. It is not the intention of the authors to claim this is the most appropriate algorithm to use, it just worked well under conditions our testing was carried out.

A.3 Redistribution of Attributes

Attributes that are labeled 'selected', but belong to a deselected concept, need to be put somewhere else in the solution, in other words, a redistribution of attributes is required. This is done by looking for a proper path to a selected concept (which should exist, as this is one of the rules for RCOS), and if multiple alternatives exist, a similar algorithm to the previous one can be used to determine the 'best' option.

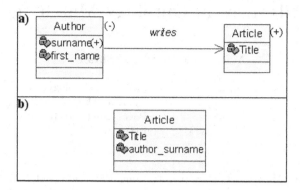

Fig. 17. Original (a) and Extracted (b) Ontology with Redistributed Attribute

Figure 17 shows an original ontology (a) where a deselected concept ("Author") has a selected attribute ("surname"). The second ontology (b) shows how the selected attribute was redistributed, and now belongs to the "Article" concept. Note that the name here has been changed to make the move of the attribute more meaningful.

Advances in Ontology Matching

Avigdor Gal and Pavel Shvaiko

[1] Technion – Israel Institute of Technology
`avigal@ie.technion.ac.il`
[2] University of Trento, Povo, Trento, Italy
`pavel@dit.unitn.it`

Abstract. Matching of concepts describing the meaning of data in heterogeneous distributed information sources, such as database schemas and other metadata models, grouped here under the heading of an ontology, is one of the basic operations of semantic heterogeneity reconciliation. The aim of this chapter is to motivate the need for ontology matching, introduce the basics of ontology matching, and then discuss several promising themes in the area as reflected in recent research works. In particular, we focus on such themes as uncertainty in ontology matching, matching ensembles, and matcher self-tuning. Finally, we outline some important directions for future research.

1   Introduction

Matching of concepts describing the meaning of data in heterogeneous distributed information sources (*e.g.*, database schemas, XML DTDs, HTML form tags) is one of the basic operations of semantic heterogeneity reconciliation. Due to the cognitive complexity of this matching process [18], it has traditionally been performed by human experts, such as web designers, database analysts, and even lay users, depending on the context of the application [79, 47]. For obvious reasons, manual concept reconciliation in dynamic environments such as the web (with or without computer-aided tools) is inefficient to the point of being infeasible, and so cannot provide a general solution for semantic reconciliation. The move from manual to semi-automatic matching has therefore been justified in the literature using arguments of scalability, especially for matching between large schemas [45], and by the need to speed-up the matching process. Researchers also argue for moving to fully-automatic, that is, unsupervised, schema matching in settings where a human expert is absent from the decision process. In particular, such situations characterize numerous emerging applications, such as agent communication, semantic web service composition, triggered by the vision of the semantic web and machine-understandable web resources [9, 82].

As integration of distributed information sources has been made more automated, the ambiguity in concept interpretation, also known as semantic heterogeneity, has become one of the main obstacles to this process. Heterogeneity is typically reduced in two steps: (*i*) matching of concepts to determine alignments and (*ii*) executing the alignment according to application needs (*e.g.*, schema integration, data integration, query answering). In this chapter, we focus only on

T.S. Dillon et al. (Eds.): Advances in Web Semantics I, LNCS 4891, pp. 176–198, 2008.

the first, *i.e.*, the matching step, automation of which still requires much research. The second step has already found a certain level of support from a number of commercial tools, such as Altova MapForce[1] and BizTalk Schema Mapper.[2]

In the context of web applications and the advent of the semantic web, a new term, in addition to schema matching, has come into existence, namely *ontology matching*. Ontologies are considered to be semantically richer than schemas in general, and therefore, techniques for schema matching can be easily adopted to ontologies but not vice versa. Therefore, in this chapter, unless explicitly referenced, we consider schema matching to be a special case of ontology matching.

Research into schema and ontology matching has been going on for more than 25 years now (see surveys [5, 79, 69, 73, 81] and various online lists, *e.g.*, OntologyMatching[3], Ziegler[4], DigiCULT[5], and SWgr[6]) first as part of a broader effort of schema integration and then as a standalone research. Recently, ontology matching has been given a book account in [30]. This work provided a uniform view on the topic with the help of several classifications of the available methods, discussed these methods in detail, *etc.* The AI-complete nature of the problem dictates that semi-automatic and automatic algorithms for schema and ontology matching will be largely of heuristic nature. Over the years, a significant body of work was devoted to the identification of automatic matchers and construction of matching systems. Examples of state of the art matching systems include COMA [21], Cupid [55], OntoBuilder [35], Autoplex [8], Similarity Flooding [58], Clio [60, 43], Glue [22], S-Match [37, 39], OLA [31], Prompt [66] and QOM [27] to name just a few. The main objective of these is to provide an alignment, namely a set of correspondences between semantically related entities of the ontologies. It is also expected that the correspondences will be effective from the user point of view, yet computationally efficient or at least not disastrously expensive. Such research has evolved in different research communities, including artificial intelligence, semantic web, databases, information retrieval, information sciences, data semantics, and others. We have striven to absorb best matching experiences of these communities and report here in a uniform manner some of the most important advances.

The aim of this chapter is to motivate the need for ontology matching (Section 2), introduce the basics of ontology matching (Section 3), and then discuss several promising directions in the area as reflected in recent research works. In particular, we focus on the following themes: uncertainty in ontology matching (Section 4), matching ensembles (Section 5), and matcher self-tuning (Section 6). Finally, we conclude with a summary and outline some directions for future research (Section 7).

---

[1] http://www.altova.com/products/mapforce/data_mapping.html

[2] http://msdn2.microsoft.com/en-us/library/ms943073.aspx

[3] http://www.ontologymatching.org/

[4] http://www.ifi.unizh.ch/~pziegler/IntegrationProjects.html

[5] http://www.digicult.info/pages/resources.php?t=10

[6] http://www.semanticweb.gr/

2   Applications

Matching ontologies is an important task in traditional applications, such as ontology integration, schema integration, and data warehouses. Typically, these applications are characterized by heterogeneous structural models that are analyzed and matched either manually or semi-automatically at design time. In such applications matching is a prerequisite of running the actual system.

A line of applications that can be characterized by their dynamics, *e.g.*, agents, peer-to-peer (P2P) systems, and web services, is emerging. Such applications, contrary to traditional ones, require (ultimately) a run time matching operation and often take advantage of more explicit conceptual models.

Below, we first discuss a motivating example and give intuition about the matching operation and its result. It is presented in the settings of the schema integration task. Then, we discuss data integration as yet another example of a traditional application. Finally, we overview a number of emergent applications, namely, P2P information systems, web service composition, and query answering on the deep web.

2.1   Motivating Example

To motivate the matching problem, let us use two simple XML schemas ($O1$ and $O2$) that are shown in Figure 1 and exemplify one of the possible situations which arise, for example, when resolving a schema integration task [80].

Let us suppose an e-commerce company needs to finalize a corporate acquisition of another company. To complete the acquisition we have to integrate databases of the two companies. The documents of both companies are stored according to XML schemas $O1$ and $O2$, respectively. Numbers in boxes are the unique identifiers of the XML elements. A first step in integrating the schemas is to identify candidates to be merged or to have taxonomic relationships under an integrated schema. This step involves ontology matching. For example, the entities with labels Office_Products in $O1$ and in $O2$ are the candidates to

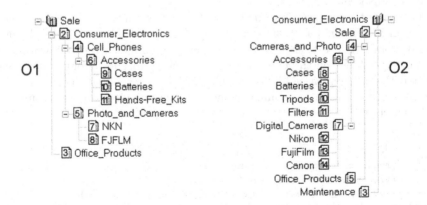

Fig. 1. Two XML schemas

be merged, while the entity with label Digital_Cameras in $O2$ should be subsumed by the entity with label Photo_and_Cameras in $O1$. Once correspondences between two schemas have been determined, the next step will generate query expressions that automatically translate data instances of these schemas under an integrated schema.

2.2 Data Integration

Data integration is a process of generating a global virtual ontology from multiple local sources without actually loading their data into a central warehouse [44]. Data integration allows interoperation across multiple local sources having access to up-to-date data.

The scenario is as follows. First, local information sources participating in the application, *e.g.*, bookstores and cultural heritage, are identified. Then, a virtual common ontology is built. Queries are posed over the virtual common ontology, and are then reformulated into queries over the local information sources. For instance, in the e-commerce example of Figure 1, integration can be achieved by generating a single global ontology to which queries will be submitted and then translated to the local ontologies. This allows users to avoid querying the local information sources one by one, and obtain a result from them just by querying a common ontology. In order to enable semantics-preserving query answering, correspondences between semantically related entities of the local information sources and the virtual ontology are to be established, which is a matching step. Query answering is then performed by using these correspondences in the settings of Local-as-View (LAV), Global-as-View (GAV), or Global-Local-as-View (GLAV) methods [53].

2.3 Peer-to-Peer Information Systems

Peer-to-peer is a distributed communication model in which parties (also called peers) have equivalent functional capabilities in providing each other with data and services [88]. P2P networks became popular through a file sharing paradigm, *e.g.*, music, video, and book sharing. These applications describe file contents by a simple schema (set of attributes, such as title of a song, its author, *etc.*) to which all the peers in the network have to subscribe. These schemas cannot be modified locally by a single peer.

Since peers are meant to be totally autonomous, they may use different terminologies and metadata models in order to represent their data, even if they refer to the same domain of interest [1, 48, 88]. Thus, in order to establish (meaningful) information exchange between peers, one of the steps is to identify and characterize relationships between their ontologies. This is a matching operation. Having identified the relationships between ontologies, they can be used for the purpose of query answering, *e.g.*, using techniques applied in data integration systems.

Such applications pose additional requirements on matching solutions. In P2P settings, an assumption that all the peers rely on one global schema, as in data integration, cannot be made because the global schema may need to be updated any time the system evolves [40]. While in the case of data integration schema matching can be performed at design time, in P2P applications peers need to coordinate their databases on-the-fly, therefore ultimately requiring run time schema matching.

Some P2P scenarios which rely on different types of peer ontologies, including relational schemas, XMLs, RDFs, or OWL ontologies are described in [10, 88, 48, 64, 76]. It is worth noting that most of the P2P data management projects, including [2] as well as Piazza [48] and Hyperion [75], focus on various issues of query answering and assume that the correspondences between peer schemas have been determined beforehand, and, hence, can be used for query propagation and rewriting.

2.4   Web Service Composition

Web services are processes that expose their interface to the web so that users can invoke them. Semantic web services provide a richer and more precise way to describe the services through the use of knowledge representation languages and ontologies. Web service discovery and integration is the process of finding a web service that can deliver a particular service and composing several services in order to achieve a particular goal, see [68, 67, 36, 32]. However, semantic web service descriptions do not necessarily reference the same ontology. Henceforth, both for finding the adequate service and for interfacing services it is necessary to establish the correspondences between the terms of the descriptions. This can be provided through matching the corresponding ontologies. For example, a browsing service may provide its output description using ontology $O1$ of Figure 1 while a purchasing service may use ontology $O2$ for describing its input. Matching ontologies is used in this context for (i) checking that what is delivered by the first service matches what is expected by the second one, (ii) verifying preconditions of the second service, and (iii) generating a mediator able to transform the output of the first service into input of the second one [30].

2.5   Query Answering on the Deep Web

In some of the above considered scenarios, *e.g.*, schema integration and data integration, it was assumed that queries were specified by using the terminology of a global schema. In the scenario under consideration, we discard this assumption, and therefore, users are free to pose queries by using their own terminology.

The so-called *deep web*, is made of web sites searchable via query interfaces (HTML forms) giving access to one or more back-end web databases. It is believed that it contains much more information [46] than the billions of static HTML pages of the surface web. For example, according to the investigations of [7] in March 2000, the size of the deep web was estimated as approximately from 400 to 550 times larger than the surface web. According to estimations of [46] in April of 2004, the deep web has expanded from 2000 by 3-7 times. At the moment, search engines are not very effective at crawling and indexing the deep web, since they cannot meaningfully handle the query interfaces. For example, according to [46], Google[7] and Yahoo[8] both manage to index 32% of the existing deep web objects. Finally, the deep web remains largely unexplored. However, it contains a huge number of on-line databases, which may be of use.

Thus, users have difficulties, first in discovering the relevant deep web resources and then in querying them. A standard use case includes, for example, buying a book with the lowest price among multiple on-line book stores. Query interfaces can be viewed as simple schemas (sets of terms). For example, in the book selling domain, the query interface of an on-line bookstore can be considered as a schema represented as a set of concept attributes, namely Author, Title, Subject, ISBN, Publisher. Thus, in order to enable query answering from multiple sources on the deep web, it is necessary to identify semantic correspondences between the attributes of the query interfaces of the web sites involved in handling user queries. This correspondences identification is a matching operation. Ultimately, these correspondences are used for on-the-fly translation of a user query between interfaces of web databases. For example, this motivating setup served in the basis of OntoBuilder [35], two holistic matching approaches presented in [45, 83], and others.

The above considered scenarios suggest that ontology matching is of great importance. Moreover, a need for matching is not limited to one particular application. In fact, it exists in any application involving more than one party. Thus, it is reasonable to consider ontology matching as a unified object of study. However, there are notable differences in the way these applications use matching. The application related differences must be clearly identified in order to provide the best suited solution in each case [30].

3   Basics

There have been different formalizations of matching and its result, see, for example, [11, 53, 49, 16, 81, 24, 30]. We provide here a general definition, synthesized from [21, 24, 80, 30]. In this chapter we focus on ontology matching and we therefore start with an informal description of what an ontology is. An *ontology* is "a specification of a conceptualization" [42], where conceptualization is an abstract view of the world represented as a set of objects. The term has been used in different research areas, including philosophy (where it was coined), artificial intelligence, information sciences, knowledge representation, object modeling, and most recently, eCommerce applications. For our purposes, an ontology can be described as a set of terms (vocabulary) associated with certain semantics and relationships. Depending on the precision of this specification, the notion of ontology encompasses several data and conceptual models, *e.g.*, classifications, database schemas, thesauri, and fully axiomatized theories. For the last model, ontologies may be represented using a Description Logic [25], where subsumption

---

[7] http://www.google.com
[8] http://www.yahoo.com

typifies the semantic relationship between terms; or Frame Logic [50], where a deductive inference system provides access to semi-structured data.

The *matching* operation determines an alignment A' (to be shortly defined) for a pair of ontologies $O1$ and $O2$. For this purpose only, we consider $O1$ and $O2$ to be finite sets of *entities*. In this general framework, we set no particular limitations on the notion of entities. Therefore, entities can be both simple and compound, compound entities should not necessarily be disjoint, *etc.*

Alignments express correspondences between entities belonging to different ontologies. A correspondence expresses the two corresponding entities and the relation that is supposed to hold between them. It is formally defined as follows:

Definition 1 (Correspondence). *Given two ontologies, a correspondence is a 5-tuple:*

$$\langle id, e_1, e_2, n, R \rangle,$$

such that

- *id is a unique identifier of the given correspondence;*
- e_1 *and* e_2 *are entities (e.g., tables, XML elements, properties, classes) of the first and the second ontology, respectively;*
- n *is a confidence measure (typically in the* $[0, 1]$ *range) holding for the correspondence between* e_1 *and* e_2*;*
- R *is a relation (e.g., equivalence* $(=)$*, more general* $(\sqsupseteq)$*, disjointness* $(\perp)$*, overlapping* $(\sqcap)$*) holding between* e_1 *and* e_2*.*

The correspondence $\langle id, e_1, e_2, n, R \rangle$ asserts that the relation R holds between the ontology entities e_1 and e_2 with confidence n. The higher the confidence, the higher is the likelihood of the relation to hold.

Let $\mathcal{O} = O1 \times O2$ be the set of all possible *entity correspondences* between $O1$ and $O2$ (as defined in Definition 1). To demonstrate the notion of a correspondence, let us consider Figure 1. Using some matching algorithm based on linguistic and structure analysis, the confidence measure (of the equivalence relation to hold) between entities with labels Photo_and_Cameras in $O1$ and Cameras_and_Photo in $O2$ could be 0.67. Let us suppose that this matching algorithm uses a threshold of 0.55 for determining the resulting alignment, *i.e.*, the algorithm considers all the pairs of entities with a confidence measure higher than 0.55 as correct correspondences. Thus, our hypothetical matching algorithm should return to the user the following correspondence:

$$\langle id_{5,4}, Photo\_and\_Cameras, Cameras\_and\_Photo, 0.67, = \rangle.$$

However, the relation between the same pair of entities, according to another matching algorithm which is able to determine that both entities mean the same thing, could be exactly the equivalence relation (without computing the confidence measure). Thus, returning

$$\langle id_{5,4}, Photo\_and\_Cameras, Cameras\_and\_Photo, n/a, = \rangle.$$

Definition 2 (Alignment). *Given two ontologies O1 and O2, an alignment is made up of a set of correspondences between pairs of entities belonging to O1 and O2, respectively. The power-set $\Sigma = 2^{\mathcal{O}}$ captures the set of all possible ontology alignments between O1 and O2.*

This definition of the matching process makes use of three matching features in addition to the input ontologies, namely: (*i*) alignment *A*, which is to be completed by the process; (*ii*) matching parameters, *p*, *e.g.*, weights and thresholds; and (*iii*) external resources used by the matching process, *r*, *e.g.*, common knowledge and domain specific thesauri.

Definition 3 (Matching process). *The matching process can be viewed as a function f which, from a pair of ontologies O1 and O2 to match, an input alignment A, a set of parameters p and a set of oracles and resources r, returns an alignment A' between these ontologies:*

$$A' = f(O1, O2, A, p, r)$$

The matching process can be schematically represented as illustrated in Figure 2. This definition of matching can be extended in a straightforward way to *multi-ontology matching*, that is, when multiple ontologies are taken as input. For simplicity of the presentation we focus here on matching between two ontologies.

In conceptual models and databases, the terms multiplicity or cardinality denote the constraints on a relation. Usual notations include $1:1$ (one-to-one), $1:m$ (one-to-many), $n:1$ (many-to-one) and $n:m$ (many-to-many). These naturally apply to the correspondences, thereby relating one or more entities of one ontology to one or more entities of another ontology.

Cardinality is only one (albeit important) example of a broader notion of alignment correctness. We introduce correctness into the matching process using a boolean function $\Gamma: \Sigma \rightarrow \{0,1\}$ that captures application-specific constraints on the process, *e.g.*, cardinality constraints and correspondence constraints. In what follows, by $\Sigma_\Gamma \subseteq \Sigma$ we denote the set of all *valid* ontology alignments in Σ, that is $\Sigma_\Gamma = \{\sigma \in \Sigma \mid \Gamma(\sigma) = 1\}$. The output of the matching process is an alignment $\sigma \in \Sigma_\Gamma$, where the process may define an (either implicit or explicit) ordering over Σ, and can provide the top ranked valid alignment. Here, we also

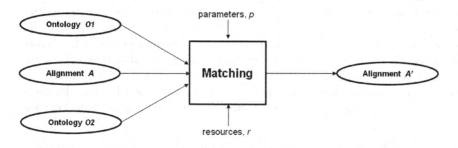

Fig. 2. The matching process

define an *exact alignment* to be a valid alignment $\sigma^* \in \Sigma_\Gamma$ that is recognized to be correct by an external observer.

In [21, 24] a 2-step method was proposed for the matching process (the rectangle in Figure 2). In the first step, a real-valued degree of similarity is automatically assigned with each correspondence. If $O1$ and $O2$ are of arity $n1$ and $n2$, respectively, then this step results in an $n1 \times n2$ *similarity matrix* M, where $M_{i,j}$ represents the degree of similarity between the i-th entity of $O1$ and j-th entity of $O2$. Various matching instantiations differ mainly in the measures of similarity they employ, yielding different similarity matrices. These measures can be arbitrarily complex, and may use various techniques for name matching, domain matching, structure matching (such as XML hierarchical representation), *etc.*

In the second step of the process, the similarity information in M is used to quantify the quality of different alignments in Σ. A single alignment is then chosen as the best alignment. The best alignment is typically considered to be the one that maximizes some *local aggregation function* (or *l-aggregator*, for short)

$$f(\sigma, M) = f(M_{1,\sigma(1)}, \ldots, M_{n,\sigma(n)}),$$

that is, a function that aggregates the degrees of similarity associated with the individual correspondences forming the alignment σ. The most common choice of *l*-aggregator turns out to be the sum (or equivalently, average) of correspondence degrees of similarity (*e.g.*, see [21, 56, 35]). In certain domains, however, other *l*-aggregators have been found appealing. For instance, an *l*-aggregator called *Dice* [21] stands for the ratio of the number of successfully matched correspondences (those that their similarity measure has passed a given threshold) and the total number of entities in both ontologies.

4   Ontology Matching Quality Models

In this section we discuss issues of quality in ontology matching. We start with a brief review of the management of imperfect information, based on [19], followed by an overview of the efforts in ontology matching quality management.

4.1   Brief Introduction to Information Imperfection

Data management tools deal regularly with imperfect information. Imperfection may be in the form of imprecision, vagueness, uncertainty, incompleteness, inconsistency, *etc.* Managing imperfections, both at the modelling (design time) level and at the querying (run time) level, can be done using tools such as probability theory, Dempster-Shafer theory, fuzzy logic, surprisal, and entropy. Over the years, several categorical classifications of the different types and sources of imperfect information have been presented. In accordance with the classifications of Bosc and Prade [15], Motro [63] and Parsons [70], imperfect information can be categorized as follows:

Uncertain information. Information for which it is not possible to determine whether it is true or false.

Imprecise information. Information that is not as specific as it should be.

Vague information. Information that include elements (*e.g.*, predicates or quantifiers) that are inherently "vague" (in the common day-to-day sense of the word cf. [63]).

Inconsistent information. Information that contains two or more assertions that cannot simultaneously hold.

Incomplete information. Information for which some data are missing.

Data management approaches to deal with *uncertainty* include the possibilistic approaches and the probabilistic approaches. With possibilistic approaches, possibility theory [87] is used, where a possibility distribution is used to model the value of an attribute which is known to be uncertain. Each possible value for the attribute is assigned a membership grade that is interpreted as the degree of uncertainty [71]. Furthermore, possibility and necessity measures are attached to each result in the result set of a query. Probabilistic approaches are based on probability theory, where each result in the result set of a query is extended with a probability, representing the probability of it belonging to the set [85].

Both approaches have their advantages and disadvantages. Probabilities represent the relative occurrence of an event and therefore provide more information than possibilities. Possibilities, however, are easier to apply because they are not restricted by a stringent normalization condition of probability theory. A probabilistic approach towards ontology matching was utilized in several works, including [8, 23], where machine learning was utilized in estimating correspondence similarity measures. For example, given a correspondence $\langle id, e_1, e_2, n, R \rangle$, the naïve Bayes method compares the probability of a set of instances of entity e_1 (e.g., brands of element NKN in $O1$, Figure 1) to serve as instances of entity e_2 (e.g., brands of entity Nikon in $O2$, Figure 1) with the probability of not serving as e_2's instances. A probability space is constructed using training data and then used for generating new correspondences.

Another work in [65], where an approach, based on combining Horn predicate logics and probability theory, was presented to harness correspondence uncertainty. A set of candidate Horn predicate rules is generated and assigned a weight. Then, a set of rules with maximum probability is selected. This work is in line with the 2-layer approach also suggested in [77] for managing uncertain data.

Imprecision of data is mostly modelled with fuzzy set theory [86] and its related possibility theory [87]. Fuzzy set theory is a generalization of regular set theory in which it is assumed that there might be elements that only partially belong to a set. Therefore, a so-called membership grade, denoting the extent to which the element belongs to the fuzzy set, is associated with each element of the universe. Two main approaches can be distinguished when modeling imprecision. First, similarity relations are utilized to model the extent to which the elements of an attribute domain may be interchanged [17]. Second, possibility distributions [71] are used, having the benefit of being suitable to cope with uncertainty (see above) and *vagueness*. In [34], a fuzzy model of ontology matching was proposed. In this model, a correspondence is assigned with a fuzzy membership degree (similar to probability, yet without a naïve assumption of correspondence independence and without constraints that stem from the need to build a probability space). Using such a model, the work continues to discuss the properties of various aggregators, transforming correspondence membership degrees into alignment similarity grades.

The treatment of *incomplete* information in databases has been widely addressed in research. A survey that gives an overview of the field is presented in [26]. The most commonly adopted technique is to model missing data with a pseudo-description, called *null*, denoting "missing" information. A more recent approach, based on possibility theory, [84] provides an explicit distinction between the cases of unknown data and inapplicable data.

4.2   Ontology Matching Evaluation

Quantitative quality measures for alignment evaluation, in works such as [20] consist of *precision*, *recall*, and a couple of their derivatives, namely *F-Measure* and *overall*. Assume that out of the $n_1 \times n_2$ correspondences $c \leq n_1 \times n_2$ are the correct correspondences, with respect to some reference alignment. Also, let $t \leq c$ be the number of correspondences, out of the correct correspondences, that were chosen by the matching algorithm and $f \leq n_1 \times n_2 - c$ be the number of incorrect such correspondences. Then, precision is computed to be $\frac{t}{t+f}$ and recall is computed as $\frac{t}{c}$. Clearly, higher values of both precision and recall are desired. Another derivative of precision and recall, dubbed *error*, was used in [61]. In many research works, precision and recall are considered to provide a form of pragmatic soundness and completeness. Towards this end, an *exact alignment* is needed, against which such soundness and completeness are measured. Notice that these measures actually derive their values from a discrete domain in $[0, 1]$.

In [41], a probabilistic interpretation was assigned with precision and recall, generating posterior distributions for them. The authors have shown the benefit of such an approach to estimate the performance of text retrieval systems. However, to the best of our knowledge, such a model has never been adopted to evaluate ontology matching results so far.

A model for representing uncertainty in schema matching was presented in [34] and will be discussed in Section 4.3. Alignments were evaluated in [6] using semantic soundness and completeness. They start from some representation language $\mathcal{L}$ (*e.g.*, a Description Logic language [4]). A schema matcher α is *semantically sound* w.r.t. F if for any correspondence $\langle id, e_1, e_2, n, r \rangle$, if $\alpha(\langle e_1, e_2, r \rangle) = T$ then $\mathcal{O} \models_{\mathcal{L}} \mathcal{T}(e_1) r \mathcal{T}(e_2)$. α is *semantically complete* w.r.t. F if for any two nodes e_1 and e_2, if $\mathcal{O} \models_{\mathcal{L}} \mathcal{T}(e_1) r \mathcal{T}(e_2)$ then $\alpha(\langle e_1, e_2, r \rangle) = T$. While providing a theoretical foundation for evaluating matchers, such correctness depends on the completeness of the ontology in use. The authors use a philosophical argument of H. Putnam [72] to say that "two agents may agree at the conceptual level, but not at the pragmatic level." That is, while a matcher may correctly identify a relationship between two concepts, it may still not entail agreement at the instance level. With such an argument at hand, tasks such as query answerability, which is one of the tasks addressed in [54] by using a formal representation language, and query rewriting, which was presented as one ultimate goal of schema matching in [35], cannot be evaluated in such a framework to be sound and complete. In particular, the use of certain answers, [3] which lies heavily on the ability to agree at the conceptual level, may be hindered.

4.3   Imperfection in Ontology Matching

Imperfection in ontology matching has been discussed both in [54] and in [6]. The former argues for the need "to incorporate *inaccurate* mappings [correspondences] and handle *uncertainty* about mappings. Inaccuracy arises because in many contexts there is no precise mapping ... mappings may be inaccurate [since] the mapping language is too restricted to express more accurate mappings." [6] went even further, arguing philosophically that even if two ontologies fully agree on the semantics and the language is rich enough, ontologies may still not convey the same meaning, due to some hidden semantics, beyond the scope of the ontologies. A similar argument was provided in [59] in the context of relational databases: "the syntactic representation of schemas and data do not completely convey the semantics of different databases." Therefore, [54] argues that "when no accurate mapping [correspondence] exists, the issue becomes choosing the *best* mapping from the viable ones." This highlights a possible benefit of specifying semantics explicitly for the purpose of efficiently pruning the search space, to allow the evaluation of valid alignments only, namely alignments that satisfy the semantic constraints of the model.

One way of modeling ontology matching as an uncertain process is to use similarity matrices as a measure of certainty. This way, a matcher needs to be measured by the fit of its estimation of a certainty of a correspondence to the real world. In [34], such a formal framework was provided, attempting to answer the question of whether there are "good" and "bad" matchers.

We have already observed that precision (denoted as $p(\sigma)$ for any alignment $\sigma \in \Gamma$) takes its values from a discrete domain in $[0, 1]$. Therefore, one can create equivalence alignment classes on Γ. Two alignments σ' and σ'' belong to a class p if $p(\sigma') = p(\sigma'') = p$, where $p \in [0, 1]$. Let us consider now two alignments, σ' and σ'', such that $p(\sigma') < p(\sigma'')$. For each of these two alignments we can compute their level of certainty, $f(\sigma', M)$ and $f(\sigma'', M)$, respectively. We say that a matcher is *monotonic* if for any two such alignments $p(\sigma') < p(\sigma'') \rightarrow f(\sigma', M) < f(\sigma'', M)$. As an example, consider once more Figure 1 and take two alignments, σ and σ', that differ on a single correspondence. In σ, NKN is matched to Nikon, while in σ', NKN is matched to FujiFilm. Clearly, the former is a correct correspondence while the latter is not. Therefore, $p(\sigma) < p(\sigma')$. If a matcher is monotonic, it should generate a similarity matrix M such that $f(\sigma, M) < f(\sigma', M)$.

A monotonic ontology matcher can easily identify the exact alignment. Let σ^* be the exact alignment, then $p(\sigma^*) = 1$. For any other alignment σ', $p(\sigma') \leq p(\sigma^*)$, since p takes its values in $[0, 1]$. Therefore, if $p(\sigma') < p(\sigma^*)$ then from monotonicity $f(\sigma', M) < f(\sigma^*, M)$. All one has to do then is to devise a method for finding an alignment σ that maximizes f.[9] In fact, this is one of the two most common methods for identifying the exact alignments nowadays [21, 34, 14]. The other common method, adopted in [56, 45] and others, is to only determine M automatically, allowing the user to identify the exact (ontology) alignment from the individual correspondences.

4.4   Imperfection as an Emergent Semantics

Imperfection can be managed and reduced using an iterative process. In such a process, initial assumptions are strengthened or discarded, and initial measures of imperfection are being refined. Such an iterative process may involve bringing together and relating information located at different places. Alternatively, one may attempt accessing a user with well-defined questions that eventually will minimize imperfection. In approaches based on possibility theory refinement can be done by composing all available fuzzy sets related to the same imperfect data. Hereby, the intersection operators for fuzzy sets (t-norms) can be used as composition operators [87].

As an example to the latter, in [33] uncertainty is refined by a comparison of K alignments, each with its own uncertainty measure (modeled as a fuzzy relation over the two ontologies). The process yields an improved ontology matching, with higher precision. For example, assume that the second-best correspondence, as generated by some heuristic, changes a correspondence of NKN with Canon to Nikon. The latter correspondence then remains unchanged for the next eight best correspondences. Therefore, in nine out of the top-10 correspondences, the correspondence of NKN with Nikon exists. If we set a threshold of 9, requiring a correspondence to appear in at least nine out of ten correspondences, then this correspondence will be included in the final alignment.

5   Matching Ensembles

Striving to increase robustness in the face of the biases and shortcomings of individual matchers, tools combine principles by which different ontology matchers judge the similarity between concepts. The idea is appealing since an ensemble of complementary matchers can potentially compensate for the weaknesses of each other. Another argument in favor of using matching ensembles was presented in [13, 74, 52]. There, using ensembles was promoted as a method for ensuring matching system extensibility. Indeed, several studies report on encouraging results when using matcher ensembles (*e.g.*, see [21, 29, 35, 55, 13, 62]).

Formally, let us consider a set of m matchers $matcher_1, \dots, matcher_m$, utilizing (possibly different) local aggregators $f^{(1)}, \dots, f^{(m)}$, respectively. Given two ontologies $O1$ and $O2$ as before, these matchers produce an $m \times n1 \times n2$ similarity cube of $n1 \times n2$ similarity matrices $M^{(1)}, \dots, M^{(m)}$. In these matrices, $M_{i,j}^{(l)}$

---

[9] In [34] it was shown that while such a method works well for fuzzy aggregators (*e.g.*, weighted average) it does not work for t-norms such as min.

captures the degree of similarity that $matcher_l$ associates with correspondence of the i-th entity of $O1$ to the j-th entity of $O2$.

Given such a set of matchers $matcher_1, \ldots, matcher_m$, we would like to aggregate the similarity measures, given the correspondences produced by the different matchers. Such a weight aggregation can be modeled using a *global aggregation function* (or *g-aggregator*, for short) $F\left(f^{(1)}(\sigma, M^{(1)}), \cdots, f^{(m)}(\sigma, M^{(m)})\right)$. For instance, a natural candidate for *g*-aggregator would be as follows:

$$F\left(f^{(1)}(\sigma, M^{(1)}), \cdots, f^{(m)}(\sigma, M^{(m)})\right) = \frac{\lambda}{m} \sum_{l=1}^{m} k_l f^{(l)}(\sigma, M^{(l)})$$

It is interpreted as a (weighted) sum (with $\lambda = m$) or a (weighted) average (with $\lambda = 1$) of the local similarity measures, where k_l are some arbitrary weighting parameters.

COMA [21], which introduced first the notion of a similarity cube reverses the roles of local and global aggregators. It first reduces the cube into a matrix, and then applies to this matrix the (common) local aggregator. Many other tools (with the exception of OntoBuilder) implicitly follow COMA's footsteps, aggregating correspondence values before determining an alignment. In [24], the limitations of replacing global and local aggregators were discussed, mainly in the scope of generating top-K alignments.

6 Matcher Self-tuning

The work in [20] specifies manual effort as a comparison criteria for measuring matchers. The discussion separates pre-match efforts from post-match efforts. The former includes training of matchers, parameter configuration, and specification of auxiliary information. The latter involves the identification of false positives and false negatives. The authors comment that "[un]fortanutely, the effort associated with such manual pre-match and post-match operations varies heavily with the background knowledge and cognitive abilities of users."

Clearly, one of the goals of ontology matching is to reduce this effort. Attempts to reduce post-match efforts focus on the generation of matchers that produce better alignments. Pre-match efforts focus on automatic parameter tuning. In this section we focus on the latter. Before dwelling into tuning, it is worthwhile mentioning here that another interesting aspect of the problem involves feature selection.

A general problem of pre-match effort was defined in [78] as follows: "Given a schema S, how to tune a matching system M so that it achieves high accuracy when we subsequently apply it to match S with other schemas." The various tuning parameters are called "knobs" in [78] and searching for the right knob values may be an intractable process. Let us first discuss a few alternatives for parameter tuning, followed by a discussion of methods to increase the efficiency of self-tuning.

An immediate approach to parameter tuning is that of machine learning. Using this approach, one provides a set of examples (positive, negative, or both) from which a tuning configuration is selected such that it optimizes a goal function. With such a configuration at hand, matching is performed. As an example, consider the LSD algorithm [23]. The algorithm uses an ensemble of learners, whose grades are combined using weighted average. To determine the weights of different learners, a linear regression is performed, aiming at minimizing the square error of the decision made by the ensemble over the test data.

Machine learning was also used in APFEL [28]. In this work users were first given the alignments for validation. Using user validation, new hypotheses were generated by APFEL and weighted using the initial feedback. User feedback was also adopted in eTuner as an additional source of information for the tuning process.

Table 1. Best alignment for two "matchmaking" web sites

www.cybersuitors.com	www.date.com
select: Country: (cboCountries)	select: Select your Country (countrycode)
select: Birthday: (cboDays)	select: Date of Birth (dob_day)
select: Birthday: (cboMonths)	select: Date of Birth (dob_month)
select: Birthday: (cboYears)	select: Date of Birth (dob_year)
checkbox: (chkAgreement2)	image: ()
checkbox: (chkAgreement1)	checkbox: Date.com - Join Now for Free! (over18)
select: State (if in USA): (cboUSstates)	select: I am a (i_am)

Another approach to tuning can be dubbed "dynamic tuning." According to this approach, knobs are not determined apriori but are rather derived from a heuristic at hand. An example of such an approach is available in [33]. For illustration purposes, we follow the example of query answering on the deep web, given in [33]. Let us consider two web sites that offer "matchmaking" services. In each of these sites, one has to fill in personal information (*e.g.*, name, country of residence, birthdate attributes). A matching algorithm called *Combined*, which is part of the toolkit of OntoBuilder [35], was applied. The algorithm returned the best alignment, containing a set of possible correspondences. A sample list of such correspondences is shown in Table 1. Each column in the table contains information about one field in a registration form in one of the web sites. The information consists of the type of field (*e.g.*, select field and checkbox), the label as appears at the web site, and the name of the field, given here in parentheses and hidden from the user. Each row in the table represents attribute correspondence, as proposed by this algorithm. The top part of the table contains four correct correspondences. The bottom part of the table contains three incorrect correspondences.

Matching algorithms face two obstacles in providing the best alignments. First, correct alignments should be identified and provided to the user. Second, incorrect alignments should be avoided. Separating correct from incorrect alignments is a hard task. When using a best alignment approach, an algorithm can discard attribute correspondences that do not reach some predefined threshold, assuming that those attribute correspondences with low similarity measures are less adequate than those with high similarity measures. By doing so, an algorithm (hopefully) increases precision, at the expense of recall. Using a threshold, however, works only in clear-cut scenarios. Moreover, tuning the threshold becomes an art in itself. As an example, let us consider Table 1. The four correct attribute correspondences received similarity measures in the range $(0.49, 0.7)$ while the other similarity measures ranged from 0 to 0.5. Any arbitrary apriori selection of a threshold may yield false negatives (if the threshold is set above 0.49) or false positives, in case the threshold is set below 0.49.

Consider now an alternative, in which the algorithm generates top-10 correspondences, that is, the best 10 correspondences between the two schemas, such that correspondence i differs from correspondences $1, 2, \ldots i - 1$ by at least one attribute correspondence. For example, the second best correspondences include: (i) checkbox: (chkAgreement2) and checkbox: Date.com - Join Now for Free! (over18) as well as (ii) checkbox: (chkAgreement1) and image: () (this last attribute is actually a button and has no associated label or field name).

Stability analysis of the method proposed in [33] assumes that such a scenario represents a "shaky" confidence in this correspondence to start with and removes it from the set of proposed attribute correspondences. Simultaneous analysis of the top-10 correspondences reveals that the four correct attribute correspondences did not change throughout the 10 correspondences, while the other attributes were matched with different attributes in different correspondences. Stability analysis suggests that the four correspondences, for which consistent attribute correspondences were observed in the top-10 correspondences, should be proposed as the "best alignment" yielding a precision of 100% without adversely affecting recall.

Tuning may be a costly effort. Exhaustive evaluation of the search space may be infeasible if tuning parameters take their values from continuous domains, and intractable even if all parameter domains are discrete. Therefore, efforts were made to reduce the search cost. Staged tuning was proposed in [78]. There, matchers were organized in an execution tree, for which the output of lower level matchers serve as input to higher level matchers. Given a K-level tree, the staged tuning starts with optimizing each matcher at the leaf level. Then, equipped with the optimal setting of the individual matchers it moves on to optimize the next level matcher, and so on and so forth.

For the tuning process to work well, there is a need of some ground truth regarding alignments. The quality of the training set has a crucial impact on the success of the tuning. In the early days of ontology matching research, the lack of an exact alignment yielded a poor validation process, in which heuristics were measured based on a few ontologies only. To alleviate this problem, two main directions were taken. The first approach, taken within the OntoBuilder project, involves a continuous effort to gather exact alignments (in the time of writing this chapter, there are over 200 exact alignments). This process is tedious and error prone, yet it provides a variety of ontologies practitioners are likely to access. The second approach, taken within the framework of the eTuner project [52] and also suggested in [51], involves the synthetic generation of a sufficient number of schema "mutations" from a few known exact alignments to allow effective learning. This approach overcomes the possible erroneous correspondences in a manually generated exact alignment. However, the quality of the learning set becomes dependent on the quality of the mutation rules. In addition, the strong correlation between mutated instances may generate biases in the learning process.

Combining the two approaches may provide a robust solution to the training set problem. In fact, a more varied training set could overcome the correlation problem, while the synthetic mutation would allow a tighter control over the learning process.

7   Conclusions

In this chapter we have introduced recent advances in ontology matching. In particular, after a brief introduction to the problem we have discussed several contemporary applications that motivate the research into automatic ontology matching as opposed to manual labor intensive effort. We have then provided a generic model of ontology matching as well as some technical details of several research directions, whose importance is highlighted by the need for automatic matching. These include the issues in matching quality, matching ensembles, and matcher self-tuning. While being far from exhaustive, we have striven to provide a good coverage of the performed efforts in these three directions. Much work is yet need to be done in these directions, including:

Ontology meta-matching: Following the model of uncertainty in ontology matching and our discussion of the usefulness of ensembles, a possible next step involves ontology meta-matching. That is a framework for composing an arbitrary ensemble of ontology matchers, and generating a list of best-ranked ontology alignments. We can formulate our task of identifying a top-K consensus ranking as an optimization problem, in which we aim at minimizing the amount of effort (in terms of time or number of iterations) the ensemble invests in identifying top alignments. Algorithms for generating a consensus ranking may adopt standard techniques for general quantitative rank aggregation and build on top of them, as proposed for example in [24].

Matcher self-tuning: This direction is still largely unexplored. In dynamic settings, such as the web, it is natural that applications are constantly changing their characteristics. Therefore, approaches that attempt to tune and adapt automatically matching solutions to the settings in which an application operates are of high importance. In particular, the challenge is to be able to perform matcher self-tuning at run time, and therefore, efficiency of the matcher configuration search strategies becomes crucial. Moreover, the configuration space can be arbitrary large, thus, searching it exhaustively may be infeasible.

Ontology matching evaluation: The evaluation of ontology matching approaches is still in its infancy. Initial steps have already been done in this direction, for example, the Ontology Alignment Evaluation Initiative (OAEI).[10] However, there are many issues to be addressed along the ontology matching evaluation lines in order to empirically prove the matching technology to be mature and reliable, including (*i*) design of extensive experiments across different domains with multiple test cases from each domain as well as new, difficult to match, and large real world test sets, (*ii*) more accurate evaluation measures, involving user-related measures, and (*iii*) automating acquisition of reference alignments, especially for large applications.

We have outlined three promising future research directions along the lines of the key themes discussed in this chapter. However, it is worth notice that ontology matching certainly requires further developments in a number of other important directions as well, including: background knowledge in ontology matching [38], social and collaborative ontology matching [89], performance and usability of matching approaches [13, 12], and infrastructures [35, 57].

Acknowledgements

Avigdor Gal has been partially supported by Technion V.P.R. Fund and the Fund for the Promotion of Research at the Technion. Pavel Shvaiko has been partly supported by the Knowledge Web European network of excellence (IST-2004-507482). We are very grateful to Fausto Giunchiglia, Mikalai Yatskevich and Jérôme Euzenat for many fruitful discussions on various ontology matching themes.

References

[1] Aberer, K.: Guest editor's introduction. SIGMOD Record 32(3), 21–22 (2003)

[2] Aberer, K., Cudré-Mauroux, P., Hauswirth, M.: Start making sense: The chatty web approach for global semantic agreements. Journal of Web Semantics 1(1), 89–114 (2003)

[3] Abiteboul, S., Duschka, O.: Complexity of answering queries using materialized views. In: Proceedings of the 17th Symposium on Principles of Database Systems (PODS), Seattle, USA, pp. 254–263 (1998)

[4] Baader, F., Calvanese, D., McGuinness, D., Nardi, D., Patel-Schneider, P. (eds.): The Description Logic Handbook: Theory, Implementation, and Applications. Cambridge University Press, Cambridge (2003)

[5] Batini, C., Lenzerini, M., Navathe, S.: A comparative analysis of methodologies for database schema integration. ACM Computing Surveys 18(4), 323–364 (1986)

[6] Benerecetti, M., Bouquet, P., Zanobini, S.: Soundness of schema matching methods. In: Gómez-Pérez, A., Euzenat, J. (eds.) ESWC 2005. LNCS, vol. 3532, pp. 211–225. Springer, Heidelberg (2005)

[7] Bergman, M.: The deep web: surfacing hidden value. The Journal of Electronic Publishing 7(1) (2001)

---

[10] http://oaei.ontologymatching.org/

[8] Berlin, J., Motro, A.: Autoplex: Automated discovery of content for virtual databases. In: Proceedings of the 9th International Conference on Cooperative Information Systems (CoopIS), Trento, Italy, pp. 108–122 (2001)

[9] Berners-Lee, T., Hendler, J., Lassila, O.: The semantic web. Scientific American 284(5), 34–43 (2001)

[10] Bernstein, P., Giunchiglia, F., Kementsietsidis, A., Mylopoulos, J., Serafini, L., Zaihrayeu, I.: Data management for peer-to-peer computing: A vision. In: Proceedings of the 5th International Workshop on the Web and Databases (WebDB), Madison, USA, pp. 89–94 (2002)

[11] Bernstein, P., Halevy, A., Pottinger, R.: A vision of management of complex models. SIGMOD Record 29(4), 55–63 (2000)

[12] Bernstein, P., Melnik, S., Churchill, J.: Incremental schema matching. In: Proceedings of the 32nd International Conference on Very Large Data Bases (VLDB), Seoul, South Korea, pp. 1167–1170 (2006)

[13] Bernstein, P., Melnik, S., Petropoulos, M., Quix, C.: Industrial-strength schema matching. SIGMOD Record 33(4), 38–43 (2004)

[14] Bilke, A., Naumann, F.: Schema matching using duplicates. In: Proceedings of the 21st International Conference on Data Engineering (ICDE), Tokyo, Japan, pp. 69–80 (2005)

[15] Bosc, P., Prade, H.: An introduction to fuzzy set and possibility theory based approaches to the treatment of uncertainty and imprecision in database management systems. In: Proceedings of the 2nd Workshop on Uncertainty Management in Information Systems: From Needs to Solutions, Santa Catalina, USA, pp. 44–70 (1993)

[16] Bouquet, P., Ehrig, M., Euzenat, J., Franconi, E., Hitzler, P., Krötzsch, M., Serafini, L., Stamou, G., Sure, Y., Tessaris, S.: Specification of a common framework for characterizing alignment. Deliverable D2.2.1, Knowledge web NoE (2004)

[17] Buckles, B., Petry, F.: Generalised database and information systems. In: Bezdek, J.C. (ed.) Analysis of fuzzy Information. CRC Press, Boca Raton (1987)

[18] Convent, B.: Unsolvable problems related to the view integration approach. In: Atzeni, P., Ausiello, G. (eds.) ICDT 1986. LNCS, vol. 243, pp. 141–156. Springer, Heidelberg (1986)

[19] Cudré-Mauroux, P.: Emergent semantics: rethinking interoperability for large scale decentralized information systems. Ph.D thesis, École Polytechnique Fédérale de Lausanne (2006)

[20] Do, H.-H., Melnik, S., Rahm, E.: Comparison of schema matching evaluations. In: Proceedings of the 2nd Workshop on Web, Web-Services, and Database Systems, Erfurt, Germany, pp. 221–237 (2002)

[21] Do, H.-H., Rahm, E.: COMA – a system for flexible combination of schema matching approaches. In: Proceedings of the 28th International Conference on Very Large Data Bases (VLDB), Hong Kong, China, pp. 610–621 (2002)

[22] Doan, A., Madhavan, J., Domingos, P., Halevy, A.: Learning to map between ontologies on the semantic web. In: Proceedings of the 11th International Conference on World Wide Web (WWW), Honolulu, USA, pp. 662–673 (2002)

[23] Doan, A.-H., Domingos, P., Halevy, A.: Reconciling schemas of disparate data sources: A machine-learning approach. In: Proceedings of the 20th International Conference on Management of Data (SIGMOD), Santa Barbara, USA, pp. 509–520 (2001)

[24] Domshlak, C., Gal, A., Roitman, H.: Rank aggregation for automatic schema matching. IEEE Transactions on Knowledge and Data Engineering 19(4), 538–553 (2007)

[25] Donini, F., Lenzerini, M., Nardi, D., Schaerf, A.: Reasoning in description logic. In: Brewka, G. (ed.) Principles on Knowledge Representation, Studies in Logic, Languages and Information, pp. 193–238. CSLI Publications (1996)

[26] Dyreson, C.: A bibliography on uncertainty management in information systems. In: Motro, A., Smets, P. (eds.) Uncertainty Management in Information Systems: From Needs to Solutions, pp. 415–458. Kluwer Academic Publishers, Boston (1996)

[27] Ehrig, M., Staab, S.: QOM – quick ontology mapping. In: McIlraith, S.A., Plexousakis, D., van Harmelen, F. (eds.) ISWC 2004. LNCS, vol. 3298, pp. 683–697. Springer, Heidelberg (2004)

[28] Ehrig, M., Staab, S., Sure, Y.: Bootstrapping ontology alignment methods with APFEL. In: Proceedings of the 4th International Semantic Web Conference (ISWC), Galway, Ireland, pp. 186–200 (2005)

[29] Embley, D., Jackman, D., Xu, L.: Attribute match discovery in information integration: Exploiting multiple facets of metadata. Journal of Brazilian Computing Society 8(2), 32–43 (2002)

[30] Euzenat, J., Shvaiko, P.: Ontology matching. Springer, Heidelberg (2007)

[31] Euzenat, J., Valtchev, P.: Similarity-based ontology alignment in OWL-lite. In: Proceedings of the 15th European Conference on Artificial Intelligence (ECAI), Valencia, Spain, pp. 333–337 (2004)

[32] Fensel, D., Lausen, H., Polleres, A., de Bruijn, J., Stollberg, M., Roman, D., Domingue, J.: Enabling Semantic Web Services: The Web Service Modeling Ontology. Springer, Heidelberg (2007)

[33] Gal, A.: Managing uncertainty in schema matching with top-K schema mappings. Journal of Data Semantics 6, 90–114 (2006)

[34] Gal, A., Anaby-Tavor, A., Trombetta, A., Montesi, D.: A framework for modeling and evaluating automatic semantic reconciliation. VLDB Journal 14(1), 50–67 (2005)

[35] Gal, A., Modica, G., Jamil, H., Eyal, A.: Automatic ontology matching using application semantics. AI Magazine 26(1), 21–32 (2005)

[36] Giunchiglia, F., McNeill, F., Yatskevich, M.: Web service composition via semantic matching of interaction specifications. Technical Report DIT-06-080, University of Trento, Italy (2006)

[37] Giunchiglia, F., Shvaiko, P., Yatskevich, M.: Semantic schema matching. In: Proceedings of the 13rd International Conference on Cooperative Information Systems (CoopIS), Agia Napa, Cyprus, pp. 347–365 (2005)

[38] Giunchiglia, F., Shvaiko, P., Yatskevich, M.: Discovering missing background knowledge in ontology matching. In: Proceedings of the 16th European Conference on Artificial Intelligence (ECAI), Riva del Garda, Italy, pp. 382–386 (2006)

[39] Giunchiglia, F., Yatskevich, M., Shvaiko, P.: Semantic matching: Algorithms and implementation. Journal on Data Semantics 9, 1–38 (2007)

[40] Giunchiglia, F., Zaihrayeu, I.: Making peer databases interact - a vision for an architecture supporting data coordination. In: Proceedings of the 6th International Workshop on Cooperative Information Agents (CIA), Madrid, Spain, pp. 18–35 (2002)

[41] Goutte, C., Gaussier, É.: A probabilistic interpretation of precision, recall and f-score, with implication for evaluation. In: Proceedings of the 27th European Conference on Advances in Information Retrieval Research (ECIR), Santiago de Compostela, Spain, pp. 345–359 (2005)

[42] Gruber, T.R.: A translation approach to portable ontology specifications. Knowledge Acquisition 5(2), 199–220 (1993)

[43] Haas, L., Hernández, M., Ho, H., Popa, L., Roth, M.: Clio grows up: from research prototype to industrial tool. In: Proceedings of the 24th International Conference on Management of Data (SIGMOD), Baltimore, USA, pp. 805–810 (2005)

[44] Halevy, A., Ashish, N., Bitton, D., Carey, M., Draper, D., Pollock, J., Rosenthal, A., Sikka, V.: Enterprise information integration: successes, challenges and controversies. In: Proceedings of the 24th International Conference on Management of Data (SIGMOD), Baltimore, USA, pp. 778–787 (2005)

[45] He, B., Chang, K.: Making holistic schema matching robust: An ensemble approach. In: Proceedings of the 11th International Conference on Knowledge Discovery and Data Mining (KDD), Chicago, USA, pp. 429–438 (2005)

[46] He, B., Patel, M., Zhang, Z., Chang, K.: Accessing the deep web: a survey. Communications of the ACM 50(5), 94–101 (2007)

[47] Hull, R.: Managing semantic heterogeneity in databases: a theoretical prospective. In: Proceedings of the 16th Symposium on Principles of Database Systems (PODS), Tucson, USA, pp. 51–61 (1997)

[48] Ives, Z., Halevy, A., Mork, P., Tatarinov, I.: Piazza: mediation and integration infrastructure for semantic web data. Jornal of Web Semantics 1(2), 155–175 (2004)

[49] Kalfoglou, Y., Schorlemmer, M.: Ontology mapping: the state of the art. The Knowledge Engineering Review 18(1), 1–31 (2003)

[50] Kifer, M., Lausen, G., Wu, J.: Logical foundation of object-oriented and frame-based languages. Journal of the ACM 42(4), 741–843 (1995)

[51] Koifman, G.: Multi-agent negotiation over database-based information goods. Master's thesis, Technion-Israel Institute of Technology (February 2004)

[52] Lee, Y., Sayyadian, M., Doan, A., Rosenthal, A.: eTuner: tuning schema matching software using synthetic scenarios. VLDB Journal 16(1), 97–122 (2007)

[53] Lenzerini, M.: Data integration: A theoretical perspective. In: Proceedings of the 21st Symposium on Principles of Database Systems (PODS), Madison, USA, pp. 233–246 (2002)

[54] Madhavan, J., Bernstein, P., Domingos, P., Halevy, A.: Representing and reasoning about mappings between domain models. In: Proceedings of the 18th National Conference on Artificial Intelligence (AAAI), Edmonton, Canada, pp. 122–133 (2002)

[55] Madhavan, J., Bernstein, P., Rahm, E.: Generic schema matching with Cupid. In: Proceedings of the 27th International Conference on Very Large Data Bases (VLDB), Rome, Italy, pp. 48–58 (2001)

[56] Melnik, S., Garcia-Molina, H., Rahm, E.: Similarity flooding: a versatile graph matching algorithm. In: Proceedings of the 18th International Conference on Data Engineering (ICDE), San Jose, USA, pp. 117–128 (2002)

[57] Melnik, S., Rahm, E., Bernstein, P.: Developing metadata-intensive applications with Rondo. Journal of Web Semantics 1(1), 47–74 (2003)

[58] Melnik, S., Rahm, E., Bernstein, P.: Rondo: A programming platform for model management. In: Proceedings of the 22nd International Conference on Management of Data (SIGMOD), San Diego, USA, pp. 193–204 (2003)

[59] Miller, R., Haas, L., Hernández, M.: Schema mapping as query discovery. In: Proceedings of the 26th International Conference on Very Large Data Bases (VLDB), Cairo, Egypt, pp. 77–88 (2000)

[60] Miller, R., Hernàndez, M., Haas, L., Yan, L.-L., Ho, C., Fagin, R., Popa, L.: The Clio project: Managing heterogeneity. SIGMOD Record 30(1), 78–83 (2001)

[61] Modica, G., Gal, A., Jamil, H.: The use of machine-generated ontologies in dynamic information seeking. In: Proceedings of the 9th International Conference on Cooperative Information Systems (CoopIS), Trento, Italy, pp. 433–448 (2001)

[62] Mork, P., Rosenthal, A., Seligman, L., Korb, J., Samuel, K.: Integration workbench: Integrating schema integration tools. In: Proceedings of the 22nd International Conference on Data Engineering (ICDE) Workshops, Atlanta, USA, p. 3 (2006)

[63] Motro, A.: Management of uncertainty in database systems. In: Kim, W. (ed.) Modern Database Systems, The object model, interoperability and beyond. Addison-Wesley, Reading (1995)

[64] Nejdl, W., Wolf, B., Qu, C., Decker, S., Sintek, M., Naeve, A., Nilsson, M., Palmér, M., Risch, T.: Edutella: A P2P networking infrastructure based on RDF. In: Proceedings of the 11th International World Wide Web Conference (WWW), Honolulu, USA, pp. 604–615 (2002)

[65] Nottelmann, H., Straccia, U.: A probabilistic, logic-based framework for automated web directory alignment. In: Ma, Z. (ed.) Soft Computing in Ontologies and the Semantic Web. Studies in Fuzziness and Soft Computing, vol. 204, pp. 47–77. Springer, Heidelberg (2006)

[66] Noy, N., Musen, M.: The PROMPT suite: interactive tools for ontology merging and mapping. International Journal of Human-Computer Studies 59(6), 983–1024 (2003)

[67] Oundhakar, S., Verma, K., Sivashanugam, K., Sheth, A., Miller, J.: Discovery of web services in a multi-ontology and federated registry environment. International Journal of Web Services Research 2(3), 1–32 (2005)

[68] Paolucci, M., Kawamura, T., Payne, T., Sycara, K.: Semantic matching of web services capabilities. In: Proceedings of the 1st International Semantic Web Conference (ISWC), Chia Laguna, Italy, pp. 333–347 (2002)

[69] Parent, C., Spaccapietra, S.: Issues and approaches of database integration. Communications of the ACM 41(5), 166–178 (1998)

[70] Parsons, S.: Current approaches to handling imperfect information in data and knowledge bases. IEEE Transactions on Knowledge and Data Engineering 8(3), 353–372 (1996)

[71] Prade, H., Testemale, C.: Generalizing database relational algebra for the treatment of incomplete or uncertain information and vague queries. Information Sciences 34, 115–143 (1984)

[72] Putnam, H. (ed.): Reason, Truth, and History. Cambridge University Press, Cambridge (1981)

[73] Rahm, E., Bernstein, P.: A survey of approaches to automatic schema matching. VLDB Journal 10(4), 334–350 (2001)

[74] Rahm, E., Do, H.-H., Maßmann, S.: Matching large XML schemas. SIGMOD Record 33(4), 26–31 (2004)

[75] Rodríguez-Gianolli, P., Garzetti, M., Jiang, L., Kementsietsidis, A., Kiringa, I., Masud, M., Miller, R.J., Mylopoulos, J.: Data sharing in the Hyperion peer database system. In: Proceedings of the 31st International Conference on Very Large Data Bases (VLDB), Seoul, South Korea, pp. 1291–1294 (2005)

[76] Rousset, M.-C., Adjiman, P., Chatalic, P., Goasdoué, F., Simon, L.: Somewhere in the semantic web. In: Wiedermann, J., Tel, G., Pokorný, J., Bieliková, M., Štuller, J. (eds.) SOFSEM 2006. LNCS, vol. 3831, pp. 84–99. Springer, Heidelberg (2006)

[77] Sarma, A.D., Benjelloun, O., Halevy, A.Y., Widom, J.: Working models for uncertain data. In: Proceedings of the 22nd International Conference on Data Engineering (ICDE), Atlanta, USA, p. 7 (2006)

[78] Sayyadian, M., Lee, Y., Doan, A.-H., Rosenthal, A.: Tuning schema matching software using synthetic scenarios. In: Proceedings of the 31st International Conference on Very Large Data Bases (VLDB), Trondheim, Norway, pp. 994–1005 (2005)

[79] Sheth, A., Larson, J.: Federated database systems for managing distributed, heterogeneous, and autonomous databases. ACM Computing Surveys 22(3), 183–236 (1990)

[80] Shvaiko, P.: Iterative Schema-based Semantic Matching. Ph.D thesis, International Doctorate School in Information and Communication Technology, University of Trento, Trento, Italy (November 2006)

[81] Shvaiko, P., Euzenat, J.: A survey of schema-based matching approaches. Journal of Data Semantics 4, 146 (2005)

[82] Srivastava, B., Koehler, J.: Web service composition - Current solutions and open problems. In: Proceedings of the Workshop on Planning for Web Services at the 13th International Conference on Automated Planning and Scheduling (ICAPS), Trento, Italy, pp. 28–35 (2003)

[83] Su, W., Wang, J., Lochovsky, F.: Holistic schema matching for web query interfaces. In: Proceedings of the 10th Conference on Extending Database Technology (EDBT), Munich, Germany, pp. 77–94 (2006)

[84] Tsichritzis, D., Klug, A.C.: The ansi/x3/sparc dbms framework report of the study group on dabatase management systems. Information Systems 3(3), 173–191 (1978)

[85] Wong, S., Xiang, Y., Nie, X.: Representation of bayesian networks as relational databases. In: Proceedings of the 5th International Conference on Information Processing and Management of Uncertainty (IPMU), Paris, France, pp. 159–165 (1994)

[86] Zadeh, L.: Fuzzy sets. Information and Control 8, 338–353 (1965)

[87] Zadeh, L.: Fuzzy sets as a basis for a theory of possibility. Fuzzy Sets and Systems 1, 3–28 (1978)

[88] Zaihrayeu, I.: Towards Peer-to-Peer Information Management Systems. Ph.D thesis, International Doctorate School in Information and Communication Technology, University of Trento, Italy (March 2006)

[89] Zhdanova, A., Shvaiko, P.: Community-driven ontology matching. In: Proceedings of the 3rd European Semantic Web Conference (ESWC), Budva, Montenegro, pp. 34–49 (2006)

Multi-site Software Engineering Ontology Instantiations Management Using Reputation Based Decision Making

Pornpit Wongthongtham, Farookh Khadeer Hussain, Elizabeth Chang, and Tharam S. Dillon

Digital Ecosystems and Business Intelligence Institute
Curtin University of Technology, G.P.O. Box U1987, Perth, WA 6845, Australia
{Pornpit.Wongthongtham,Farookh.Hussain,Elizabeth.Chang,
Tharam.Dillon}@cbs.curtin.edu.au

Abstract. In this paper we explore the development of systems for software engineering ontology instantiations management in the methodology for multi-site distributed software development. Ultimately the systems facilitate collaboration of teams in multi-site distributed software development. In multi-site distributed environment, team members in the software engineering projects have naturally an interaction with each other and share lots of project data/agreement amongst themselves. Since they are not always residing at the same place and face-to-face meetings hardly happen, there is a need for methodology and tools that facilitate effective communication for efficient collaboration. Whist multi-site distributed teams collaborate, there are a lot of shared project data updated or created. In a large volume of project data, systematic management is of importance. Software engineering knowledge is represented in the software engineering ontology whose instantiations, which are undergoing evolution, need a good management system. Software engineering ontology instantiations signify project information which is shared and has evolved to reflect project development, changes in the software requirements or in the design process, to incorporate additional functionality to systems or to allow incremental improvement, etc.

Keywords: Multi-Site Software Development, Software Engineering Ontology, Reputation, Trust.

1 Introduction

The term "Ontology" is derived from its usage in philosophy where it means the study of being or existence as well as the basic categories [1]. Therefore, it is used to refer to what exists in a system model.

Definition 1. An ontology, in the area of computer science, is the effort to formulate an exhaustive and rigorous conceptual schema within a given domain, typically a hierarchical data structure containing all the relevant elements and their relationships and rules (regulations) within the domain [2].

T.S. Dillon et al. (Eds.): Advances in Web Semantics I, LNCS 4891, pp. 199–218, 2008.
© IFIP International Federation for Information Processing 2008

Definition 2. An ontology, in the artificial intelligence study, is an explicit specification of a conceptualisation [3, 4]. In such an ontology, definitions associate the names of concepts in the universe of discourse e.g. classes, relations, functions) with describing what the concepts mean, and formal axioms that constrain the interpretation and well-formed used of these terms [5].

For example, by default, all computer programs have a fundamental ontology consisting of a standard library in a programming language, or files in accessible file systems, or some other list of 'what exists'. However, the representations are poor for some certain problem domains, so more specialised schema must be created to make the information useful and for that we use an ontology.

To represent the software engineering knowledge, the whole set of software engineering concept representing generic software engineering knowledge is captured as domain knowledge in ontology. A particular project or a particular software development probably uses only part of the whole sets of software engineering concepts. For example, if a project uses purely object oriented methodology, and then the concept of a data flow diagram may not be necessarily included. Instead, it includes concepts like class diagram, activity diagram and so on. The specific software engineering concepts used for the particular software development project representing specific software engineering knowledge are captured as sub domain knowledge in ontology. The generic software engineering knowledge represents all software engineering concepts, while specific software engineering knowledge represents some concepts of software engineering for the particular project needs. In each project, there exists project information or actual data including project agreement and project understanding. The project information especially meets a particular project need and is needed with the software engineering knowledge to define instance knowledge in ontology. Note that the domain knowledge is separate from instance knowledge. The instance knowledge varies depending on its use for a particular project. The domain knowledge is quite definite, while the instance knowledge is particular to a project. Once all domain knowledge, sub domain knowledge and instance knowledge are captured in ontology, it is available for sharing among software engineers through the Internet. All team members, regardless of where they are, can query the semantic linked project information and use it as the common communication and knowledge basis of raising discussion matters, questions, analysing problems, proposing revisions or designing solutions, etc.

Ontology is machine-understandable. A machine, in the form of application, can use the software engineering knowledge represented in the ontology and carry out software engineering knowledge management. Software engineering knowledge management refers to the ways that project information, a concrete state of the conceptual structures of software engineering domain, are gathered, managed, and maintained for remote teams' collaboration in multi-site software development. In order to make such project information assets available for transfer across sites, a range of specific software development processes is identified and captured as knowledge, know-how, and expertise [6] which is in the form of ontology. Software engineering knowledge management is specifically tied to the objective of providing unified semantic knowledge sharing and improved understanding within distributed teams, transparent and understandable task accomplishment for remote teams, aware of current project state, etc. In order to explore software engineering knowledge management as part of the solution, in the following sections we clarify instantiations retrieval and instantiations manipulation.

Software engineering knowledge, formed in the ontology, represents abstraction of software engineering domain concepts and instantiations. The abstraction is divided into generic software engineering knowledge and specific software engineering knowledge. The abstraction of generic knowledge represents the whole software engineering concepts, while the abstraction of specific knowledge represents the software engineering concepts used for some particular projects. The instantiations knowledge simply represents the project information. The abstraction of the specific software engineering knowledge has its instantiations being used to store data instances of the multi-site projects. Each abstraction of the specific software engineering knowledge can have multiple instantiations in different circumstances of the projects. The corresponding concrete data instances are stored as instantiations. The abstraction and its instantiations are made explicit and enable retrieval by remote team members.

Sowa [7] stated that natural languages are too ambiguous. In order to eliminate ambiguity, project information is captured corresponding to its abstraction. This makes project information assumptions explicit as its abstraction is made explicit in the ontology. Explicit instantiations of project information knowledge share meaning and understanding amongst distributed teams. It leverages remote team perspective. Project information is formulated corresponding with software engineering concept structures in the ontology. This makes project information easy to navigate and be retrieved punctually. A navigational view and integrated view of the retrieved instantiations of knowledge assist team members with team collaboration.

In order to understand systems of software engineering ontology instantiations management and its architecture, in the next section we analyse instantiations in the software engineering ontology and define instantiations transformation. Accordingly, management systems of safeguard, ontology, and decision maker are introduced and discussed.

2 Software Engineering Ontology Instantiations

Software engineering ontology instantiations are derived as a result of populating software engineering project information and are referred to as ontology instances of ontology classes. Instantiations are also known as instance knowledge of the software engineering ontology. In other words, once the software engineering ontology is designed and created, it needs to be populated with data relating to the project. This process is usually accomplished by mapping various project data and project agreement to the concepts defined in the software engineering ontology. Once mappings have been created, project information, including project data, project agreement, and project understanding, is in a semantically rich form and management is needed to maintain the instantiations.

2.1 Instantiations Analysis

The software engineering ontology contains abstractions of the software engineering domain concepts and instantiations. There are two types of the abstraction which are the generic software engineering and the specific software engineering. The abstraction of the generic one represents the concepts that are common to a whole set of software engineering concepts, while the abstraction of the specific one represents the set of software engineering concepts that are specifically used for some categories of particular projects. The instantiations, also known as population, are simply the project data. The abstraction of the specific software engineering ontology has its instantiations ultilised for storing data instances of the projects. Each abstraction can have multiple instantiations in different circumstances of projects. The corresponding concrete data instances are stored as instantiations. In this study, the software engineering ontology integrates abstractions and instantiations together, rather than separating them by storing instances in a traditional relational database style linked to the knowledge base. The latter, SQL queries, can help with the large volume of concept and instance management and maintenance. Nevertheless, in the software engineering ontology, the data volume is not very large and coherent integration of abstraction and instantiations are important in the software engineering projects especially in multi-site software development environment. Putting them together instead of separately would be more suitable for this study. For example, each project contains a different narrow domain (specific software engineering ontology) and limited numbers of data instances. The domain specific ontologies are locally defined; that is, they are derived from the generic software engineering ontology so they are not created with respect to some global declarations. This indicates that abstractions and instantiations are better stored together instead of separately. In conclusion, ontology instantiations for software engineering knowledge management actually means management of the instantiations.

In reality, in software engineering projects, the project data over a period of time needs to be modified to reflect project development, changes in the software requirements or in the design process, in order to incorporate additional functionality to systems or to allow incremental improvement. Since changes are inevitable during software engineering project development, the instantiations of the software engineering ontology is continuously confronted with the evolution problem. If such changes are not properly traced or maintained, this would impede the use of the software engineering ontology. Due to the complexity of the changes to be made, at least a semi-automatic process becomes increasingly necessary to facilitate updating tasks and to ensure reliability. Note that this is not ontology evolution because it does not change the original concepts and relations in the ontology, rather instantiations of the ontology change or that conform to the ontology change.

Thus, software engineering domain changes that are produced by new concepts, and change in the conceptualisation as the semantics of existing terms are modified with time, are all outside the scope of this study.

When there are changes made to the instantiations of the ontology, they are all recorded by a logger object. Basically, instantiations can be updated by three basic operations: add, delete and modify. The add operation extends the existing instantiations of the ontology with new instantiations. The delete operation removes some instantiations from the ontology. The modify operation modifies some instantiations of the ontology but it still keeps its original construct. Generally, any update to the instantiations of ontology can be described by a sequence of the three operations. For example, a delete operation followed by an add operation can be considered as a replacement operation. Notice that the replacement operation loses its original construct while the modify operation still maintains its construct.

2.2 Instantiations Transformation

In this section, we particularly report on how software engineering project data are transformed or mapped into concepts formed in the software engineering ontology as instance knowledge. Conversely, the instance knowledge can be transformed back to more presentable and semantic project data e.g. diagram-like project data. Once transformed, instance knowledge is available for sharing among multi-site teams. Manipulation of semantics such as instance knowledge can be carried out by users or remote members. These are shown in Figure 1.

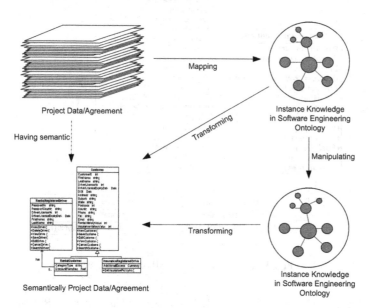

Fig. 1. Instantiations transformation

An example of transformation is given in Figure 2 which shows a class diagrams ontology model. Figure 3 shows an example UML class diagram that will be transformed into the ontology model in Figure 2 as instance knowledge.

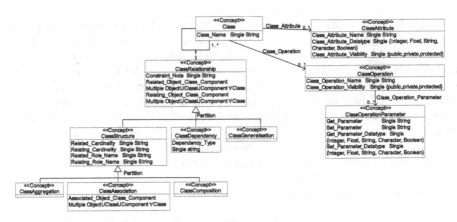

Fig. 2. Class diagrams ontology relations

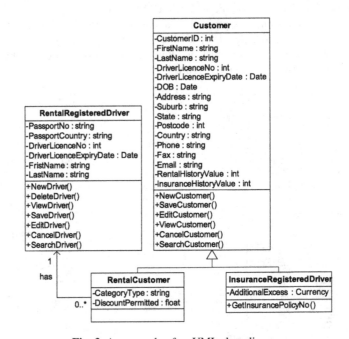

Fig. 3. An example of an UML class diagram

As from Figure 3, UML classes Customer, RentalCustomer, InsuranceRegisteredDriver, and RentalRegisterredDriver apply as instances of the ontology concept Class in class diagrams ontology. Explicit domain knowledge from concept Class elicit that class consists of its properties, its operations and its relationships. This is by referring respectively, in the class diagrams ontology model, to relations Class_Attribute, Class_Operation, and association ontology class ClassRelationship. The concept Class instance Customer has relation has_Attribute with concept ClassAttribute instances CustomerID, FirstName, LastName,

DriverLicenceNo, etc. For example, the concept instance DriverLicenceNo has relations Class_Attribute_Datatype with xsd:string of 'Integer' and has relations Class_Attribute_Visibility with xsd:string of 'Private'. These are shown below in Figure 4.

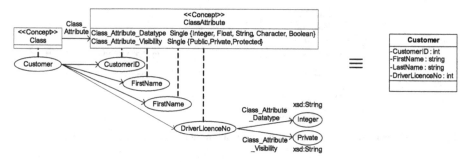

Fig. 4. Transformation of UML class Customer and its attributes to class diagrams ontology

For a particular UML class Customer, operation NewCustomer() applies as an instance of concept ClassOperation in the class diagrams ontology model. The concept Class instance Customer has relation Class_Operation with concept ClassOperation instance NewCustomer. The concept instance NewCustomer has relations Class_Operation_Visibility with xsd:string of 'Public.' These are shown below in Figure 5.

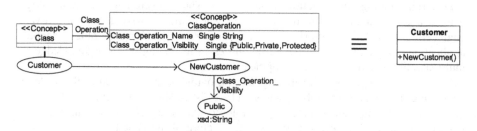

Fig. 5. Transformation of UML class Customer and its operation to class diagrams ontology

For the particular class diagram shown in Figure 3, the concepts of generalisation relationship and association relationship are applied. An instance of concept ClassGeneralisation has relations Related_Object_Class_Component with concept Class instances RentalCustomer and InsuranceRegisteredDriver and has relations Relating_ Object_Class_Component with concept Class instance Customer. Instance of concept ClassAssociation has relations Related_Object_Class_Component with concept Class instance RentalRegisteredDriver, has relations Relating_ Object_Class_Component with concept Class instance RentalCustomer, has relations Related_Cardinality with xsd:String of '1', has relations Relating_Cardinality with xsd:String of '0..*', and has relations Related_Role_Name with xsd:String of 'has'. These are shown below in Figure 6.

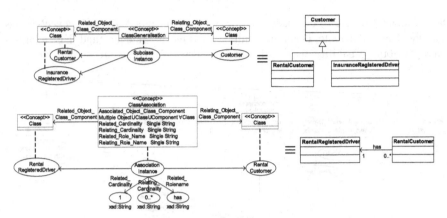

Fig. 6. Transformation of generalisation and association relationships to class diagrams ontology

All project data/agreements which are instantiations of the software engineering ontology need management to promote the use of semantic project data for multi-site distributed software development.

3   Development of Systems for Software Engineering Ontology Instantiations Management

In this paper we present development of applications in systems to facilitate software engineering ontology instantiations management. The software engineering ontology is made available to any application to deploy. The ability to make use of the software engineering knowledge, described in the software engineering ontology, enables applications in the systems to have capabilities in managing instance knowledge in multi-site distributed software development.

Management tasks for software engineering ontology are assigned to the systems containing a number of sub systems. There is a set of systems to facilitate management of software engineering ontology named safeguard system, ontology system, and decision maker system. The architecture of the systems in the multi-site environment is shown in Figure 7.

Team members, regardless of where they are, connect to the web server via a web browser. This will enable team members to directly use the system without having to download any software or install any application. Each team member is served by the intelligent systems tool as the communication media. This allows direct communication between different team members using a messaging system and allows monitoring and recording of the activities of the team members as well. Each team member is provided with a particular set of access privileges that are dependent of the role of that team member in the project. The set of sub systems within the intelligent support systems architecture is comprised of: safeguard system, ontology system and decision maker system. The safeguard system represents system authentication for user authorisation and determination of the access level. The

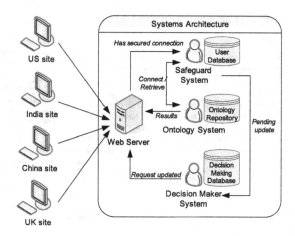

Fig. 7. Model of Management Systems

safeguard system communicates with the ontology system if the user / team member wants to query or update the software engineering ontology. The ontology system manipulates and maintains the software engineering ontology repository. The decision maker system operates tasks if an operation needs to be certified. The decision maker system is responsible for decision making on the matter of updating the software engineering ontology including acknowledgement of the decision made to all involved team members. As can be seen from the model of management systems, only the safeguard system has any connection with the user database. This means that the safeguard system performs all recording of user activities as well. All other systems call the safeguard system and pass the information to log all the events that the system carried out. Thus, tracking can be accomplished by the safeguard system if needed. Not only does it allow tracing; the safeguard system can determine bottlenecks, if there are any occurrences, with the use of the timestamp. The ontology system is the only one manipulating the software engineering ontology. Thus, it is the only one that has access to the ontology repository. All other systems contact the ontology system in the cases of wanting to view, query, or update the ontology. The decision maker system has its own database to store data for decision making occurring in the systems.

In detail, the functionalities of each system can be observed in Figure 8. The safeguard system functionalities include system authentication, access level allocation, solution proposal management and monitoring of the user activities. Functionalities in the ontology system include navigating, querying and manipulating software engineering ontology. For the decision maker system, a method of reputation based voting is used in the system.

Every team member can navigate the software engineering ontology but no changes allow (number 3). A team member can log into systems. Once logged in, system authentication (number 1) in the safeguard system verifies user access from user database. Once authorised, the member will be provided the access privileges from the safeguard system (number 2). The member can now navigate, query, make

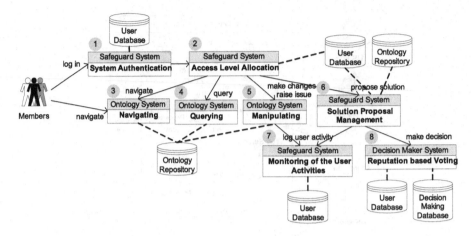

Fig. 8. Functionalities provided in each system

changes, raise issue, or propose solution. Navigation, query, and manipulation of instantiations in software engineering ontology are functioned by the ontology system (number 3-5). Manipulation of instantiations is functioned by the ontology system and is recorded into user database by the safeguard system (number 7). Solution proposal is managed by safeguard system (number 6). Any decision is made by decision maker system (number 8).

4   Safeguard of Software Engineering Ontology

To implement security features into the systems, it has been decided to appoint the safeguard system. The safeguard system implements and enforces the systems' authentication, the access control policies and member activity log. All these operations have different logic involved whose details are given in the next following sections.

4.1   Software Engineering Ontology Access Authentication

Once users log into the system, the user identification will be verified with the user database handled by relational database. Once authorised, the user can access, modify or update the ontology depending on the access privileges held by the user whose details are given in the next section. There are a few security levels, mainly:

- Software engineers – they can be:
  - Software engineers, who have no access to updating, but can only look up, or
  - Software engineers who have an access to update project data, or
- Team leader / project manager – they have unlimited access to all functions such as updates, backups, ontology maintenance and database access.

It should be noted that only the team leader / project manager has access to the database server. Team leaders, who do not have any of the access levels stated, will not have access to the specific section of the ontology.

4.2 Determination of the Software Engineering Ontology Access Level

A user logged into the systems after authentication will be provided with different services according to different access levels. All team members are provided with the service to query the software engineering ontology. The list below shows the different access levels:

- Querying level – only querying service that allows no changing
- Add and modifying level – restricted access to add and modify service of the software engineering ontology instances (project data). At this level, some operations may be required to be hold through making decision system e.g. request for revision of project design model. Simple updates like the status of the project or documentation update, would immediately be updated to the software engineering ontology
- Full access level – unrestricted access to all services provided

These access levels are given according to the different status of the team members. The hierarchy of the software engineering sub-ontology is used to categorise hierarchy to in order to determine the access to the status of team members. The team leader is the one who assigns proper access privileges to each member in the team. For example, the sub-ontology 'software design' would require a designer team or the sub-ontology 'software requirement' would demand an analyst team to access, add, and modify their project data. Nevertheless, the team designer can look up project requirements through the sub-ontology 'software requirement' but no changes have been allowed, which means they are on the querying level. The team leader will have full access including monitoring team member activities. The process of viewing, querying and manipulating software engineering ontology is done by the ontology system. The safeguard system will only verify the authorisation, the access level and log activities and then pass the request to the ontology system.

4.3 Monitoring of the User Activities

Every single completed activity will be recorded by the logger application resided in the safeguard system. This allows monitoring of all the user activities. The safeguard system is requested by any other systems i.e. ontology system and decision maker system to record team member activities. Because the safeguard system is the only system that connects to the user database; thus, if there is any action to log into the database, it will be done by the safeguard system.

4.4 Solution Proposal Management

In a software engineering project, once issues arrive, one can raise these issues with the team and the suggested solutions can be proposed by any one in the project team. All team members then can vote for either the selected solution, or they can support the original. This is like a communication tool that allows a project team member to voice opinions or suggestions on the particular issue that has arisen. Multiple solutions can be proposed for a single issue. After voting, the decision maker system whose details are in the later section, operates decision making showing the proposed solution that has been chosen and acknowledged by the team. Functionalities of a solution proposal management therefore include raising an issue, proposing possible solutions, voting for a final solution and retrieving all parts of an issue and its pending proposed solutions.

5 Software Engineering Ontology Management

The purpose of having an ontology system is to manage connections with the software engineering ontology. The ontology system is built on top of Jena [8] which we would like to gratefully acknowledge. Jena developed by the Hewlett-Packard Company is a Java framework having capacity of manipulating ontologies [9]. The version of Jena used is Jena 2.1. The ontology system provides navigating, querying, and manipulating software engineering ontology. The design philosophy of the ontology system is to use the in-memory storage model and serialise it into a physical document stored in the ontology repository. It is an attempt to minimise the query response time. Note that this is not like a knowledge base system that uses the data based model to query the ontology and instance data.

There are three different services here in the ontology system: navigating, querying and manipulating services which are given in the next sections.

5.1 Navigation of Software Engineering Ontology

In this section, we deal with the accessing of information held in the software engineering model. Software engineering concept structures are formulated so that it can easily be navigated. A team member can navigate in the software engineering ontology for clarification or classification certain concepts. The information provided is in hierarchical form so upper level concepts or lower level concepts or adjacent concepts can easily be navigated.

Technically for this function, the ontology system focuses on the software engineering ontology model, the set of statements that comprises the abstraction and instantiations. To navigate the software engineering ontology, the ontology system reads OWL software engineering ontology into a model and then accesses the individual elements.

5.2 Query of Software Engineering Ontology

The previous section dealt with the case of navigating the software engineering model. This section deals with searching the software engineering model. As stated in the earlier section, the in-memory storage model is used hence a query primitive supports. The query facilities of RDQL [10] which is a data based model held in a persistent store, is not within the scope of this study.

It serves as a searching tool to help narrow down the vast number of concepts in the ontology. Through the use of the ontology search function, the team member can re-classify concepts to match their project needs. This leads to the specific ontology. Note that the information provided by this function is all in XML format, which means that it can be easily managed to display only a certain part of the information retrieved or be able to provide a different display interface with the same set of information retrieved.

5.3 Manipulation of Software Engineering Ontology

This section deals with manipulating the software engineering model. In the specific software engineering ontology which contains project data, a team member can add, delete, and update the project data. However, the ontology system will only allow direct updates for the minor changes/updates. The changes will be recorded and team members will be advised of the changes. An example of minor changes is enumerated types where the changes allowed are already fixed and team members cannot put in other values. Another example of minor changes is a changing of status of a document with the option of, for example, 'verified' or 'processing'. By default, any updating apart from the minor changes will be done by the decision maker system and be recorded. Even the ontology system considers whether they are minor changes or major changes, though there is an option for a team member to select whether or not these changes will go through the decision maker system. In the decision maker system, the changes will not be updated immediately to the specific ontology. They need to be voted by members of the community and therefore need to be stored in the decision making database. The process of decision making is handled by the decision maker system whose details are given in the next section. The ontology system simply checks whether the update request had been authorised before being updated. Basically, for major changes, the ontology system will pass the request of changes to the decision maker system to proceed further with processes of, for example, gathering information, consulting the ontologies in ontology repository etc. Once it has passed through the decision maker system, the updating can be done by the ontology system. Every activity will be recorded and the results of the processes are sent to the user that made the enquiry and to every team member involved.

5.4 Software Engineering Ontology System Model Packages

The architecture of the ontology system consists of three packages: 'generic', 'specific' and 'ontology'. The 'generic' package defines the interfaces of the data structures of generic software engineering ontology and generic software engineering ontology objects. Likewise, the 'specific' package defines the data structures of specific software engineering ontology interfaces and specific software engineering ontology objects, such as class and its instances. Both 'generic' and 'specific' packages provide an in-memory implementation of the data models of generic and specific ontologies respectively. The 'ontology' package provides the utilities for the ontologies defined in 'generic' and 'specific' packages.

Generic Package. Generic software engineering ontology can be accessed by anyone without system authentication. It is used for a search of concepts relating to the software engineering domain. Unlike specific software engineering ontology, it is meant to be used for the projects, and therefore system authentication is required. Generic search allows searching of any concept within the software engineering ontology. Search results display the contents of the concept the user specifies including its subclasses, its properties and restrictions. The output is in XML format in order for it to be displayed easily on the web browser. Basically, the display of the hierarchy of subclasses can be accomplished using a recursive function. The function will find out the entire sub concepts of a concept by recursively calling the function itself over and over again until no more sub concepts can be found.

Specific Package. A specific package provides a set of functionalities that helps the project team to have a mutual understanding through the use of specific software engineering ontology. Not only can the project members update project data, but also by withdrawing partial relevant knowledge from the software engineering ontology, issues can be discussed and solutions proposed. The set of functionalities includes: view, query, add, delete and modify instances or simply project data and properties.

To retrieve instances of a concept or ontology class, a function retrieves all direct instances related to the class or the concept. From here, users can browse this instance information. All information associated with the instance is like its definition, its properties (both object properties and data type properties) or its relationships, value inside those properties and its restriction.   The main purpose of retrieving its relationships are firstly to help the team members understand its underlying concept; secondly, to help discussion on the issues or solutions; and lastly, to help update project data to be completed according to its domain concept. It is easy to become confused if discussion takes place with words only, especially when there are many ambiguous words in the software engineering domain. Therefore, by retrieving the relationships associated with instances, it can help team members illustrate what they truly mean. This is done even better with the Java drawing toolkit which can be used to draw a relationship diagram.

Manipulating specific software engineering ontology is an essential tool to help maintain a project because all project data is stored as instance. In reality, project data are always updated from time to time. When project data need to be updated or added, a function even helps to check essential parameters needed in order to retrieve from its associated relationships, its restrictions etc. Updating is divided into two types of update: minor and major. Any significant changes, such as those that annihilate certain information or add an entirely new instance to a project, are considered as a major update. Additionally, all object properties are considered as a major update because they reflect the changes of relationships. Requirements that satisfy the condition of being a minor update are firstly, any changes made by members in their field of expertise or simply in their team. For example, a designer making changes in the domain of project design will have the right to do so, therefore they are considered as minor updates. However, the designer making changes to the domain of project implementation will then not have valid rights and the changes will be considered as major updates instead. All data type properties are also classified as minor updates.

Ontology Package. The 'ontology' package is a compilation of functions that provides the utilities for the ontologies defined in 'generic' and 'specific' packages. It does not belong to any category and does not have enough information to create its own category either. The functions in the package are mainly like (i) getting ontology name space, (ii) search engine for the ontology system, (iii) ontology class or concept restrictions and property characteristics checker to specify the range or restrict the values of input, (iv) converting information into XML format for output and (v) parsing and serialising ontologies in OWL language.

Firstly, getting ontology namespace is needed to extract the namespace of the ontology. The OWL file stores all the information of the different URL and namespace in the header of the file. When OWL file is loaded into the ontology model by calling the Jena modelfactory [11], all the URLs and namespace for the ontology can be retrieved. Since the ontology model loads all URLs first then loads the namespace of the ontology, the namespace is always the last element. By using Java's StringTokeniser [11] the namespace of the ontology can be retrieved with ease.

Secondly, a function in the package serves as a search engine for the ontology system. It especially includes finding any close match of ontology class or concept. This is useful when the search does not return any exact match to the user.

Ontology restrictions include quantifier restrictions and cardinality restrictions and ontology property characteristics include functional, inverse functional, symmetric and transitive properties. Functions of checking all ontology restrictions and property characteristics are all in the package to restrict the conditions. A function checks whether there is a minimum cardinality restriction implemented on the concept or ontology class. Minimum cardinality restriction refers to the minimum number of properties that must be input in order to satisfy the condition. For maximum cardinality restriction, a function checks for a maximum number before a new property value is added. If the maximum cardinality number is reached, adding of a new property value is disallowed, or else adding of a new property value is allowed. If there is a cardinality restriction present, it means there can be no more and no less cardinality than the cardinality specified. The quantifier restrictions of allValueFrom means that all the values of this property to whom this restriction applies, must have all values falling within its range. Likewise, someValueFrom restrictions, some values of the property that this restriction applies to must have some values falling within its range. Therefore, the aim of a function for this is to check whether the new value which is going to be added falls into the category (if so return true; if it does not fall into the category, return false). These kinds of restrictions are only for adding new object properties.

Fourthly, a function is to convert information retrieved into XML format for output. This function is used often to display instance information including its restrictions information retrieved. All restrictions follow the same output format with XML tag as the name of the restriction and its value.

Lastly, parsing and serialising ontologies in OWL language are needed for format translation. A format or syntax translator requires the ability to parse, represent the results of the parsing into an in-memory ontology model, and then serialise. Manipulation capabilities for example would also be required, in between the parsing and serialising processes, in order to allow construction and editing of ontologies. In the implementation view, parsing is taking the OWL file and converting it to an in-memory ontology model. Conversely, analogous to the parsing, serialising produces an OWL concrete syntax in the form of a syntactic OWL file from an in-memory ontology model.

6 Decision Support

As the name itself suggests, the job of the decision maker system is to make a decision on an issue such as a major update request. In this study, we have developed a combination of two techniques to implement the decision maker system. The decision making is based on members in the teams agreeing to vote, along with the reputation of each individual member involved in the software engineering project. In the following sub sections, details of the voting and reputation techniques are presented. We illustrate the combination of both techniques in the last sub section.

6.1 Voting System

The voting system provides a means for making changes to the reflected project data or instance knowledge in the software engineering ontology based on votes from each member of the project teams. Every member in the teams involved in a given software engineering project has a right to vote for proposed solutions. Everybody's vote is worth points. Below is a list of requirements for the voting system.

- A member can work on a project or multiple projects at the same time
- A member can work in a team or in multiple teams
- A member can work in different teams in different projects
- A project involves multiple teams and multiple members

The vote cast by each team member is mathematically weighted by the factor of which 'members who actually work on a task have the best understanding of that task'. In other words, if a member votes on an issue which arises within the area he/she is working on, presumably this falls within his/her area of expertise, then his/her vote carries more weight than that of a member who does not have expertise in the issue area, or who does not really work on it. There are four areas of expertise categorised by following four software processes in the software engineering domain. These are software requirement, software design, software construction, and software testing. Members classified in these areas of expertise are analysts, designers, programmers or implementers, and tester respectively. Figure 9 gives an example of this classification. A member, for example, may work in the design team for a particular project and may also work in the requirements team and construction team in another project. It is assumed, for example, that the designers of a project who

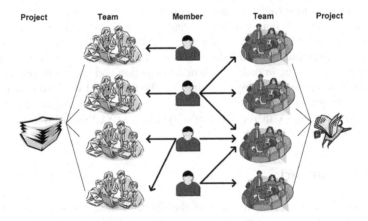

Fig. 9. An example of members working in project teams

work on the project design, have expertise in project design, or know more than others do about this aspect of a project. Thus, if the project issue relates to project design, the votes of members in the design team carry more weight than others.

Table 1 shows an example of three possible solutions named A, B, and C in the issue of project design. Let us assume that solution A received a single vote from a designer, solution B received a single vote from an analyst, and solution C received a single vote from a programmer. The vote of each of these people is weighted by their expertise in the area of the issue. From the above description, solution A then would receive the maximum vote points since it has been voted by a member in the design team who, it is assumed, has some degree of expertise in the project design because he/she actually works on it.

Table 1. An example of the three possible solutions in the issue of project design

Solution	Requirement (y)	Design (x)	Construction (y)	Testing (y)	Design Issue Voting Points $(x>y)$
A		√			1 x x
B	√				1 x y
C			√		1 x y

6.2 Reputation System

The reputation based system provides a means for making the changes to the reflected project data in the software engineering ontology based on the reputation of the team members involved in the software engineering project. Below is a list of requirements for the voting system.

- A member has a reputation value for a particular area or domain in a given project
- A member can have a different reputation value for a different area or domain in a different project
- The reputation value of the team member continues to increase if the team member votes for the chosen (or correct) solution and vice versa
- The reputation value of the team member decreases if the team member did not vote for the chosen (or correct) solution and vice versa.

The reputation value of members may change with time. In other words, at a given time and in a particular area, reputation value may increase, decrease or remain the same. Figure 10 shows as example of different reputation values in the different areas of a member working on different projects. Using the Markov Model [12], the change in the reputation value of each team member in a given phase is tracked. Additionally, using the Markov Model, we consider what could be the most probable future reputation value of a given team member in the category of the issue at a time in which the decision has to be made.

The calculation of a user's reputation value, which is a value of either 1 or 2, is based on the past reputation points for different domains. In order to calculate a user's reputation value, the first step is to calculate the current state value (CSV) which is

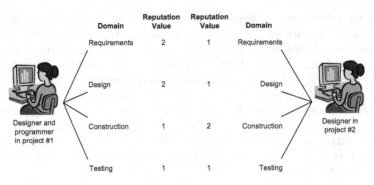

	Reputation Value	Reputation Value	
Domain			**Domain**
Requirements	2	1	Requirements
Design	2	1	Design
Construction	1	2	Construction
Testing	1	1	Testing

Fig. 10. An example of different reputation value in the different area of expertises of members working in different project

the latest up-to-date reputation value; second, calculate the Markov matrix; and third, multiply CSV with the Markov matrix in order to arrive at the determined reputation value of the user [12]. Along with the explanation, we will provide an example for further clarity. If the last reputation value is a 2, the CSV is a matrix [0 1]. If the last reputation value is a 1, the CSV is a matrix [1 0]. There can be only these two possibilities. For example, a set of reputation value list is c. As can be seen, the last value is a 2 then the CSV for that member is a matrix [0 1]. Once CSV has been found, the Markov matrix is calculated next. The transition states matrix is needed. Since there are only 1s and 2s, there are four states of transition namely: 1-1 state, 1-2 state, 2-1 state and 2-2 state. By counting the numbers of each state, we form the transition states matrix. As from the example, there is none of 1-1 state; there is one 1-2 state; there is one 2-1 state and there are three 2-2 states as shown below.

$$\text{Transition States Matrix} = \begin{bmatrix} 0 & 1 \\ 1 & 3 \end{bmatrix} \tag{1}$$

By counting a frequency of transition from state 1, in the above example state 1 transition frequency is one and state 2 transition frequency is four. This is used to calculate the percentage of whether it changes state or stays at the state. For the above example, 1-1 state has 0/1 that means 0% or 0, 1-2 state has 1/1 that means 100% or 1, 2-1 state has 1/4 that means 25% or 0.25 and 2-2 state has 3/4 which means 75% or 0.75. From here the Markov matrix is as followed.

$$\text{Markov Matrix} = \begin{bmatrix} 0 & 1 \\ 0.25 & 0.75 \end{bmatrix} \tag{2}$$

By multiplying the Markov matrix with the CSV, we will be able to obtain the reputation value of the user. The probability of the reputation value is given below.

$$\text{Reputation Value Probability} = \begin{bmatrix} 0 & 1 \\ 0.25 & 0.75 \end{bmatrix} \times \begin{bmatrix} 0 & 1 \end{bmatrix} = \begin{bmatrix} 0.25 & 0.75 \end{bmatrix} \tag{3}$$

The probability of the reputation value is in the form of [a b]. If the value a is greater than b, it means the most probable reputation value will be a 1 and if value b is greater than a, it means the most probable reputation value will be a 2. Therefore, in this example, the reputation value probability of [0.25 0.75] where 0.75 is greater than 0.25, shows that the user's reputation value is worth 2.

6.3 Reputation Based Voting for Making Decision

Whenever issues arise, such as a major update request of project data, the decision making system sends a message to every team member asking him/her for an opinion. Subsequently, it then gathers and stores the possible solutions for that particular issue. Of all the possible solutions, one solution is chosen by asking all team members to vote for one of the proposed solutions. The reputation based decision will then determine the total number of votes for each solution and, as mentioned earlier, each vote is weighted by the expertise of the person casting it in the area of the problem. Additionally, the reputation value of individual member who votes is weighted.

For example, assume that a weighting value for member who his/her expertise is not in the area of the issue is 0.2 and a weighting value for member who his/her expertise is in the area of the issue is 0.8. Let us follow the previous example of the three possible solutions named A, B, and C on the project design issue. Let assume that for the design area, the reputation point of a member who votes for solution A is 1. Since this member's expertise is in the area of design, which is the area where the issue is raised, (i.e. project design), this member's vote would have a value of 0.8 (1 multiplied by 0.8). If the reputation value of a member who votes for solution B is 2, then this member's vote would have of value 0.4 (2 multiplied by 0.2) because this member's expertise area is requirement (this member is an analyst) while the issue is about project design. Similarly, if the reputation value of member who votes for solution C is 1, then this member's vote would have a value of 0.2 (1 multiplied by 0.2) because this member's expertise area is construction (this member is a programmer) while the issue is about project design. 0 shows the calculation of the voting points.

Table 2. An example of voting point calculation

Solution	Requirement	Design	Construction	Testing	Design Issue Voting Points
A		0.8 x 1			0.8
B	0.2 x 2				0.4
C			0.2 x 1		0.2

For a particular issue, whichever solution has the highest vote value will be chosen. Therefore, from the example, solution A that has the highest vote value is chosen as a final solution. Once a solution has been chosen and finalised, the project data in the software engineering ontology will be updated along the lines of the chosen solution. The system advises all team members of the final decision and records the event. The users' reputation points are also updated for future use.

7 Conclusion

Ultimately, the systems facilitate collaboration of teams in multi-site distributed software development. We have explored the development of systems for management of software engineering knowledge formed in the software engineering ontology. We have analysed instantiations in the software engineering ontology. Instantiations signify project information which is shared and evolved to reflect project development, changes in the software requirements or in the design process, to incorporate additional functionality to systems or to allow incremental improvement, etc. Accordingly, management systems have been introduced and discussed. Detailed specific management systems of safeguard, ontology, and decision maker have been given.

References

1. Witmer, G.: Dictionary of Philosophy of Mind - Ontology (2004) (cited May 11, 2004), http://www.artsci.wustl.edu/~philos/MindDict/ontology.html
2. Smith, B.: Ontology. In: Floridi, L. (ed.) Blackwell Guide to the Philosophy of Computing and Information, pp. 155–166. Blackwell, Oxford (2003)
3. Gruber, T.R.: A translation approach to portable ontology specification. In: Knowledge Acquisition (1993)
4. Gruber, T.R.: Toward principles for the design of ontologies used for knowledge sharing. In: International Workshop on Formal Ontology in Conceptual Analysis and Knowledge Representation, Deventer, The Netherlands. Kluwer Academic Publishers, Padova (1993)
5. Beuster, G.: Ontologies Talk given at Czech Academy of Sciences (2002)
6. Rumizen, M.C.: Complete Idiot's Guide to Knowledge Management. Alpha (2001)
7. Sowa, J.F.: Conceptual Structures: Information Processing in Mind and Machine. Addison Wesley, Reading (1984)
8. Carroll, J.J., et al.: Jena: Implementing the Semantic Web Recommendations, Digital Media Systems Laboratory, HP Laboratories Bristol (2004)
9. McBride, B.: Jena: Implementing the RDF Model and Syntax Specification. In: Semantic Web Workshop, WWW 2001 (2001)
10. Seaborne, A.: Jena Tutorial: A Programmer's Introduction to RDQL (Updated February 2004)
11. McCarthy, P.: Introduction to Jena: use RDF models in your Java applications with the Jena Semantic Web Framework, SmartStream Technologies, IBM developerWorks (2004)
12. Chang, E., Dillon, T., Hussain, F.K.: Trust and Reputation for Service Oriented Environment: Technologies For Building Business Intelligence And Consumer Confidence. John Wiley and Sons, Chichester (2006)

Representing and Validating
Digital Business Processes

Lianne Bodenstaff[1], Paolo Ceravolo[2], Ernesto Damiani[2], Cristiano Fugazza[2],
Karl Reed[3], and Andreas Wombacher[4]

[1] Information Systems Group, Dept. of Computer Science,
University of Twente (Nl)
l.bodenstaff@utwente.nl
[2] Dept. of Information Technologies,
Università degli Studi, Milan (It)
{ceravolo,damiani,fugazza}@dti.unimi.it
[3] Computer Science Dept.,
LaTrobe University, Melbourne, (Aus)
k.reed@latrobe.edu.au
[4] School of Computer and Communication Sciences,
École Polytechnique Fédérale de Lausanne (Ch)
andreas.wombacher@epfl.ch

1 Introduction

Today, the term *extended enterprise* (EE) is typically meant to designate any collection of organizations sharing a common set of goals. In this broad sense, an enterprise can be a whole corporation, a government organization, or a network of geographically distributed entities. EE applications support digitalization of traditional business processes, adding new processes enabled by e-business technologies (e.g. large scale Customer Relationship Management). Often, they span company boundaries, describing a network of relationships between not only a company and its employees, but also partners, customers, suppliers, and markets. In this scenario, *Business Process Modeling* (BPM) techniques are becoming increasingly important. Less than a decade ago, BPM was known as *workflow design* and was aimed at describing human-based processes within a corporate department. Today, BPM is used to design the orchestration mechanisms driving the interaction of complex systems, including communication with processes defined and executed by third parties according to well-defined protocols. Also, it can be used to check compatibility and consistency of the individual business processes that are defined by collaborating business entities. A large number of methodologies, languages, and software tools have been proposed to support *digital* BPM; nonetheless, much work remains to be done for assessing a business process model validity with respect to an existing organizational structure or w.r.t. external constraints, like the ones imposed by security compliance regulations. In particular, Web-based business coalitions and other inter-organizational transactions pose a number of research problems. OMG's *Model Driven Architecture* (MDA) [Aßmann et al., 2005] provides a framework

T.S. Dillon et al. (Eds.): Advances in Web Semantics I, LNCS 4891, pp. 219–246, 2008.

for representing processes at different levels of abstraction. In this paper we rely on a MDA-driven notion of business process model, constituted by three distinct components:

- A *static domain model*, including the domain entities (actors, resources, etc.);
- a *workflow model*, providing a specification of process activities;
- a *value model*, describing the value exchange between parties.

In the modeling of static models, we shall focus on expressive formalisms constituted by controlled fragments of natural languages, introducing their translation into logics-based static domain models, and describing their relations with Semantic Web (SW) metadata formats [W3C]. In fact, the latter allow to assign a specific semantics to entities in the domain model; particularly, we are interested in the entities that, for a number of reasons, may result in under-specified descriptions. This distinction will prove of foremost importance with regard to the derivation of implicit knowledge. The static model can also be used to provide a comprehensive description of the entities that interact with each other in the workflow model and the resources that are exchanged during workflow execution. Visual languages are typically used to produce business process descriptions that regulate the interaction between the different actors in the EE scenario. Logic-based formalismsmodels can be easily derived from business process descriptions and can therefore be integrated with the static model for checking consistency and computing a wide range of business metrics that take into account dynamic aspects of the business environment. Finally, although the process model represents the main source of information driving the actual implementation of business activities, it may not be the focus for business analysts that are required to evaluate the net outcome of transactions in terms of the value exchange between the interacting parties. Consequently, the overall model will be completed by providing the *value model* underlying business processes. As for process models, these are also typically expressed by means of visual languages and can be translated into logic-based to obtain data structures that are amenable to automated processing.

The Chapter is structured as follows: in Section 2, we introduce rule-based business modeling and its translation into a logic-based formalism. Section 2.2 addresses two distinct semantics that can be applied to the knowledge base that is derived from business rules, highlighting the need for integration of both paradigms into a *hybrid* deduction system. Furthermore, Section 2.4 introduces the issues related with the different *modal* interpretations of business rules that are required. Section 3 provides an overview of formalisms for modeling process workflows and then focuses on a practical example of BPMN diagram describing the orchestration of independent processes. Section 4 completes the picture with a value model to be associated with the entities that have been introduced in the static model and have been instantiated in the workflow model in order to define processes. Section 5 is addressing the relations between the three models that have been introduced by indicating some of the possible cross-checking mechanisms that can bind the distinct layers in the actual implementation. Finally, Section 6 draws the conclusions and highlights the main open issues.

2   Rule-Based Structural Description

This Section introduces the *business rules* (BR) approach to business domain modeling [Ross, 2003]. BR allows for a thorough specification of the entities that populate a specific state of affairs and the mutual relations between them. The high expressivity that is required by rules has led business analysts toward the adoption of natural language as the encoding formalism for BR. This strategy clearly fulfils the needs of knowledge sharing between humans, but inevitably complicates any sort of automated processing on rules. A tradeoff between expressivity and formal specification of statements is constituted by *controlled natural languages* and, among them, *controlled English* (CE): these formalizations are derived from natural languages by constraining the admissible sentential forms to a subset that is both unambiguous and expressive. A widely acknowledged example of CE is the *Attempto Controlled English* (ACE) [Fuchs et al., 1999], a general-purpose controlled natural language supporting specification of complex data structures, such as ontologies. As an example, a simple rule in ACE that may contribute to the definition of a business domain is the following:

$$\text{A customer provides a credit card to a retailer.} \tag{1}$$

General-purpose controlled languages can be provided with a formal (e.g., logics-based) semantics; however, they fall short of being capable to model all the aspects of a business domain. With regard to the expressive power required by BR, rule-based languages may need to cover higher order logics and also, as explained in Section 2.4, may specify the modal interpretation to be associated with a statement. Also, as we discussed in Section 3, the static description provided by BR needs to integrate with process descriptions and, possibly, originate object-oriented data structures that software developers may use to flesh out applications. The recognition of these requirements was a major driver of OMG's *Semantics of Business Vocabulary and Business Rules* (SBVR) proposal [OMG, 2006], aimed at specifying a business semantics definition layer on top of its software-oriented layers. In the OMG vision, BR can then be integrated with the development process represented by OMG's own *Model Driven Architecture* (MDA) and, consequently, extend the applicability of object-oriented modeling not only to software product development but also to business process modeling and maintenance. SBVR provides business analysts with a very general controlled language, whose syntax visually separates the different tokens in a sentence (nouns, verbs, reserved keywords) with different styles and colors[1]. As an example, the rule in (1) corresponds to the following SBVR *fact type*.

$$\text{Obligaton: a } \underline{customer} \; provides \; \text{a } \underline{credit \; card} \text{ to a } \underline{retailer} \tag{2}$$

---

[1] Here we shall not deal with *color markup* of rules, which is primarily intended to ease the reading of a large rule corpus.

For the sake of clarity, in the remainder of this Section we are not going to stick to any specific Controlled English (CE) formalism for expressing rules. However, we stress the importance of carefully evaluating the expressivity of candidate CE languages, prior to encoding business intelligence into one of these formalisms, because it may not necessarily meet the requirements of more comprehensive frameworks for corporate data reuse. ACE and SBVR represent only two of the many available CE formalisms, which may vary according to *i)* the syntactic restrictions that are applied to the corresponding natural language, *ii)* the fragment of *first-order logic* (FOL) that can be conveyed by statements, and *iii)* the applicability of automated reasoning. The reader can refer to [CLT] for a more complete survey of controlled natural language formalisms. Here, we investigate the feasibility of automated deductions over business rules, particularly in the EE scenario where independent business entities are required to integrate.

2.1   Formal Grounding of Business Rules

By using CE formalisms for expressing BR, it is possible to apply translation mechanisms that lead to a univocal logic formulation of statements. As an example, the grounding in formal logic provided by ACE is constituted by Discourse Representation Structures (DRS) [Fuchs and Schwertel, 2003] that represent a subset of FOL and provide a pathway to executable logic formulations [2]. More importantly, an Attempto Parsing Engine (APE) is available either as a standard Web interface and as a webservice, so that the engine can be remotely queried by programming logic developed by third parties. As an example, the DRS corresponding to the simple rule in (1) is the following:

$[A, B, C, D]$
$object(A, atomic, customer, person, cardinality, count\_unit, eq, 1) - 1$
$object(B, atomic, credit\_card, object, cardinality, count\_unit, eq, 1) - 1$
$object(C, atomic, retailer, person, cardinality, count\_unit, eq, 1) - 1$
$predicate(D, unspecified, provide\_to, A, B, C) - 1$

APE also provides a mapping between a subset of ACE and the OWL DL ontology language [W3C, 2004]; it can therefore take advantage of DL *reasoners* [Haarslev and Moller, 2001, Parsia et al., 2003] to infer implied knowledge. As will be shown in the following of this Section, DL represents only a small fragment of FOL; particularly, it is also limited to expressing binary relations between entities. As a consequence of this, even the simple ternary relation binding *customers*, *resellers*, and *credit cards* in (1) cannot be expressed without "objectifying" the relation by means of a newly introduced concept definition. This amounts to expressing predicate D in the corresponding DRS as a concept

---

[2] Recall that full FOL is proven to be undecidable; therefore, deriving the DRS corresponding to an ACE statement does not imply that such logic formulation can also be executed by programs.

definition that is the domain of three binary relations whose ranges (i.e., the categories of entities the relations map to) are concepts *customer, reseller*, and *credit card*, respectively. For a more traditional processing of ACE rules, DRS can also be translated into RuleML [Boley et al., 2001] to be processed by *rule engines*, such as the Jena framework [Jena]. Note that, in this case, the term "rule" is not indicating a BR, but instead the Horn fragment of FOL which guarantees a sound and complete reasoning on rules by applying either forward- or backward-chaining derivations. One of the main challenges of drawing inferences based on a BR model is the capability of applying the so-called *hybrid reasoning* on the knowledge base. The distinct inference paradigms which are associated, respectively, with OWL DL reasoning and Horn rules execution need to be integrated. It is not possible to adopt a single inference technique, because the entities in the business domain may have a different semantics associated with them. Information under full control of the stakeholder (e.g., a company) writing the model (e.g., the notion of *employee*) can be modeled as in traditional database design. In this case, BR simply provide a lingua franca by means of which business analysts and software developers can more easily translate company data requirements into real-world implementations. Other knowledge, however, needs to be introduced in order to compete and cooperate in the EE scenario (e.g., the notion of *competitor*); this knowledge is not under the modeler's full control, and may therefore be incomplete[3]. We indicate by *closed-world assumption* (CWA) the approach implemented by applications that only process complete data under the full control of their owner. In this case, failing to retrieve answers to a query (say, 'retrieve the credit card data associated with customer John Smith') automatically implies that such data do not exist. Consequently, customer John Smith constitutes a valid answer to the query 'retrieve customers that do **not** have a credit card associated with them' because incomplete knowledge amounts to false (i.e., negative) knowledge. This notion of negation (generally referred to as *negation as failure*) leads to the *non-monotonic reasoning* that provides the correct interpretation of closed systems. In the context of logic inference, the term 'non-monotonic' essentially means that conclusions (e.g., that *John Smith* is a valid answer to the previously defined query) may be contradicted by adding information to the knowledge base (e.g., the assertion John Smith provides VISA-041). On the contrary, we indicate by *open-world assumption* (OWA) the monotonic approach to inference that should be applied to heterogeneous data sources, such as those collected by individual systems in the EE scenario, and also (according to business analysts) to proprietary descriptions expressed by business rules, wherever not explicitly stated otherwise. OWA represents a fundamental requirement for Semantic Web (SW) languages [W3C] and, consequently, SW applications may also process data structures expressed by BR that cannot be considered as complete knowledge.

---

[3] This kind of incomplete descriptions may also express proprietary entities from within the business model. In fact, the complexity of business descriptions that need to be expressed by BR may not make it possible to exhaustively express the business domain.

2.2   Interpreting Entities in the Business Domain

The semantics of BR is typically described by providing a mapping from the rules syntax to some well-known logic formalism. Three (potentially conflicting) basic requirements of such formal models have been identified:

1. *High expressive power.* A basic requirement for the underlying logic is to match the high expressive power of the specific CE without further constraining the sentential forms that can be interpreted. Business analysts are accustomed to using plain English and would not accept any too severe limitation to the expressive power of the modeling language.
2. *Tractability.* In order to automate rule checking and execution, the underlying logics's expressive power has to be carefully balanced against tractability.
3. *Non-falsifiability.* BR semantics should rely on *monotonic reasoning*, i.e. on inference paradigms whose conclusions cannot be contradicted by simply adding new knowledge to the system.

These three requirements have been emphasized by business analysts and researchers as guidelines toward finding the correct logical interpretation of business models, but are NOT satisfied by current BR modeling. Furthermore, as anticipated above, managing this category of descriptions in the EE may pose novel requirements. Specifically, aggregating heterogeneous data sources that are not under full control of each system participating in the EE demands more attention when data is evaluated. The first set of entities that will be described belong to this category of *open* descriptions and we will show that only some of them can lead to automated deductions in such a way that their full semantics is preserved. The business rules that follow are meant to describe a generic *product* that is made available in a market as a consequence of cooperation among *manufacturers*, *distributors*, and *resellers*. These entities are to be considered external to the system that will process the information; you may suppose that a company is doing this in order to monitor markets that are interested by their business. Consequently, we are going to consider each of these entities as open; that is, incomplete with regard to their formal definition and also with regard to existing instance data associated with them. As an example, consider the following business rules:

a product is produced by exactly one manufacturer   (3)

a product is distributed by at least one distributor   (4)

a product is reselled by at least one reseller   (5)

Clearly, the rules above define constraints that instances of concept *product* must satisfy. Furthermore, they refer to attributes of a product instance (*produced by*, *distributed by*, and *reselled by*) that relate product instances with (possibly complex) data structures expressing *manufacturers*, *distributors*, and *resellers*. They may also indicate datatype attributes, such as *price*, *weight*, etc., that are required by metrics based on numeric calculations. Considering state of the art paradigms for data storage, constraints (3)-(5) cannot take the form of mandatory attributes in a relational schema (or more expressive trigger-based constraints) because we assume that instance data may be incomplete, e.g. both the following tuples do indicate, in the knowledge base, valid product instances:

product	manufacturer	distributor	reseller
ITEM-01	COMP-01	-	-
ITEM-02	-	-	SHOP-01

On the contrary, rules (3)-(5) can be easily translated into the following FOL statements[4]:

$$\text{Product}(x) \rightarrow \exists! y.\text{Manufacturer}(y) \land \text{producedBy}(x, y) \tag{6}$$

$$\text{Product}(x) \rightarrow \exists y.\text{Distributor}(y) \land \text{distributedBy}(x, y) \tag{7}$$

$$\text{Product}(x) \rightarrow \exists y.\text{Reseller}(y) \land \text{reselledBy}(x, y) \tag{8}$$

Unfortunately, existing FOL reasoners cannot process statements (6)-(8) because they do not comply with the Horn fragment: Specifically, all variables in a Horn rule consequent (*head*) must match variables in the rule antecedent (*body*), while variable y in statements (6)-(8) is not bound to any variable in the rule body. When applying reasoning, enforcing these rules clearly amount to asserting the *existence* of hypothetical class instances related with a product instance by the three properties. Because of the CWA approach of rule reasoners, the semantics of (6)-(8) cannot be expressed.

Instead, languages that are specifically designed for modeling incomplete data sources (like the ones used in the SW) can express these constraints without requiring instances of *concept* Product to actually refer to instances of concepts Manufacturer, Distributor, and Reseller:

$$\text{Product} \sqsubseteq \; = 1 \; \text{producedBy.Manufacturer}$$

$$\text{Product} \sqsubseteq \; \geq 1 \; \text{distributedBy.Distributor}$$

$$\text{Product} \sqsubseteq \; \geq 1 \; \text{reselledBy.Reseller}$$

For the sake of clarity, here we express OWL structures by means of the corresponding Description Logics (DL) syntax. However, there is a one-to-one correspondence between OWL constructs and DL assertions[5]. Inference procedures that are associated with SW languages allow to evaluate data structures according to the OWA, while querying a (structurally equivalent) relational data model may not derive all possible conclusions. In fact, as for the Horn fragment of FOL, databases are bound to consider only existing data instances when executing queries. Consider for example the following query:

retrieve all instances that have a reseller associated with them

---

[4] In *knowledge representation* (KR), concept definitions are typically indicated by a leading uppercase letter; instead, property definitions start with a lowercase letter.

[5] The OWL Lite and OWL DL sub-languages are isomorphic to the $\mathcal{SHIF}(\mathbf{D})$ and $\mathcal{SHOIN}(\mathbf{D})$ DLs [Baader et al., 2003], respectively, where $(\mathbf{D})$ indicates support for XML Schema datatypes.

Clearly, a database query would return ITEM-02 as the only individual satisfying the query because no reseller is associated with the other tuple. Instead, DL reasoning paradigms may derive that, because of rule (5), ITEM-01 must have a reseller, even if its identity is not known to the system at the moment. Consequently, the relational data model grounding mainstream dataware housing cannot be as expressive as SW formalisms when modeling data structures that are, by definition, incomplete. This not a minor difference, as incompleteness is the most common feature of information exchanged in an inter-organizational business process; it may also be an explicit decision that is taken to avoid defining aspects that are not relevant to the model and, nevertheless, cannot be considered as false knowledge. Unfortunately, the restricted set of constructs that are provided by SW languages, such as OWL DL, can express only a limited subset of the FOL structures that may stem from BR formalization. Firstly, although OWL is very good at expressing constraints on concept and property definitions, business rules often need to take into consideration data instances for their enforcement. Secondly, the model-theoretic approach to OWL reasoning services has dramatic consequences on computational complexity and this inevitably narrows the set of logic structures that can be expressed. As an example, the following definition cannot be modeled with OWL DL:

$$\text{a direct distributor is a manufacturer of a product} \tag{9}$$
$$\text{that is also a distributor of the product}$$

In fact, translating this rule amounts to comparing the fillers of properties producedBy and distributedBy (i.e., the instances at the other end of these relations) for any given product, in order to determine if the product's manufacturer is also a distributor for the same product. Instead, (9) can be easily translated into a Horn rule of the following form:

$$\text{Product}(x) \land \text{producedBy}(x,y) \land \text{distributedBy}(x,y) \rightarrow \text{DirectDistributor}(y) \tag{10}$$

In order to straightforwardly integrate Horn rules with the structural component of the knowledge base, rules are expressed in the SWRL formalism [Boley et al., 2004]. Since the entities on the left-hand side of the formula are to be considered open with regard to inference, it is possible that evaluating them under the CWA (the only possible interpretation for Horn rules) may not reflect the actual semantics of (9). Moreover, constraints expressed by BR on n-ary relations may not be expressed with OWL by objectifying the relation; consider the following rules introducing the notion of *market*.

$$\text{a product is distributed in at least one market} \tag{11}$$
$$\text{a product that is distributed in a market is distributed} \tag{12}$$
$$\text{by exactly one distributor in the market}$$

While it is possible to express (11) as a DL concept definition, the constraint expressed by (12) cannot. In fact, the triples market-product-distributor should first be grouped according to the market, then according to the product, and only then it may be checked whether the constraint holds. This degree of complexity cannot be expressed as DL concept and property definitions. The second category of entities that can populate the business domain is constituted by *closed* entities, i.e. data structures that are under full control of the system. These data can be expressed with the wide range of SWRL constructs and can be evaluated according to CWA, with no loss in the semantics being expressed. Let us introduce in the business vocabulary the notion of *article* to indicate, among products in a market, those that are produced by the company under consideration:

$$\text{an article is a product that is produced by the company} \qquad (13)$$

Whereas *article* represents a closed entity (i.e., the company is supposed to exhaustively enumerate its products in the knowledge base), rule (13) defines it as a specialization of *product*, which is an open entity. This is a major motivation for the conjunct evaluation of both categories of constructs. Moreover, the open/closed status of an entity may also be implicitly derived by those of the entities defining the former. Consider the following definition of *target market*:

$$\text{a target market is a market and an article is distributed in the market}$$

Since knowledge on articles is, by definition, complete, the rule identifies a closed specialization of the open concept *market* whose instances are of direct interest to the company because some of its products are distributed in that market. Even if, singularly taken, articles and products can be expressed in their full semantics by SWRL rules and OWL constructs, mixing them up may not preserve this property. This is due to information interchange between the distinct reasoning engines that are processing, respectively, closed and open constructs.

2.3   Interpretation Issues

So far, we have been using the term 'interpretation' in the broader sense of 'the act of interpreting'. Now, in order to explain the differences between CWA and OWA reasoning, we must shift to the precise notion of 'interpretation' used in *model theory*, that is a 'mapping from the language to a world'. In the barebones knowledge base introduced in this Section, interpretations can be roughly assimilated to assignments of individuals to variables in the logic structures derived from BR. In order to exemplify this, we further specialize concept *reseller* with concepts *shop retailer* and *web-enabled retailer*. These concepts distinguish resellers that are capable of selling goods online from those that don't.

$$\text{a reseller is a shop retailer or a web-enabled retailer} \qquad (14)$$

Note that rule (14) also implies that, in our simple example, a reseller has to be *either* a shop retailer *or* a web-enabled retailer. These new open entities may be straightforwardly expressed with OWL DL through the union operator:

$$\text{Retailer} \equiv \text{ShopRetailer} \sqcup \text{WebEnabledRetailer}$$

Now suppose that two distinct rules are created (it may be by different people and for different purposes) associating the newly introduced entities with discount rates that can be applied to them.

$$\text{a shop retailer has a discount of 10\%} \tag{15}$$

$$\text{a web-enabled retailer has a discount of 15\%} \tag{16}$$

These kind of associations generally require SWRL definitions because rules (15) and (16) amount to declaring new property instances relating individuals to literals '10%' or '15%'.

$$\text{ShopRetailer}(x) \rightarrow \text{hasDiscount}(x, \text{`10\%'})$$

$$\text{WebEnabledRetailer}(x) \rightarrow \text{hasDiscount}(x, \text{`15\%'})$$

Finally, suppose that individual SHOP-01 in the knowledge base is known to be a reseller, but it is not known whether it is a shop or a web-enabled retailer (this can be formalized with the assertion Reseller(SHOP-01)). Now, in order to show that it may not be straightforward to derive all possible answers to a query, it is sufficient to query the knowledge base for individuals that have a discount rate associated with them. Intuitively SHOP-01 should be returned because, by rule (14), either of the rules should be applicable to the individual (albeit it is not known, at the moment, which one). On the contrary, this is the typical situation where the model-theoretic approach of OWL reasoning cannot be integrated with the single-model approach of SWRL reasoning without losing information. Specifically, the former will consider SHOP-01 as either instance of ShopRetailer and WebEnabledRetailer in each of the interpretations that are computed[6]; instead, SWRL reasoning would not consider either assignment to hold in the interpretation computed on the basis of facts that are explicitly known to the system.

2.4   Modal Evaluation of Business Rules

Business rules determine which states and state transitions are possible or permitted for a given business domain. Modal BR can be of *alethic* or *deontic* modality. Alethic rules are used to model necessities (e.g., implied by physical laws) which cannot be violated, even in principle. For example, an alethic rule may state that an employee must be born on at most one date. Deontic rules are used to model obligations (e.g., resulting from company policy) which ought to be obeyed, but may be violated in real world scenarios. For example, a deontic rule may state that it is forbidden that any person smokes inside any company building. It is important to remark that widespread domain modeling languages such as the Unified Modeling Language (UML) typically express alethic statements only. When drawing a UML class diagram, for instance, the modeler is

---

[6] Actually, also as instances of both concepts at the same time, because this is not explicitly prohibited by definition (14).

stating that domain objects belonging to each UML class MUST have the attribute list reported in the class definition, implicitly taking an alethic approach to domain modeling. In business practice, however, many statements are deontic, and it is often important (e.g., for computing metrics) to know if and how often they are violated. Much research work has been done to provide a logics-based model for BR including modalities. Indeed, supporting modalities does not mean that it is mandatory to map the BR to a modal logic. For instance, work by the BR OMG team and specifically by Terry Halpin (including his package NORMA [Curland and Halpin, 2007], an open-source tool which supports deontic and alethic rules) addresses logical formalization for SBVR by mapping BR's deontic modalities into modal operators *obligatory (O)*, *permitted (P)* (used when no modality is specified in the rule), and *forbidden* (F). Deontic modal operators have the following rules w.r.t. negation:

$$\sim Op \equiv P \sim p$$
$$\sim Pp \equiv O \sim p \equiv F\ p$$

Other modal operators used for mapping BR alethic rules are *necessary* ($\Box$), i.e. true in all possible states of the business domain, *possible* ($\Diamond$), i.e. true in some state of the business domain, and *impossible* ($\sim \Diamond$). Alethic operators' negation rules are as follows:

$$\sim \Diamond p \equiv \Box \sim p$$
$$\sim \Box p \equiv \Diamond \sim p$$

Terry Halpin's NORMA approach represents BR as rules where the only modal operator is the main rule operator, thus avoiding the need for a modal logics model. Some allowed BR formulations that violate this restriction may be transformed into an equivalent NORMA expression by applying modal negation rules, the *Barcan formulae*, and their converses:

$$\forall p \Box Fp \equiv \Box \forall p Fp$$
$$\exists p \Diamond Fp \equiv \Diamond \exists p Fp$$

For instance, the BR *For each customer, it is necessary that he provides a credit card* is transformed into *It is necessary that each customer provides a credit card*[7]. However, BR rules emerging from business modeling cannot always be transformed into rules where the only modal operator is the main operator. To support such cases, in principle there is no alternative but to adopt a semantics based on a modal logic; but the choice of the "right" modal logic is by no means a trivial exercise, due to tractability and expressive power problems [Linehan, 2006]. Modal logics engines do exist; for instance, MOLOG [Fariñas del Cerro, 1986]

---

[7] Another transformation that could be used in this context is the one based on Barcan formulae's *deontic variations*, i.e. $\forall p OFp \equiv O \forall p Fp$. We shall not discuss here the application of these transformations to normalizing modal logic formulas; the interested reader can refer to [Linehan, 2006].

has been developed by the Applied Logic Group at IRIT from 1985 on, initially supported by the ESPRIT project ALPES. MOLOG is a general inference machine for building a large class of meta-interpreters. It handles Prolog clauses (with conjunctions and implications) qualified by modal operators. The language used by MOLOG can be multi-modal (i.e., contain several modal operators at the same time). Classical resolution is extended with modal resolution rules defining the operations that can be performed on the modal operators, according to the modal logics that are chosen. However, instead of choosing a modal logics and applying MOLOG-style modal reasoning, most current approaches to BR semantics are based on Horn rules, which is the basis of Logic Programming and counts on many robust implementations. Actually, DLs also provide all the necessary constructs to evaluate multi-modal formulas (by mapping operators to universal and existential quantifiers) but, for DL as for Horn FOL, processing modal formulations in conjunction with the OWL and SWRL constructs introduced so far can easily lead to undecidability. Consequently, much work has to be done to translate BR into models that can be safely executed by reasoners preserving most of the semantics of the original definitions.

3 Declarative Process Flow Description

An important component of a successful business strategy is related with the organization of process work flows. To this purpose, a business process is viewed as the sequence of activities and decisions arranged with the purpose of delivering a service, assuring security and effectiveness, in accordance to the service life cycle. Due the the procedural nature of notations typically used for these descriptions, process flows are usually validated against process termination, verifying the absence of interferences and procedural inconsistencies. Violation metrics based on this validation can be easily devised; still, they do not exhaust the possible metrics that can be calculated on a flow. This Section discusses process flow description, underlining the role played by declarative representations in supporting model cross-checking or sharing a process description among a community of users.

3.1 Workflow Languages

Graphical notations aimed at describing process flows are one of the most widespread tools for supporting business process modeling. Their popularity is due to the capability of supporting both immediate reading and rigorous formalization. Another important advantage is that graphical notations are understandable by all business users (e.g., business analysts designing a process, technical developers implementing it, business people monitoring the process, etc.). A broad range of standards allowing formalization of process flows exist. A partial list of more relevant standards includes: UML Activity Diagram, UML EDOC Business Processes, IDEF, ebXML BPSS, Activity-Decision Flow (ADF) Diagram, RosettaNet, LOVeM, and Event-Process Chains

(EPCs)[van der Aalst et al., 2003] In May 2004, OMG proposed a standard aimed at reducing the fragmentation of notations and methodologies. This standard was named Business Process Modeling Notation (BPMN) [Bauer et al.] and was designed as a tradeoff between simplicity of notation and expressivity. Another very diffused standard is constituted by UML Activity Diagrams. Currently, these two standards are gaining a large diffusion: on the one hand, the first is more popular in the business analysts community; on the other hand, the latter is more popular in the software analysts community. A recent OMG initiative [BMI] is aimed at reconciling UML AD with BPMN by means of an integrated metamodel.

In general, the properties of a flow of transaction cannot be captured by a declarative formalization. This is primarily due to the dynamic nature of transactions that describe dependencies among events and may require the specification of complex processes with the presence of event-driven behaviors, loops, real-time evaluation of actions, and parallelism. Model checkers for declarative theories require a finite state space whereas dynamic process, in general, have an infinite state space. Nowadays, Petri nets are widely adopted for workflow modelling and they have a formal semantics by means of which model checkers can be implemented [Grahlmann, 1997]. Another widely adopted formalization is π-calculus [H Smith, 2003] and it is a process algebra describing mobile systems. Key notions of this formalization are *communication* and *change*. Distinct π-calculus processes may communicate by referring to other processes trough links and pointers. By doing this, the development of a process can be inserted into another and generate a new development cycle. Activities in a workflow are conceptually mapped to independent π-calculus processes. This way, processes use events as the form of communication to determine the behavior of the workflow. Another option is to use formalizations based on higher-order logic, such as situation calculus or temporal logic. Situation calculus was introduced by John McCarthy in 1963, it is a logic formalism designed for reasoning about dynamic domains. Recently, it was used as a base for designing a programming language named ConGolog [De Giacomo et al., 2000]. Temporal logic is a logic aimed at reasoning about propositions qualified in terms of time. Traditionally, temporal Logic formalizes only one of the two paradigms that are required in order to deal with dynamic and concurrent systems. In fact, the information to be derived from the formalization of a dynamic system can involve either the properties a state must satisfy and also the temporal dependencies between events. Some early works, such as [Nicola, 1995], proposed a formalization including both states and events. In [Gnesi and Mazzanti, 2003], such a formalization is applied to system modeling, i.e. UML diagrams.

3.2   A Temporal Logic for UML Statecharts

The main problem in applying model checking to business processes is the state space explosion: for real-life case studies, the state space is usually too large to be efficiently mapped. One solution is to encode the state space symbolically, using predicates, rather than enumerating it. This way, we may work on a more abstract representation while preserving the structure of the dynamic model. The most common way to adopt a predicate-based model is the adoption of temporal logics. In particular we can mention Linear Temporal Logic (LTL) and Computation Tree Logic (CTL)[Jain et al., 2005]. In order to deal with the infinite states generated by loops, special model checkers have been implemented. These model checkers develop algorithms for strong fairness. A strong fairness constraint is aimed at excluding loops. If p and q are properties, we state that if p is true infinitely often, then q must be true infinitely often as well. Intuitively, a property p can only be true infinitely often if there is some kind of loop in the model in which p is made true. Consequently, the strong fairness constraint on (p, q) says that if there is some loop which makes p true infinitely often, then q must be made true infinitely often by the loop as well. If this is not the case, the loop is not strongly fair and the loop must be exited after a finite number of iterations.

3.3   Declarative Representation

Despite the limitations in describing dynamic and concurrent systems, declarative formalizations are not irrelevant to dynamic and concurrent systems, and can be exploited for some important tasks related to validation such as:

- consistency checking;
- data exportation;
- annotation.

Consistency Checking. This is a task where declarative formalizations play the main role. Traditional formalisms are aimed at verifying performance properties of workflow models. For instance, a typical problem is to identify if a path is terminating or which tasks are in dependency with others. Declarative formalizations cannot support this kind of controls but act very well for evaluating the consistency of the objects acting in the transaction or the data objects exchanged in the transaction, as discussed in [Haarslev and Moller, 2001].

Data Exportation. Basically any notation used for representing process flow rely on an XML format used for exporting data. This is a declarative description of the flow, usually limited to a syntactic description of the elements in the notation, that in principle could be queried for consistency checking purposes. This approach is not straightforward, because it requires reconstructing the semantics of the notation directly in the query step. Since the usual approach is to provide a semantic mapping between XMI and the individual format to be queried [Fox and Borenstein, 2005].

Annotation. In [Melnik and Decker, 2000], a RDF format has been provided for representing UML diagrams. RDF is a language for data annotation that allows to attach complex assertions (in the form of triples subject-predicate-object) to URIs, i.e. any type of resource. Typically this language is used in systems for cooperative design, such as described in [Ceravolo et al., 2007]. The final output of this approach allows to share process flow annotations and cooperatively update the description of a process.

3.4   Consistency Checking

Here we propose an example of consistency check between the structural part of the model and the process flow. Fig. 1 shows a BPMN diagram describing process coordination between distinct actors, represented by different *swim lanes* in the diagram. A declarative description of the flow can describe business transactions in terms of the actors involved in the transaction plus input and output data required for executing the transaction. As an example, the static model may

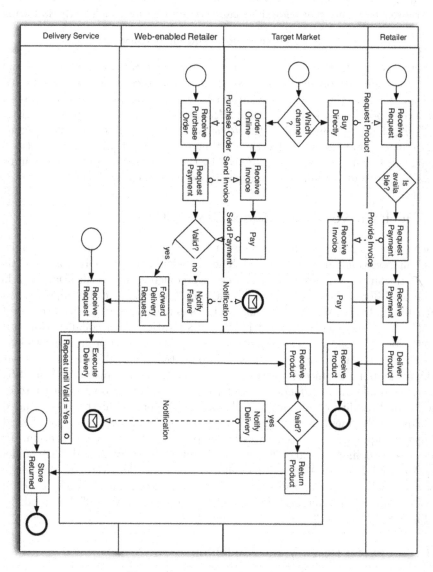

Fig. 1. An example of process flow

feature concept definitions for *Event, Activity, Gateway*, and all the other BPMN constructs. Entities in Fig. 1, such as the tasks activated by the *Retailer* in the first lane, can then be expressed in terms of assertions (i.e., subject-predicate-object triples), such as the following:

ReceivePayment rdf:type Task

RequestPayment rdf:type Task

Instances of these concepts may be related with each other in order to express general requirements that should be satisfied by processes, such as the following:

ReceivePayment follows RequestPayment

Clearly, a diagram that contradicts the requirement can be spotted at design time. Another possible usage of the static model is to constrain the instances of concept *message* that can be exchanged between *Tasks*. As an example, reference to a specific instance of concept *Message* named *invoice* can be restricted to *Tasks* that are contained in the *Retailer* lane. More interestingly, dynamic requirements that are related with run-time execution of processes can also be expressed. In this case, logfiles produced by the execution of processes are interpreted as concept instances and properties relating them with each other.

4   Declarative Value Model Description

Before implementing and executing a business collaboration, models describing this collaboration can be developed. These models help to analyze a priori the collaboration with different stakeholders. Agreements and clarifications can be made on different levels of the collaboration. A model especially important for describing inter-organizational collaborations is a *value model*, estimating profitability for every actor involved in the collaboration. In a collaboration each stakeholder is *profit and loss responsible*. Analyzing a priori profit opportunities as well as agreeing on the exchanges of value between the different stakeholders is highly important. In this Section, we use a running example for illustration. In our example, a *manufacturer* develops a global value model in order to estimate profitability of his business, before implementing his business, and to *monitor* his business during the life cycle of the collaboration. Several modeling techniques can be used to estimate profitability of a collaboration, e.g. Business Modeling Ontology [Osterwalder and Pigneur, 2002] and REA modeling [McCarthy, 1982]. Although modeling techniques differ, value models depict always which *entities of value* are exchanged between stakeholders. Here, we refer to entities of value as *value transfers*. We discuss two different modeling techniques by means of our running example.

4.1   Graphical Based Value Modelling

REA modelling [McCarthy, 1982] is a widely acknowledged business modeling technique using a graphical representation of the actors and their relations. Here,

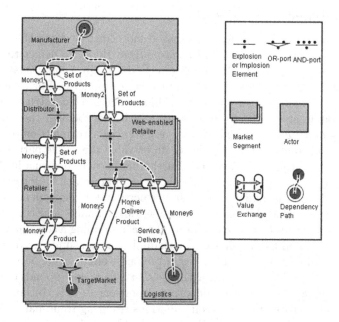

Fig. 2. Business Case as Value Model

Table 1. Estimations

	Money1	Money2	Money3	Money4	Money5	Money6
Total Value	2000	850	450	50	48	3
# of Products	50	20	10	1	1	1
Euros per Product	40	42.50	45	50	48	3

we use another modeling technique which also has a graphical representation, e³-value modeling [Gordijn and Akkermans, 2003]. Figure 2 depicts our example as an e³-value model. The manufacturer sells products to *distributors* and to *web-enabled retailers*. Distributors sell products to *retailers* who, in turn sell products to clients in the *target market*. Web-enabled retailers sell products directly to clients in the target market. Furthermore, they need *logistics* to deliver products to clients. A *consumer need* is fulfilled by one or more value transfers. Value transfers dependent on each other are connected through a *dependency path*. Quantifications of constructs in a dependency path influence the quantification of other constructs in that path. For example, the manufacturer gets higher profits from selling to a target market constituted by two hundred clients than to a market of one hundred only. This model also supports quantification of relations (not shown in the picture). Furthermore, the manufacturer gets higher profits from selling to web-enabled retailers than from selling via distributors. Table 1 depicts estimations the manufacturer made on the number of sold products and the value of each product. These estimations, together with the graphical model, provide an estimation of profitability.

4.2   Logic Based Value Modelling

Another modeling technique for value based modeling, closer to the other ones used in this chapter, is using a *logical formalism*. Here, we depict the running example as a value model in Prolog style in FOL. Again, *value transfers* as well as *dependencies* between these value transfers are modeled. Each construct in e^3-value can be mapped to a *predicate* in this model. Each construct has several arguments, at least denoting the *actor* to which they belong, their *name* and estimated *quantification*, e.g. *number of occurrences*. A connection element has a unique *id* and connects two constructs. In e^3-value this is the part of a dependency path connecting two constructs. Furthermore, each construct in the model captures an *equation*. We demonstrate the general approach by formalizing two predicates. The first is modeling a consumer need and its dependencies on other constructs. In e^3-value this is modeled as a start stimulus and corresponds to predicate *Start*. The second is modeling a *choice* in fulfilling a consumer need and dependencies related to this choice. For example, the consumer need is to buy a product which might be fulfilled by either buying it online or by purchasing it in a store. In e^3-value this is modeled with an *OR-port* and corresponds to predicate *Or_Split*. To create the equation denoting dependencies of a consumer need in a value model, a distinction between equations resulting from market segments (*Equation_StartMS*) and actors (*Equation_StartActor*) is made. To derive the equation, first the estimated number of *occurrences* of the start stimulus (consumer need) is required. When the equation depicts a market segment, also the number of customers in the market segment (*count*) is necessary. Furthermore, the *id* of the connection element, indicating dependency on other constructs, is necessary. For the equation of a choice, again the *id* of the connection element is needed. Furthermore, the estimated number of times a choice is made (*fraction*) is needed.

$Equation\_StartMS(id, occurrences, count)$ $\leftarrow$
   $Start(actor, name, occurrences)$ $\wedge$
   $Connection(id, name, name_2)$ $\wedge$
   $Market\_Segment(actor, count)$

$Equation\_StartActor(id, occurrences)$ $\leftarrow$
   $Start(actor, name, occurrences)$ $\wedge$
   $Connection(id, name, name_2)$ $\wedge$
   $Actor(actor)$

$Equation\_Split(id_1, fraction_1, fraction_2, id_2)$ $\leftarrow$
   $Or\_Split(actor, name_1, fraction\_up, fraction\_down)$ $\wedge$
   $Connection(id_1, name_1, name_2)$ $\wedge$
   $Connection(id_2, name_3, name_1)$ $\wedge$
   $((fraction_1 = fraction\_up \ \wedge \ fraction_2 = fraction\_down)$
   $\vee$
   $(fraction_2 = fraction\_up \ \wedge \ fraction_1 = fraction\_down))$

Now, we illustrate the creation of the equations based on the gathered data. There are two sets of equations calculated. The first set is the set of equations *without* instantiations. The second set of equations consists of equations which are instantiated with the values as estimated. In our example these are the estimations made by the manufacturer in Table 1. Predicate *Value* contains two arguments representing these estimations. The first argument is the corresponding name in the equation and the second argument is the estimated value. Next, the derivations of both sets of equations are depicted.

$x = y \quad \leftarrow$
  Equation_StartActor(x, y)
  $\vee$
  (Equation_StartActor$(x, occurrences)$ $\wedge$ Value$(occurrences, y)$)

$x = y \cdot z \quad \leftarrow$
  Equation_StartMS(x, y, z)
  $\vee$
  (Equation_StartMS$(x, occurrences, count)$ $\wedge$ Value$(name, y)$ $\wedge$
  Value$(count, z)$)

$y = \frac{r}{r+s} \cdot x \quad \leftarrow$
  Equation_Split(y, r, s, x)
  $\vee$
  (Equation_Split$(id_1, fraction_1, fraction_2, id_2)$ $\wedge$ Value$(fraction_1, r)$ $\wedge$
  Value$(fraction_2, s)$)

Next, formalization for the market segment *TargetMarket* is depicted, illustrating formalizing our running example. When formalizing a value model, all market segments and actors, as well as each construct and connection element, are described. Furthermore, the estimations made by the manufacturer are added with the *Value* predicate. Predicate *Value_Transfer* represents the actual value transfer where the last argument is the estimated average value which are also represented in Table 1.

Market_Segment$(TargetMarket, Count_1)$
Start$(TargetMarket, Start_1, Occurrences_1)$
Or_Split$(TargetMarket, Or_1, S, R)$
Value_Transfer$(Targetmarket, Transfer_1, Value_1)$
Value_Tranfser$(Targetmarket, Transfer_2, Value_2)$
Connection$(Id_1, Start_1, Or_1)$
Connection$(Id_2, Or_1, Transfer_1)$
Connection$(Id_3, Or_1, Transfer_2)$
Value$(Count_1, 100)$
Value$(Occurrences_1, 5)$
Value$(S, 2)$
Value$(R, 1)$
Value$(Value_1, 50)$
Value$(Value_2, 48)$

Expressing the value model in a logical formalism and quantifying it with estimations enables profitability calculations on the collaboration. These calculations are used for decision making. In this Section we showed different representations of value models. Both representations, graphical as well as logic based, enable reasoning on the profitability of the collaboration. Although these are two distinct representations, both model explicitly *actors* and *value exchanges* between those actors. Essentially, this is the important part for a priori evaluating profitability of a collaboration.

5   Relations between Models

Models are per se an abstraction of the real world focusing on a particular aspect like, for example, the flow or the value aspect. Dependent on the model, the analyses that can be performed differ: in case of the process flow model it can be checked whether there are dead ends in the execution or whether deadlocks can occur; on the other hand, in the value model, profitability can be investigated. However, all these models describe the same system and therefore the different models can be related with each other. The aim would be to have a description of the system according to the different aspects. To check whether different models fulfill this property, relations between the different models have to be investigated. Well known relations are equivalence checking, whether two models describe exactly the same, and consistency checking, whether two systems are contradicting each other. Equivalence of two different model types (like for example process flow and value model) will never be given since the different models focus on different aspects while neglecting other aspects. With regard to process flows and value models, the former focus on the order of message exchanges, while the latter disregards these aspects and focuses on the occurrences and values of exchanged value objects. As a consequence, consistency[8] seems more appropriate, since it focuses on contradiction free models which can be checked on the communalities of the involved models.

There are plenty of potential pitfalls that could make models inconsistent, like for example different understandings of the architects of the different models on how the actual system really works (in case it has been already implemented), modifications of a single model without maintaining the remaining models, or the discrepancy between the modeled behavior and the behavior of customers in concrete business situations. Furthermore, there are plenty of cases where consistency between two models cannot be checked at all like, for example, in case models are representing the system on a different level of granularity. In case of the process flow and the value model, this means that the value model considers the modeling at the granularity of companies, while the process flow model is based on the notion of business units within companies. However, the investigation of relations between the models, as well as the actual behavior of

---

[8] In some literature the term compatibility or soundness is used instead of consistency but addressing the same problem.

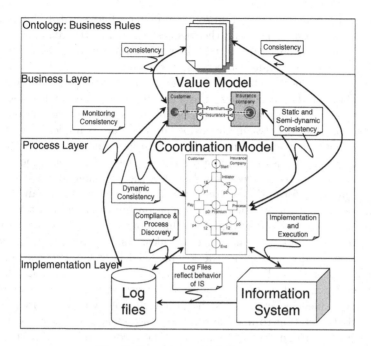

Fig. 3. Consistency Relations

the system, are valuable to improve, correct, or adapt the models according to a continuously changing environment[9].

In the following of this Section, different relations between the models introduced in Sections 2-4 (see Figure 3) are mentioned and some references are provided. Afterwards, an overview of two sample consistency relations are presented.

5.1   Overview

Figure 3 is structured in four layers according to the usual information system life cycle: specifically, the business rules, business, process, and implementation layers. First, understanding how a particular business works and what the dependencies between the different parties and business objects are is important. Second, a business idea containing a business model showing the profitability of the business is constructed. Third, it is specified how you want to do business and how to coordinate the different parties involved. Finally, the information system is implemented deployed and and becomes operational, resulting in log files representing the execution of the information system to a specific level of detail.

---

[9] We want to point out that, in the following, the focus is on relations of different model types. There exist quite some work on relations between models of the same type.

The relations depicted in Figure 3 are defined between the different layers. One set of relations is between the business rules and the remaining layers. In particular, the business rules terminology is used in the value model to describe the different actors or value objects. In our examples we used terms like "target market" or "product" in business rules and value model. Also, the process flow model relies on the terminology provided in the business rules, for example, to name swim lanes or to express what business objects are related to which messages exchanged between the swim lanes. As an example, "target market" and "product" represent, respectively, a swim lane and a business object. This common terminology is essential for direct comparison between value and process flow models. Since the process flow describes the exchange of messages sent and received by an information system, the log file generated by an information system uses the same messages. This relation indirectly allows to relate some messages to business objects in the value model, which in there turn can be related to terminology in the business rules. As a consequence, value model, process flow model, as well as the log files can be related to the terminology in the business rules. This set of relations allows a first checking whether the dependencies in each underlying layer is consistent with the rules specified in the business rules declaration. Further, the terminology provided by the business rules and the relation to a value exchange, a message exchange, and a log file entry as well as to an actor, a swim lane, and a communication partner enables checking of relations between the different models.

Figure 3 shows three consistency relations between value and process flow model: static, semi-static and dynamic consistency.

- **Static consistency.** Checks whether the set of exchanged values in the value model corresponds to the business objects exchanged in the process flow model represented by message exchanges. Here the aim is to relate the value objects and business objects exchanged, as well as the dependencies between value exchanges and business object exchanges. A more detailed description can be found in Section 5.2.
- **Semi-dynamic consistency.** Checks whether the forecast of a value model can be accomplished by the process flow model. This is mainly focusing on investigating expected changes in a business scenario, like for example the increase of the amount of expected customers using the system or an expected change of user behavior for example because of price changes for a particular modeled product group compared to other modeled product groups.
- **Dynamic consistency.** Checks whether the execution of the process flow model is consistent with the expected behavior represented in the value model. This relation cannot be evaluated directly since the process flow is not executed directly. However, the information system implementing the process flow model is executed directly. Therefore, dynamic consistency can be derived from investigating the three relations between value model, log files (hence implicitly the information system) and the process flow model depicted in Figure 3: first this is *monitoring consistency*, second the relation

between log file and information system representing the *behavior of the information system* in Figure 3, and third the relation between process flow model and information system describing the *implementation and execution* of the process flow model by the information system. The *dynamic consistency* relation is essential since it gives an a-posteriori evaluation of whether the process flow and the value model are indeed describing the implemented information system. In case there are inconsistencies, the models have to be adapted and re-evaluated with regard to the expected profitability and behavior respectively. A more detailed description of the dynamic monitoring relation can be found in Section 5.3.

The **compliance and process discovery** relation has not been mentioned so far. One aspect of this relation is also known as process mining [van der Aalst and Weijters, 2005], where the aim is to derive a process model from log file data. This relation can be applied for example in case of exceptional behavior in information systems, which is not reflected in the process flow model. It is closely related to the dynamic monitoring relation between value model and log files.

5.2   A Priori Model Evaluation: Static Consistency

During the modeling phase it can be checked whether the value model and the process flow model are still consistent, that is, are not contradicting each other. This check is performed on the commonalities between the two models, which are the actors and swim lanes, the value exchanges and message exchanges, and the dependencies between exchanges in the corresponding model. Relating swim lanes and actors to each other is done using business rules. Considering the *Delivery Service* swim lane in the process flow model (see Figure 1) and the *Logistics* actor in the value model (see Figure 2) different terms are used describing the same concept. A business rule describing that a logistics company provides a delivery service is necessary to indicate that the *Delivery Service* swim lane and the *Logistics* actor are not contradicting. The same requirement of a business rule applies to the message exchange *payment* and the value exchange *money*. Obviously money is used to perform a payment.

Based on this concept relations of value models and process flow models, a consistency checking can be performed. A value model and a process flow model are consistent if for every set of message exchanges representing an execution sequence in the process flow model there exist a set of value exchanges representing a single dependency path in the value model and vice versa. However, there exist value exchanges which do not result in a message exchange, like an experience or a knowledge gain. Further, there exist message exchanges in the process flow which do not have a corresponding value exchange. In the example depicted in Figure 1, the message exchanges *request product* and *provide invoice* do not have a corresponding value exchange since these message exchanges are for coordination purposes only. A more detailed description of the approach has been discussed in [Zlatev and Wombacher, 2005].

242   L. Bodenstaff et al.

5.3   A Posteriori Model Evaluation: Monitoring Consistency

After the modeling phase, the system is implemented and executed. It is important to check during the life-cycle of the collaboration whether estimations in the
a priori value model are met by the running system. One approach for checking
this dynamic consistency is to *monitor* the running system. This is the *monitoring consistency* relation in Figure 3 between log file and value model. Here,
we show monitoring of the running system based on log files produced by the
information systems. This approach has been formally introduced in previous
work by one of the authors of this chapter [Bodenstaff et al., 2007].

In our running example, the manufacturer monitors the collaboration. His
main interest is monitoring the *ratio* between products sold to distributors and
web-enabled retailers, since he gets higher profits from selling to web-enabled retailers than from selling to distributors. The tool used for the monitoring shows
the realized number of value exchanges with distributors as well as web-enabled
retailers. The manufacturer can now compare these realized values with estimations made in the original value model. Furthermore, the tool enables reasoning
over the relation between the realized values and the different constructs in the
value model. The equations derived from the value model are added to the monitoring tool, showing all quantified values and their relations. Figure 4 depicts

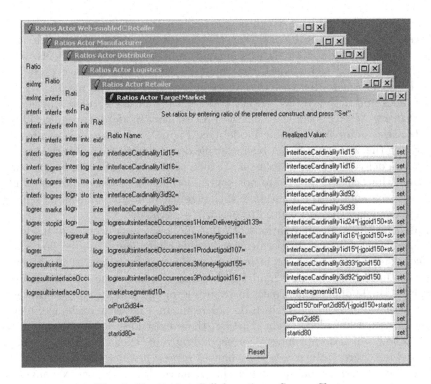

Fig. 4. Monitoring Collaboration - Screen Shot

a screen shot of the tool where the realized values and their relation to the different constructs are shown. Showing these relations enables the manufacturer to reason over effects the constructs have on the value exchanges and on which constructs are influenced by the realized values.

6   Conclusions

In the last few years, the concept of the business process model has become increasingly popular. When designing a new business process, modeling accuracy is likely to be a crucial factor for being able to assess its strength and weaknesses. The Web infrastructure and the availability of Semantic Web-style metadata have enabled new types of inter-organizational business processes, which in turn need new business process models. Despite the surge of interest in representing business process models, reasoning on process descriptions still faces many open issues, mostly due to lack of a shared notion of business process representation. Many approaches focus on producing taxonomies or categorizations, or on stating what aspects of business process models should be included in (or excluded from) modeling. The capability of representing multiple facets is indeed an important requirement for business process models. In this chapter, we discussed the different facets such a representation must possess, starting from standard ones like workflow representation and arriving to the capability of defining the structure of a companys value chain and describing the position of the process actors within the value network. Also, evidence has shown that initial business process models are often unsuccessful and need to keep being modified until a viable model is found; therefore, a multi-faceted representation of the process needs *i)* to support a priori evaluation of feasibility *ii)* to be able to evolve based on run-time evidence. While we believe a visual approach to modeling to be preferable from the modelers point of view, in this chapter we focused on the logics-based models underlying each facet of the process representation. Namely, we developed on the idea that the different facets of a business process model require different formalizations, involving different inference techniques, some of them Semantic Web-style, other closer to classical Prolog reasoning. Using a simple business process model, we discussed how well our multi-faceted representation can be integrated by a hybrid approach to reasoning, transferring knowledge and metrics obtained reasoning on each part of the model to the other parts. While we believe results to be encouraging, more work is needed. Specifically, we wish to highlight the following open issues;

- *Critical areas detection*: A multi-faceted business process model needs to include a way of finding necessary changes, when the original model does not work as envisaged. This might be done by including more options in the original model, plus decision points on when to start using these options based on constraints expressed un the value-model.
- *Representing pragmatics*: Reasons for stakeholders: actions should be part of the model. Our sample model shows what is meant to happen, but not why. Research on agent-based systems has shown that reasons for actions

are difficult to express using logics-based modeling. Also, current business process models focus on the point of view of the business owner. It might be useful to enlarge the model including the point of view of potential customers, e.g. based on market research data.

Acknowledgments

This work was partly funded by the Italian Ministry of Research under FIRB contract n. RBNE05FKZ2_004, TEKNE and by the Netherlands Organisation for Scientific Research (NWO) under contract number 612.063.409. The authors wish to thank Claudio Pizzi for his valuable comments on modal logic reasoning.

References

[Aßmann et al., 2005] Aßmann, U., Aksit, M., Rensink, A. (eds.): MDAFA 2003. LNCS, vol. 3599. Springer, Heidelberg (2005)

[Baader et al., 2003] Baader, F., Calvanese, D., McGuinness, D.L., Nardi, D., Patel-Schneider, P.F. (eds.): The Description Logic Handbook: Theory, Implementation, and Applications. Cambridge University Press, Cambridge (2003)

[Bauer et al.] Bauer, B., Müller, J., Roser, S.: A model-driven approach to designing cross-enterprise business processes. In: Meersman, R., Tari, Z., Corsaro, A. (eds.) OTM-WS 2004. LNCS, vol. 3292, pp. 544–555. Springer, Heidelberg (2004)

[BMI] BMI. Business modeling & integration domain task force, http://bmi.omg.org/

[Bodenstaff et al., 2007] Bodenstaff, L., Wombacher, A., Reichert, M., Wieringa, R.: Monitoring collaboration from a value perspective. In: Proceedings of 2007 Inaugural IEEE International Conference on Digital Ecosystems and Technologies IEEE-DEST 2007, pp. 134–140 (2007)

[Boley et al., 2004] Boley, H., Dean, M., Grosof, B., Horrocks, I., Patel-Schneider, P.F., Tabet, S.: SWRL: A Semantic Web Rule Language Combining OWL and RuleML (2004)

[Boley et al., 2001] Boley, H., Tabet, S., Wagner, G.: Design rationale of RuleML: A markup language for semantic web rules (2001), http://citeseer.ist.psu.edu/boley01design.html

[Ceravolo et al., 2007] Ceravolo, P., Damiani, E., Viviani, M.: Bottom-up extraction and trust-based refinement of ontology metadata. IEEE Transactions on Knowledge and Data Engineering 19(2), 149–163 (2007)

[CLT] CLT. Controlled Natural Languages, http://www.ics.mq.edu.au/~rolfs/controlled-natural-languages/

[Curland and Halpin, 2007] Curland, M., Halpin, T.A.: Model driven development with norma. In: HICSS, p. 286 (2007)

[De Giacomo et al., 2000] De Giacomo, G., Lesperance, Y., Levesque, H.J.: ConGolog, a concurrent programming language based on the situation calculus. Artificial Intelligence (2000)

[Fariñas del Cerro, 1986] Fariñas del Cerro, L.: Molog: A system that extends prolog with modal logic. New Gen. Comput. 4(1), 35–50 (1986)

[Fox and Borenstein, 2005] Fox, J., Borenstein, J.: XMI and the many metamodels of enterprise metadata. In: XML conference and exhibition (2005)

[Fuchs and Schwertel, 2003] Fuchs, N.E., Schwertel, U.: Reasoning in Attempto Controlled English. In: Bry, F., Henze, N., Małuszyński, J. (eds.) PPSWR 2003. LNCS, vol. 2901, pp. 174–188. Springer, Heidelberg (2003)

[Fuchs et al., 1999] Fuchs, N.E., Schwertel, U., Schwitter, R.: Attempto Controlled English (ACE) Language Manual, Version 3.0 (1999),
http://attempto.ifi.unizh.ch/site/pubs/papers/ace3manual.pdf

[Gnesi and Mazzanti, 2003] Gnesi, S., Mazzanti, F.: A mu calculus for temporal logic. In: ACM Specifying and Verifying and Reasoning about Programs (2003)

[Gordijn and Akkermans, 2003] Gordijn, J., Akkermans, J.M.: Value-based requirements engineering: Exploring innovative e-commerce ideas. Requirements Engineering 8(2), 114–134 (2003)

[Grahlmann, 1997] Grahlmann, B.: The PEP tool. In: Proceedings of CAV (1997)

[H Smith, 2003] Smith, H., Fingar, P.: Business Process Management The Third Wave. Meghan-Kiffer Press (2003)

[Haarslev and Moller, 2001] Haarslev, V., Moller, R.: Description of the RACER system and its applications. In: International Workshop on Description Logics (2001)

[Jain et al., 2005] Jain, H., Kroening, D., Sharygina, N., Clarke, E.: Word level predicate abstraction and refinement for verifying rtl verilog. In: DAC 2005: Proceedings of the 42nd annual conference on Design automation, pp. 445–450. ACM Press, New York (2005)

[Jena] Jena. Jena, A Semantic Web Framework for Java,
http://jena.sourceforge.net/

[Linehan, 2006] Linehan, M.: Semantics in model-driven business design. In: Proc. of 2nd International Semantic Web Policy Workshop, SWPW 2006 (2006)

[McCarthy, 1982] McCarthy, W.E.: The REA Accounting Model: a Generalized Framework for Accounting Systems in a Shared Data Environment. Accounting Review 57, 554–578 (1982)

[Melnik and Decker, 2000] Melnik, S., Decker, S.: A Layered Approach to Information Modeling and Interoperability on the Web. In: Semantic Web Workshop (2000)

[Nicola, 1995] Nicola, R.D.: Three logics for branching bisimulation. Journal of the ACM (1995)

[OMG, 2006] OMG. Semantics of Business Vocabulary and Business Rules Specification (2006), http://www.omg.org/cgi-bin/apps/doc?dtc/06-08-05.pdf

[Osterwalder and Pigneur, 2002] Osterwalder, A., Pigneur, Y.: An e-business model ontology for modeling e-business. In: Proceedings of the 15th Bled E-Commerce Conference - Constructing the eEconomy (2002)

[Parsia et al., 2003] Parsia, B., Sivrin, E., Grove, M., Alford, R.: Pellet OWL Reasoner. Maryland Information and Networks Dynamics Lab (2003),
http://www.mindswap.org/2003/pellet/

[Ross, 2003] Ross, R.G.: Principles of the Business Rule Approach. Addison-Wesley Longman Publishing Co., Inc., Boston (2003)

[van der Aalst et al., 2003] van der Aalst, W., Hofstede, A., Weske, M.: Business process management: A survey. In: van der Aalst, W.M.P., ter Hofstede, A.H.M., Weske, M. (eds.) BPM 2003. LNCS, vol. 2678, pp. 1–12. Springer, Heidelberg (2003)

[van der Aalst and Weijters, 2005] van der Aalst, W., Weijters, A.: Process-Aware Information Systems: Bridging People and Software through Process Technology. In: Process Mining, pp. 235–255. Wiley & Sons, Chichester (2005)

[W3C] W3C. Semantic web activity, http://www.w3.org/2001/sw/

[W3C, 2004] W3C. OWL Web Ontology Language Overview (2004), http://www.w3.org/TR/owl-features/

[Zlatev and Wombacher, 2005] Zlatev, Z., Wombacher, A.: Consistency between e^3-value models and activity diagrams in a multi-perspective development method. In: OTM Conferences, vol. (1), pp. 520–538 (2005)

Towards Automated Privacy Compliance in the Information Life Cycle

Rema Ananthanarayanan, Ajay Gupta, and Mukesh Mohania

IBM India Research Lab, 4-C, ISID, Vasant Kunj,
New Delhi - 110 070, India

1 Introduction

Management of data is an increasingly challenging task for enterprises because of the increasingly diverse and complex requirements that come to bear from various fronts. One issue is the protection of the privacy of the information held, while ensuring compliance with various rules and regulations relating to data management. Privacy may be defined as the claim of individuals, groups and institutions to determine for themselves when, how and to what extent information about them is communicated to others. [1] In its broadest scope this is a problem that has challenged civil rights advocates the world over, at different times. However, with the growth of the web, the issue gains special significance in view of the following.

1. **Volumes of data involved:** Enterprises today hold large amounts of confidential information, in the form of transactional data, customer preferences gathered from web sites and other business data. Data warehouses on the petabyte scale are expected to be increasingly common, and large parts of this data would involve sensitive information about customers and employees. This could be in structured format, in relational tables, or in unstructured format, in the form of spreadsheets, emails and other reports.
2. **Intentional and unintentional exposure:** [1] cites many instances of accidental and intentional violation of privacy resulting from privacy accidents, ethically questionnable behaviour and lax security measures with respect to such data. The resulting litigation and adverse publicity have made companies increasingly sensitive to how such data is handled. Any exposure on this front directly affects the bottomline of companies and hence an increasing concern of any CIO is the effective protection of privacy data.
3. **Increasing rules and regulations:** A few high-profile cases of privacy abuse or neglect have led to increased awareness alround of the dangers of inadequate privacy protection. Consumers are increasingly demanding control over how the privacy information is used and safe-guarded. Regulations have come from both the indutry and governments. The former is an effort to build customer trust and goodwill, while the latter is to provide a measure of protection where self-regulation cannot be depended upon.

---

[1] This definition is attributed to Professor Alan Westin, Columbia University, 1967

T.S. Dillon et al. (Eds.): Advances in Web Semantics I, LNCS 4891, pp. 247–259, 2008.

Initial solutions to the problem of privacy data management were based at the level of each individual application, and mostly introduced manual processes or at best semi-automated processes. However, these solutions are inefficient for handling the high volumes of data that are increasingly becoming common. Further, they do not scale across different applications in the enterprise. Hence it becomes very urgent to drive the search for automated solutions, which would further be application-neutral. These automated solutions need to protect the data at every stage independent of which application accesses the data.

In this work, we look at some of the issues associated with the problem of privacy protection, and present some directions of work that would help automate the solution to the problem. Our work is organised as follows. Section 2 discusses the background to the problem in terms of how customer requirements have resulted in regulations necessitating right business processes for efficient handling. In section 3 we discuss some of the IT-related standards that have evolved in response to the demands of privacy protection of information from enterprises, and some client-side solutions to the problem. In section 4 we discuss some server-side solutions to the problem. Some of these architectures are currently in use in industrial products, while others are proposed for new privacy rule structures that may become more common. We conclude in section 5 highlighting some current open issues and future directions of work.

2   Background

Personal identifiable information (PII) may be defined as any information such as an identification number, name and phone number or address that helps to identify the individual to whom the information pertains. Online transactions have increased the amount of personal identifiable information that is available to enterprises, from their customers. This repository of information is very useful for mining applications that mine for patterns and trends, to study individual preferences and customise offerings to individuals based on their purchasing pattern or preferences and related applications. However, increasingly customers are unwilling to share personal information unless they are assured that some safeguards are in place for protecting this information and controlling its use. One of the earliest and well-known studies to assess user attitudes about online privacy is [2]. One of the main findings of this report was that users registered a high level of concern about privacy in general and on the Internet. The study findings reported that 17% of the web users were privacy fundamentalists who were extremely concerned about any use of their data and generally unwilling to provide data even when protection measures were in place. 56% were observed as a pragmatic majority, who had specific concerns which were generally reduced by privacy protection measures such as privacy policies on Web sites. The balance 27% were observed as being marginally concerned and generally willing to provide data to Web sites under almost any condition, although they expressed a mild general concern about privacy. In a different study specifically focusing on health information, [3] some key findings were that customers rate personal healthcare and financial information the most sensitive types of consumer personal information. While 80% of the customers visit health sites for information, they express high concerns about privacy and security in their surfing. Further, because of their privacy concerns, many consumers visiting these sites do not share their personal data and hence fail to take full advantage of these sites.

In view of the above, enterprises started publishing privacy policies that stated how the information gathered on their site would be used or would not be used. This served as a self-regulatory exercise to convince users that the enterprise cared about protecting user privacy. Most of these privacy statements drew broadly on the fair information practices (first articulated in the US Department of Health, Education and Welfare report, also known as the HEW report [4]), whose highlights are descibed in table 1.

Table 1. Fair information practice principles

Notice/Awareness	Web sites should provide full disclosure of what personal information is collected and how it is used.
Choice/Consent	Consumers at a web site should be given choice about how their personal information is used.
Access/Participation	Once consumers have disclosed personal information, they should have access to it.
Integrity/Security	Personal information disclosed to web sites should be secured to ensure the information stays private.
Enforcement/Redress	Consumers should have a way to resolve problems that may arise regarding sites' use and disclosure of their personal information.

Apart from these, privacy policies in general have also drawn from the OECD guidelines. These are guidelines on international policy for the protection of privacy and transborder flows of personal information [5] drawn up by the Organization for Economic Co-operation and Development (OECD). These guidelines represent a consensus on the general guidance concerning the collection and management of personal information. The main driving principles laid down by the OECD for data collection are collection limitation, data quality, purpose specification, use limitation, security safeguards, openness, individual participation and accountability. In the next section we discuss some specific policy frameworks that draw from these guidelines.

3   Policy Frameworks

3.1   Frameworks

The Platform for Privacy Preferences (P3P) is a standard established by the WorldWide Web Consortium (see http://www.w3.org/P3P) that served as a first step in the front-end for building customer trust and goodwill. P3P enables web site owners to define and publish the privacy policy followed by the web site, in manual and machine-readable formats. This allows user-agents to read and interpret privacy policies automatically, on behalf of users. Users may then decide what information to share and what not to disclose, based on the site's privacy policy. Though P3P does not state how the policies will be enforced, leaving it to the individual enterprises, P3P has been widely adopted as a front-end measure by companies hoping to establish trust and build confidence in the users. These standards have served as a basis for client-side implementations of privacy policies. For instance, browsers such as Netscape 7.0 and Internet Explorer 6.0 support the standard by giving users more control over cookies and allowing the user to easily view the web site's privacy policy.

However, publishing good privacy polices does not automatically translate to processes that enforce these policies. One practice that has developed has been for individual web sites to obtain privacy seals from third-party seal providers, who are established and reputed companies in this line. Privacy seal companies such as TRUSTe (http://www.truste.org) and BBBOnline (http://www.bbbonline.com) provide privacy seals which are images displayed on the Web sites of companies that register for these seals. These seals basically ascertain that the companies care about user privacy and hence have registered with the seal providers who now have the responsibility to

1. Review their privacy policies periodically and ensure that it is in line with the latest applicable privacy legislation;
2. Ensure that the Web site accurately discloses its data collection activity;
3. Audit the privacy-related business processes periodically to ensure that the stated policies are being followed.

Another standard which defines how privacy policies may be written, in order to be easily enforceable is XACML. XACML or extensible access control language [6] is an XML-based language for access control, standardized by the Organization for the Advancement of Structured Information Standards (OASIS). XACML describes both the access control policy language (to represent access control policies of who can do what and when) and a request/response language to express queries about whether a particular access should be allowed (request) and describes answers to those queries (responses).

3.2   Privacy Compliance at the Back-End

At the back end, works such as the Hippocratic data bases [1] draw upon and extend the OECD guidelines, to define a set of standards in terms of data access and protection, that would comprise the Hippocratic database. Here it is envisaged that in addition to the existing capabilities of databases today, such as the ability to manage persistent data and the ability to access a large amount of data efficiently, data repositories in future would also need to ensure that they adhere to a certain set of standards in terms of data access and protection. These privacy principles, also borrowing from the OECD guidelines, include broadly purpose specification, consent, limited collection, limited use, limited disclosure, limited retention, accuracy, safety, openness and compliance. In the next section we discuss the impact of some of the above on the implementation of privacy-compliant solutions for data management. Traditionally Role-Based Access Control has been widely used for controlling access to information [7]. Users are assigned roles based on their job functions, and access or deny rights to data is based on the user roles. [8] discusses language constructus and implementation designs for fine grained access control that can be applied to current database management systems, to transform them to their privacy equivalents. These constructs include column restrictions, row restrictions and cell restrictions; the work also describes how privacy rules in P3P can be translated into proposed constructs. The usage control or UCON model [9] encompasses traditional access control along with trust management and digital management and presents a future direction for access control over the traditional model. UCON enables fine-grained control over digital objects, for instance, print once as opposed to unlimited prints. Subjects, objects and rights are the core components while other components are involved in the authorization process. Obligations are defined as mandatory requirements that a subject has to perform after obtaining or exercising rights on an object. An example is, a consumer may have to accept metered payment agreements before obtaining the rights for the usage of certain digital information. By including the notion of obligation with authorization, the model provides for better enforcement on exercising usage rights for both provider and consumer subjects. UCON does not state how the obligations are enforced.

While these schemes apply to data residing in individual applications, subsequent directions have been to design solution architectures that apply at an enterprise level, to ensure compliance with the various privacy rules, for data accessed by different applications. One approach is to define middleware that abstracts privacy and data-handling rules from applications and applies it across the various systems centrally. In the subsequent sections, we discuss these schemes.

4   Middleware for Privacy Protection

Standards such as P3P provide for privacy compliance at the client side, ensuring adherence in terms of privacy seals and audits. Privacy enforcement at the database level enforces privacy restrictions on the server side. We now discuss some schemes for enforcing privacy through middleware designed for this purpose. The middleware abstracts privacy and data-handling rules from applications and applies it across the various systems centrally.

4.1   Information Life Cycle

Privacy policies mandate what information is collected, how it is used and how long it is retained. As such, the policy could impact the data at different stages of the information life cycle. Hence we describe the approach in terms of the various stages in the life cycle of the data item. Table 2 gives examples of different legal directives that impact data retention at different points the life cycle of the data.

Table 2. Instances of regulations affecting data management at different stages in the lifecyle of the data item

Requirement	Instance of regulation
Mandated disclosure	Sections of the Sarbanes Oxley Act mandate disclosure of different classes of information within different time periods which may immediately, the next business day, or on a quarterly basis.
Mandated nondisclosure	The Gramm-Leach-Bliley Financial Privacy Act limits instances in which financial institutions may disclose non-public personal information about a customer to non-affiliated third parties.
Mandated retention	As stipulated in the Health Insurance Portability and Accountability Act 1996, (HIPAA) hospitals and healthcare providers must maintain medical records as well as billing records on medicare, medicaid and maternal and child health for at least 6 years.
Disposal	For operational efficiency, most enterprises define disposal schedules for disposable information such as mass communications, draft versions of documents and duplicates of documents. For instance, the rule may state that disposable information can be disposed within 90 days or sooner, unless it is specifically required for business purposes, when it may be retained for longer periods, but not exceeding 2 years.
Mandated non-collection	The Children's Online Privacy Act mandates that web sites collecting personal information from children under the age of 13 may do so only if they have obtained verifiable parental consent before the collection, use or disclosure of any personal information. Else such information may not be collected.

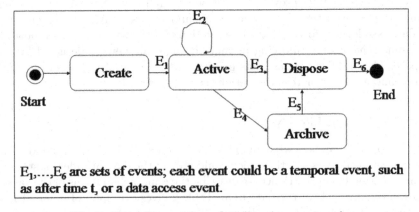

$E_1, ..., E_6$ are sets of events; each event could be a temporal event, such as after time t, or a data access event.

Fig. 1. State diagram for a data item in an enterprise

Figure 1 shows a simplified view of the life cycle of a data item or information resource, in the enterprise. This figure defines four states, in addition to a 'Start' state and an 'End' state. Information may be added to different repositories when people register and submit information, when sales transactions happen and at other points in time. The creation of such information may itself be within the scope of the privacy policy. Transitions from one state to another may be effected based on user tasks such as reading or updating documents and temporal events such as after a certain time period, or after m accesses by users. The above figure is only one possible state diagram. The actual state diagram would be defined by the record-keeping and data-handling rules of the enterprise. Further, some of the states in the figure, such as the 'Active' state, could themselves be viewed as comprising additional states.

4.2   Artefacts of a Privacy Policy

In this section we define in detail the various artefacts of a typical privacy policy. We have been guided by a policy definition similar to the XACML standard. Each privacy rule defines whether a data access is allowed or denied, based on the kind of data being accessed, the user role, the intent and intended action. In addition, each rule may be permitted only if certain conditions in the environment are met, and further, each rule may also mandate one or more obligations on the system. Figure 2 shows the artefacts of a typical policy and the various components are defined below.

Definition 1. *A privacy policy comprises a set of privacy rules.*

A rule in a privacy policy specifies whether a specific request is to be allowed or denied, in the context of some conditions in the environment evaluating to true; the rule may optionally also mandate one or more obligations. Formally, we define a rule as under.

Definition 2. *A rule is four tuple, comprising an event, a condition, a permission and optionally, one or more obligations.*

Each of these components is now defined below. An event relates to those components of a user request that are used to determine whether the request is allowed or denied.

Definition 3. *An event is a 4-tuple, comprising the user category, the data category, the action and the purpose associated with a user request.*

An event E is represented as $E = <u,d,a, p>$ where u is the user category to which the user (who is requesting the data) belongs, d is a data category to which the data item belongs, a is the action that the user wishes to perform on the resource, and p is the purpose for which the resource is being accessed. The following are some of the other terms that we will use in the rest of the paper.

User category: Each user of the data in the system is assigned to one or more user categories, based on the user roles. The user categories could be hierarchical.

Data category: Each data item that is to be privacy-protected is assigned to a specific data category. The data categories could be hierarchical.

Action: Users may access the data items to perform one or more of the specified set of actions. A representative set of actions could be {'View', 'Modify', 'Delete'}.

Purpose: Users may access the data item for a specific purpose, from one of the set of purposes specified. A representative set of purposes could be {'Marketing', 'Research', 'LegalPurposes', 'Administrative'}.

Data item: A data item is an entity that is to be privacy protected, ie., whose access is controlled by one or more rules in the privacy policy. Each data item belongs to some data category. Examples of data items include emails, facsimile documents, customer records and scanned images (of xrays, for instance, when the domain is the health industry).

The privacy rules come into effect when a user makes a request on the system.

Definition 4. *A user request is a 4-tuple comprising the user id, the data item being requested, the intended action and purpose.*

Note that each user request maps internally to an event, where the user is mapped to the user category and the data item is mapped the data category. Some more terms that we use are:

Condition: This is a clause associated with some rules, which needs to be true, for the rule to 'Allow' access. Examples of conditions include *If the user is above 18 years of age, If the user has consented, ...*

Permission: This is a binary predicate that states whether the user request is allowed or denied. It can take one of the two values, 'Allow' and 'Deny'.

Obligation: An obligation is a task optionally associated with some rules, that specifies one or more actions that need to be mandatorily performed after the user request is fulfilled. Some examples are *Log all accesses to this data* or *Email the customer if this information is used for marketing.*

Figure 2 is an object model representation of the key objects of interest, and their relationships, in the context of a privacy policy as we have defined above. Note that in theory a large number of events are possible, but in practice the system is interested only in events which are to be specifically permitted or specifically denied. All other events are not usually catalogued in the policy, but assumed to take on a default permission of 'Allow' or 'Deny'. Hence in practice the number of events that figure in the policy are much smaller than the total number of events based on the possible combinations of user groups, data items and other entities that define an event. Further, as we can see, a user request is only a specific instance of an event.

Figure 3 shows a view of the data management in terms of the privacy policy enforcement. As we can see from the figure, the relevant rule in the privacy policy is associated with the data item at the time of the data creation, which could be at the time of the data collection, for instance. If any rule applies to this data item, all subsequent accesses to this data item are monitored. A rule is

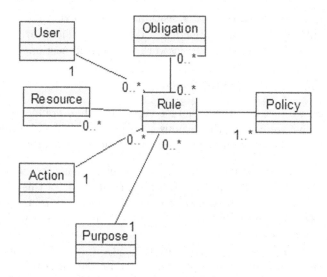

Fig. 2. Object model of the rules in a policy

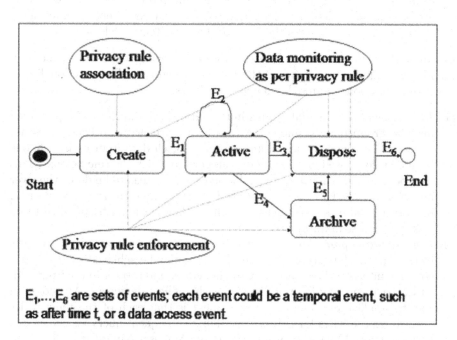

$E_1,...,E_6$ are sets of events; each event could be a temporal event, such as after time t, or a data access event.

Fig. 3. States and applicable policy management

fired on the occurrence of an event. An event could be temporal, such as at time t or after time t_1. An event could be an access event such as on the first access of a data item, on every access of a data item, or on the n'th access of a data item. On occurrence of an event the relevant execution rule is fired, and the data item stays in its current state or is transitioned to the next state, depending on the privacy rule that is being applied. We define the middleware necessary to enforce rule compliance in such a framework, in the next section.

4.3   Components of the Middleware

In this section we describe the various components that would be required to perform data management at the various stages of the information life cycle, as shown in figure 3. These components and their functionality are:

1. A policy translator that translates the high-level language policy to a machine-readable format, for example, XML. (The functionality of this component is outside the scope of the current work and hence we do not discuss this further.)
2. A policy association module that associates each data item with the applicable rule/s in the policy.
3. A request interceptor, that monitors the data items for the occurrence of events. An event could be temporal or based on data access.
4. An event handler that is invoked by the request monitor whenever an event of interest occurs. The event handler basically reports whether the specific access request is allowed or denied, in the case of access events. If applicable, the event handler updates the state of the data item as relevant.

Figure 4 presents a possible architecture for privacy compliance end-to-end with the above components, extending the notions in figure 3. The functionality of the various components may be defined as follows.

1. *Rule association:* The data items in the various repositories to be protected may be grouped into different categories for which the privacy rules are defined. The rule association step associates each data item with zero, one or more rules, based on the specific category to which the rule belongs. This may be done at setup time, and subsequently may need to be done whenever new rules or new data items are added to the system. If the rules are specified in a structured or semi-structured format, then it should be possible to do this step in a semi-automated manner.
2. *Request interceptor:* The request interceptor intercepts any user requests to the data in the various applications. This may be achieved by building a common interface that is accessed by the request interceptor in the front-end. At the back-end, this interface invokes each application-specific API, for each specific application. The only requirement is that each of the repositories need to expose APIs for access control and subsequent querying, which can then be invoked by the request interceptor. Any user request is made to the request interceptor. The request interceptor than passes on the necessary information to the event handler (described subsequently) and based on the reply from the event handler, determines whether the access is to be allowed or not. If yes, then the request is made to the API of the specific application and the results returned to the user.

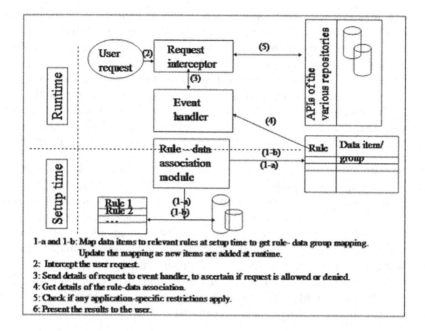

Fig. 4. Architecture of a simple privacy rule enforcement system

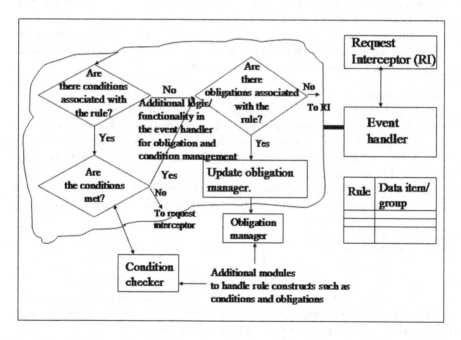

Fig. 5. Architecture of a rule enforcement system with associated conditions and obligations

3. *Event handler:* The event handler determines at runtime whether any rule is applicable for a data access. The request interceptor intercepts the user request and sends the details to the event handler. The event handler determines if one or more rules are applicable for the access event defined by this data request. The event handler performs this step based on the information available after rule association in the first step above. Based on this, the event handler returns an Allow or Deny ruling to the request interceptor. The system may have a top-level policy rule that states how access requests for events that are not specifically allowed or denied are to be handled.

In this architecture, we have considered simple privacy rules that are associated with an Allow or a Deny ruling, and are not associated with any conditions or obligations. In this case, it is possible to combine the functionality of the request interceptor and the event handler. However, if we consider privacy rules that are associated with conditions and obligations, then we can achieve enforcement of the same by including additional modifications in the event handler. This is shown in figure 5. The functionality of the event handler is extended to check for the validity of the condition. An additional component, the obligation manager, is also required to handle obligations. Here, when an event occurs, the event handler has to perform the following additional tasks.

1. If there are any conditions associated with the rule, then it needs to be checked whether the condition is met, and the data access allowed only if the condition is met.
2. If there are any obligations associated with the rule, then the obligation manager needs to be informed of the data access event and other needed information to enforce the obligation execution.

These additional details are shown in figure 5.

In this section we have discussed some possible architectures for privacy protection of data, using a middleware approach. The advantage of such as approach is that it enables the architecture to be uniformly applied across multiple content repositories, irrespective of their underlying format. The repositories only need to expose APIs for requesting data, and for the repository-specific access control that may be in place over and above the privacy policy. A further advantage is that no changes need to be made to the way the data is currently stored.

5   Conclusions

Protection of privacy information is a very important challenge across enterprises. Most solutions that exist today are application-specific or content-specific and hence do not scale readily or lend themselves to automation readily. However automated solutions are a very important requirement in view of the increasing volumes of data that enterprises handle, and the increasing exposure that enterprises face on the risk and compliance front. In view of these, we have presented a possible architecture for privacy protection at the back-end. Our initial architecture is for a system comprising simple privacy rules that specify which user groups can access which data items or data categories. The advantage of this architecture is that it can scale readily across different applications in the enterprise and is independent of the type of content or repository. Further it lends itself readily to automation with some minor additions to the architecture shown. Subsequently, we have presented extensions to the same architecture for privacy rules that may be more complex, for instance, including conditions and obligations. These would be essential constructs of privacy rules in future, in view of the increasing emphasis on risk and compliance management. Further, our architecture also aims to integrate the privacy management along with the information life-cycle management, since our contention is that future data management systems would be closely integrated with the information life cycle.

A number of open issues remain in related areas. At the top-level, the rules are framed in a business context, in a high-level language. However, for full automation, we need a way of translating the rules to a machine-understandable format. Currently this is a manual step in the process, which would need to be done for introduction of any new rule in the system. Further, conditions and obligations are increasingly a part of the policy framework. Hence the automated system needs to handle automated checking of conditions and automated obligation execution, for full scaleability.

References

1. Agrawal, R., Kieman, J., Xu, R.Y.: Hippocratic databases. In: Proceedings of the 28th VLDB Conference (2002)
2. Cranor, L.F., Ackerman, M.S.: Beyond concern: Understanding the net user's attitude about online privacy. Technical report TR 99.4.3, AT&T Labs Research (1999)
3. Westin, A.F.: How the public views health privacy: Survey findings from (1978–2005), http://www.pandab.org/HealthSrvyRpt.pdf
4. Records, computers and the rights of citizens: report of the secretary's advisory committee on automated personal data systems (1973), http://www
5. Oecd guidelines on the protection of privacy and transborder flows of personal data (1980), http://www.oecd.org
6. Xacml 2.0 specification set, http://www.oasis-open.org/committees/xacml/
7. Sandhu, R.S., Coyne, E.J., Feinstein, H.L., Youman, C.E.: Role-based access control models, vol. 29(2), pp. 38–47 (1996)
8. Agrawal, R., Bird, P., Grandison, T., Kiernan, J., Logan, S., Rjaibi, W.: Extending relational database systems to automatically enforce privacy policies. In: 21st International Conference on Data Engineering (ICDE 2005), pp. 1013–1022 (2005)
9. Park, J., Sandhu, R.: Towards usage control models: Beyond traditional access control. In: Proceedings of the 7th Symposium on Access Control Models and Technologies (2002)

Web Semantics for Intelligent and Dynamic Information Retrieval Illustrated within the Mental Health Domain

Maja Hadzic and Elizabeth Chang

Digital Ecosystems and Business Intelligence Institute, Curtin University of Technology
G.P.O. Box U1987, Perth 6845, Australia
{m.hadzic,e.change}@curtin.edu.au

Abstract. Much of the available information covering various knowledge domains is distributed across various information resources. The issues of distributed and heterogeneous information that often come without semantics, lack an underlying knowledge base or autonomy, and are dynamic in nature of are hindering efficient and effective information retrieval. Current search engines that are based on syntactic keywords are fundamental for retrieving information often leaving the search results meaningless. Increasing semantics of web content (web semantics) would solve this problem and enable the web to reach its full potential. This would allow machines to access web content, understand it, retrieve and process the data automatically rather then only display it. In this paper we illustrate how the ontology and multi-agent system technologies which underpin this vision can be effectively implemented within the mental health domain. The full realisation of this vision is still to come.

Keywords: web semantics, ontology, multi-agent systems, ontology-based multi-agent system, mental health, intelligent information retrieval, dynamic information retrieval.

1 Introduction

In 2007, the internet is frequently used as an information source for a multiplicity of knowledge domains. The general public uses Google predominately to obtain information covering these domains. The users inevitably have different access and understanding of the results they obtained from their search. As Google is not built to separate authoritative from dubious information sources, the users may have to use specialised search engines.

The accumulation of published information is an additional problem that complicates the search. For example, biomedical researchers would use PubMed which is a service of the U.S. National Library of Medicine that includes over 16 million citations from life science journals for biomedical articles back to the 1950s. Using the PubMed search engine, the user receives a list of journals related to the specified keyword. It is now left to the user to read each journal individually and try to establish links within this information. This would be easy if the journal list consisted of only a few journals. However, the journal list usually consists of

T.S. Dillon et al. (Eds.): Advances in Web Semantics I, LNCS 4891, pp. 260–275, 2008.

thousands of journals and the medical researchers usually do not have time to go through these results thoroughly. There is a high chance that some important information will be overlooked.

Specific and targeted searches are very difficult with current search engines. For example, a search for "genetic causes of bipolar disorder" using Google provides 960,000 hits consisting of a large assortment of well meaning general information sites with few interspersed evidence-based resources. The information provided by the government sites is not necessarily returned on the first page of a 'google' search. A similar search on Medline Plus retrieves all information about bipolar disorder plus information on other mental illnesses. The main problem of the current search engines is that they match specific strings of letters within the text rather than searching by meaning.

There is a need to design an intelligent search engine that would perform a search based not on keywords but on the meaning of the information. The search engine would go through the available information, understand this information and select highly relevant information as well as link this information and present it in a format meaningful to the user.

In this paper, we will briefly introduce the technologies underpinning such a search engine. We will address current issues related to information access and retrieval from the web. We will introduce the meaning of web semantics and the role of ontologies and agent technologies in the creation of a semantically rich environment. We will also illustrate this on an example from the mental health domain.

2 Information Variety

The number of active users and information resources are increasing each day. Adding content to the web is quite an easy task and access to the web is uncomplicated and fast.

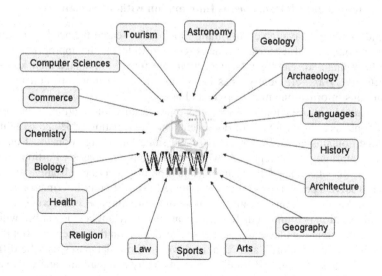

Fig. 1. Information on the web covers various knowledge domains

The available information covers various knowledge domains and various disciplines such as astronomy, geology, biology, computer sciences, history, arts, religion and so forth (Figure 1).

Finding relevant information among all the information resources is a difficult task. The sheer volume of currently available information limits access and retrieval of required information. Current use of web content is based on browsing and syntactic key-searches. Browsing makes use of hyperlinks to navigate from one web page to the next. In syntactic key-word searches, the key-words, made of patterns of characters, are matched to strings occurring in resources on the web. At present, these Google-like search engines are fundamental for retrieving information but the search results are often still meaningless and completely outside the domain that users are interested in.

3 Information Retrieval Hindrances

The usefulness of available data needs to be maximised. This is becoming more important each day. The organisation and management of the available information and existing information resources are the main focus.

We identified the following underlying issues with difficult data access, retrieval and knowledge sharing:

1. The increasing body of distributed and heterogeneous information;
2. There is no underlying knowledge base available to enable shared understanding of concepts to facilitate information retrieval and knowledge sharing; and
3. Databases and information resources are autonomous, heterogeneous and dynamic (content is continuously updated). There is no efficient tool to help coordinate them.

3.1 Distributed and Heterogeneous Information without Semantics

The rapid increase in available information makes the search for needed information even more complicated. Extensive time is required to locate the appropriate information by browsing and searching. Organisations are forced to manage their knowledge more effectively and efficiently. It is becoming more and more difficult to manage and share this information amongst potential users.

The available information is distributed over various information resources. The internet contains a huge number of documents and information which covers different knowledge domains and different areas of the same knowledge domain. Lots of the available information is repeated within numerous databases. Portions of this information may be related to each other and/or may overlap. Usually, knowledge required to solve a problem is spatially distributed over different information resources as shown in Figure 2. In this situation, sharing of information becomes crucial.

The information is found in different formats that may not be compatible with each other. Two different databases may contain information on the same topic but the way this information is structured and organised within those databases may be different. In these situations, it is quite difficult to automatically compare and analyse available information. Moreover, the available data is structured in a way to suit a particular

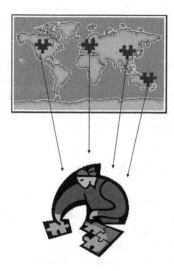

Fig. 2. Information distribution

purpose of the database and associated community. Important information in one database may be found to be insignificant for another database and be removed regardless of its importance within a larger scope. The heterogeneity of the information is becoming one of the main problems in efficient information use.

3.2 Underlying Knowledge Base Is Not Available

Knowledge integration methodologies are becoming more important as technology advances and interconnections between organisations grow. Efficient storage, acquisition and dissemination of knowledge require structured and standardised organisation of data. As some information needs to be reused and shared, so to do the basic concepts used to describe the information that needs to be shared. These basic concepts may be organised differently for specialised systems but the fundamental concepts must be common to all applications. A unifying framework that represents shared understanding of domain knowledge would enable intelligent management of the available information.

3.3 Autonomous, Heterogeneous and Dynamic Information Resources

Currently, information resources exist autonomously [1]. In most cases, the resources were designed for a single purpose and independently from others. Each separate knowledge domain generated its own data and, therefore, its own information sources. The integration of those autonomous information resources has been promoted during the last few years.

The information resources are quite heterogeneous in their content, data structure, organisation, information management and the like. A range of analysis tools are available but their capabilities are limited because each is typically associated with a particular database format. As each information source usually has its own structure, content and query language; available analysis tools often work on only a limited subset of data. Heterogeneity of the information resources makes their integration difficult.

One of the main issues is the dynamic nature of the internet environment. Content within the information resources is changeable as it is continuously updated and modified. The data and information can be added hourly or daily and may be removed the next minute. A solution needs to be developed that can address the issues associated with dynamic information resources.

4   Information Retrieval in Science

Information and computer sciences are becoming vital as they impact on a number of disciplines. Computers are used to store and manage the increasing amount of information. Also, computers are predominantly used for data analysis performed after (large-scale) experiments. Various sciences are experiencing those kinds of changes. As a result, the new generation of scientists are regarded as computational scientists as well.

We show an example from the biomedical domain in Figure 3. The researchers formulate specific hypotheses through the study of the contents of huge databases (arrow 1). Laboratory experiments need to be performed in order to test the hypotheses (arrow 2). This results in a collection of data. Computers are used to organise the data, perform needed calculations and present the data in a meaningful format for the users (arrow 3). Computers will also store this data and let the users access it when needed.

There is a need to create systems that can apply all available knowledge to scientific data. For any knowledge domain, the size of an existing domain knowledge base is too large for any human to assimilate. Computers are able to capture and store

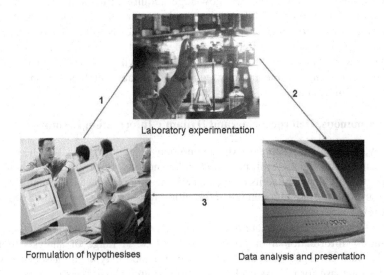

Fig. 3. Use of computers in the biomedical domain

all this information but are unable to link this information and derive some useful knowledge from it. Better predictions can be made using the bigger knowledge base. But the predictions are still being made on a relatively small subset of the available knowledge due to the:

1. size of the available information;
2. autonomous, distributed and heterogeneous nature of the information resources; and
3. lack of tools to analyse the available information and derive useful knowledge from it.

As a result, some important information is being neglected. Sometimes, this can impact on the outcomes significantly. We need a system that can effectively and efficiently use all the available information.

Scientists would be able to progress much faster if using all the available knowledge efficiently. Usually, one particular problem can be solved by various research teams. Only when found in context with the other available information, the information provided by each individual team shows its real value. Each research team provides a piece of information that together with the information provided by other research teams forms a complete picture.

Cooperation between different groups and professionals is also important. Different groups have different capabilities, skills and roles. Even if they all have common goals, they may be operating and executing their tasks on different knowledge levels. In most cases, the solution to a problem involves coordination of the efforts of different individuals and/or groups. There is an obvious need for these people to complement and coordinate with each other in order to provide the definition and/or solution to their common problem. Complexity of the information retrieval process is due to the:

1. huge body of information available;
2. nature and characteristics of information resources; and the
3. fact that many users still use collections of stand-alone resources to formulate and execute queries. This places a burden on the user and limits the uses that can be made of the information. Users need mechanisms to capture, store and diffuse domain-specific knowledge.

Use of internet for information retrieval is still a complex task. Most problems need to be decomposed into smaller interdependent sub-problems. A sequence of small tasks needs to be performed in order to solve the overall problem where each task may require querying of different information resources. Following the problem decomposition, the user needs to [2]:

• identify and locate appropriate information resources
• identify the content of those resources
• target components of a query to appropriate resources and in optimal order
• communicate with the information resources
• transform data between different formats
• merge results from different information resources

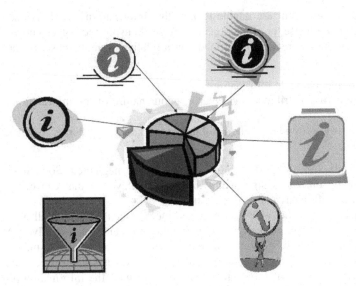

Fig. 4. Integration of heterogeneous information

One of the biggest issues in the information integration is heterogeneity of different resources. Usually, a number of different information resources need to be queried in order to solve a problem effectively. Each information resource has its unique data structure, organisation, management and so forth. This is represented in Figure 4 by different colours. It is difficult to integrate the information when found in different formats. Additional difficulties encountered by the users include:

a) Medical information alone totals several petabyte [1]. In 2007, a researcher setting a genetic human trial for 'bipolar disorder' would have needed to sift through a multitude of information from various sources (for example, Entrez Gene) to have found that loci 2p13-16 are potential positive sites for this disorder (the information originating from the research reported by Liu and his team [3]).

b) As research continues, new papers or journals are frequently published and added to the databases and more and more of this published information is available via the internet. Problematically, no collaborative framework currently exists to help inform researchers of the latest research and where and when it will become available.

c) Huge volumes of mental health information exist in different databases mostly having their own unique structure. This is equivalent to the situation in the Australian libraries before Libraries Australia (http://librariesaustralia.nla.gov.au/apps/kss) and the development of universal cataloguing standards. No tool is available to help manage, search, interpret, categorise and index the information in these disparate databases.

d) Most published results cover one specific topic. Usually, one must combine and analyse information regarding various topics in order to get an overall picture. Currently, no tool exists that allows examination and analysis of different factors simultaneously.

e) Portions of information or data on the internet may relate to each other, portions of the information may overlap and some may be semi-complementary. No knowledge based middleware is available to help identify these issues. 'Common knowledge', for example, may reduce the possibility of undertaking the same experiments such as examining the same region of DNA sequence by different research groups thus saving time and resources. This helps to create a cooperative environment making big research tasks coherent between different research teams. Gap analysis may also be easier.

5 Search Engines

We previously mentioned a number of factors responsible for complexity of the information retrieval process and these include:

1. the sheer volume of information and its continual increase;
2. nature and characteristics of information resources; and
3. many users still use collections of stand-alone resources to formulate and execute queries.

It is possible to design a search engine which will not be affected by the three above-mentioned problems. The sheer volume of the information and its continual increase can be handled by putting this information in a universally agreed and shared standardised format which can be 'understood' by computers and applications. The same is true for nature and characteristics of information resources. If their information content is formatted according to the universally agreed and shared format, their nature and characteristics will be transformed to support efficient and effective information retrieval. The user will not need to use a collection of stand-alone resources; all information resources together will 'act' as one big information resource that contains uniformly formatted information from various knowledge domains.

Currently, the fourth and main problem associated with complex and inefficient information retrieval is the nature and characteristics of the current search engines. Google-like search engines based on syntactic keywords are fundamental for retrieving information but the search results are often still meaningless. For example, consider the concept 'plate'. This concept has different meanings in different domains such as architecture, biology, geology, photography, sports etc (http://www.answers. com/topic/plate) (Figure 5). Doing a search for 'plate' using current search engines, you may get varying results from various knowledge domains. If you try to specify a search by typing in 'plate, sports', you may still get something like: 'Everyone needs to bring a plate with food after the football on Sunday night'.

We need search engines to look for word meanings within a specific context. The content of the web pages needs to be described in a way that search engines look for the meaning of the word within a specific context rather then the word itself. The content of the most current information resources is described using 'presentation-oriented' tags and the associated search engines work on this principle. In order to bring a positive transformation on this issue, the content of the information resources needs to be described using 'meaning-oriented' tags so that the search engines can be designed to search for the meaning of the information rather than simply its

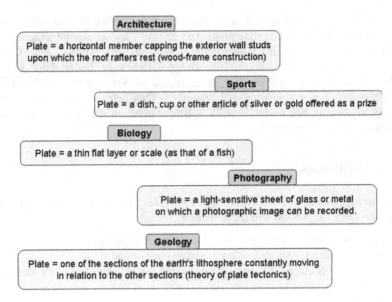

Fig. 5. The term 'plate' has different meanings within different knowledge domains

appearance in the text. This is one of the major concerns and goals in creating and increasing Web Semantics. We will discuss this more in the following section.

6 Web Semantics

Semantic Continuum moves from the semantically poor and easily understood by humans towards semantically rich and easily 'understood' by computer expressions. Semantics that exist only in the minds of humans who communicate and build web applications is implicit. Semantics may also be explicit and informal such as found in thesauri and dictionaries where terms are described and defined. Semantics may also be explicit and formal for humans such as in computer programs created by people. There is less ambiguity the further we move along the semantic continuum and it is more likely to have robust correctly functioning web applications. Here, the semantics is explicit and formal for machines. This Semantic Continuum is shown in Figure 6.

The current web can reach its full potential through enabling machines to be more actively involved in the web 'life-processes'. Machines need to be able to 'understand' the web content, access it, retrieve and process the data automatically rather then only displaying it. They need to be able to meaningfully collect information from various information sources, process it and exchange the results with other machines and human users (Figure 7). The web information that is semantically rich and easily 'understood' by computers can serve this purpose. Namely, the current web can reach its full potential through increasing semantics of its content; commonly referred to as Web Semantics.

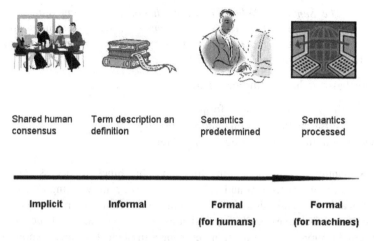

Shared human   Term description an   Semantics   Semantics
consensus   definition   predetermined   processed

Implicit	Informal	Formal	Formal
		(for humans)	(for machines)

Fig. 6. Semantic Continuum

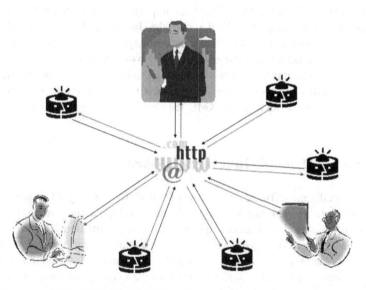

Fig. 7. Machines and humans exchange information

The main and the biggest project currently associated with establishing and increasing Web Semantics is the Semantic Web project. Tim Berners-Lee of the World Wide Web Consortium is leading this project. He aims to transform the current web into a semantic one where its content is described in computer-understandable and computer-processable meaning.

> '*The Semantic Web is an extension of the current Web in which information is given well-defined meaning, better enabling computers and people to work in cooperation.*' (Berners-Lee et al. 2001, pp. 34-43)

'*The Semantic Web is a vision: the idea of having data on the Web defined and linked in a way that it can be* used by machines *not just for display purposes, but for automation, integration and reuse of data across various applications.*' (www.w3c.org)

The emphasis in these definitions is on:

- well-defined meaning of the information
- machines use the information (automation)
- cooperation through data integration and reuse

The Semantic Web project aims to enable machines to access the available information in a more efficient and effective way. They are working on two different options towards the realisation of their vision. In the first option, the machine reads and processes a machine-sensible specification of the semantics of the information. In the second option, web application developers implant the domain knowledge into the software which will enable the machines to perform assigned tasks correctly.

The first two languages that played an important role in Web Semantics are: eXtensible Markup Language (XML) and the Resource Description Framework (RDF). XML allows the users to create their own tags and, in this way, enables them to add arbitrary structure to their documents. However, it says nothing about the meaning of this structure. On the contrary, meaning is expressed by RDF. The meaning is encoded in sets of three: things, their properties and their values. These sets of three can be written using XML tags and form webs of information about related things.

7 Agents for Dynamic Information Retrieval

Agents are intelligent software objects used for decision-making. They are capable of autonomous actions and are sociable, reactive and proactive in an information environment [4]. Agents can answer queries, retrieve information from servers, make decisions and communicate with systems, other agents or with users. Use of agent-based systems enables us to model, design and build complex information systems.

The Semantic Web project aims to transform the current web into a semantic one where agents can understand web content. This concept is presented in Figure 8. In this environment, an information resource will contain data as well as metadata which describe what the data are about. This allows agents to access, recognise, retrieve and process relevant information. They are also free to exchange results and to communicate with each other. In such environments, agents are able to access and process the information effectively and efficiently.

The web then becomes more than just a collection of web pages. Agents can 'understand' and 'interpret' meaning of the available information, perform complex tasks and communicate with each other and with humans. Humans can use one or several such agents simultaneously.

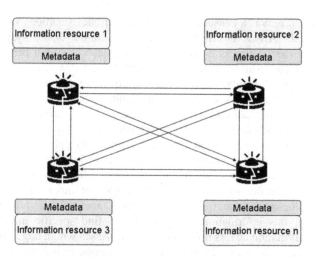

Fig. 8. Annotation of information resources enables agents to work cooperatively

8   Ontologies for Intelligent Information Retrieval

Ontology is an enriched contextual form for representing domain knowledge. Ontology provides a shared common understanding of a domain and means to facilitate knowledge reuse by different applications, software systems and human resources [5, 6].

Berners-Lee et al. emphasise the interdependence of agent technology and ontologies [7]. Intelligent agents must be able to locate meaningful information on the web and understand the meaning of this information, to advertise their capabilities and to know the capabilities of other agents. All this can be achieved through use of ontologies; firstly, through ontology sharing between agents and secondly, through the annotation of information resources.

Agents need to communicate and interact with other agents. This requires them to share a common understanding of terms used in communication. The specification of the used terms and exact meaning of those terms is commonly referred to as the agent's ontology [8]. As this meaning must be available for other software agents, it needs to be encoded in a formal language. Agents can then use automated reasoning and accurately determine the meaning of other agents' terms.

Ontology as 'a formal and explicit specification of a conceptualisation' allows the information providers to annotate their information [7]. The annotation phase is a crucial step in creating a semantically rich environment which can be intelligently explored by users' agents. The annotation provides background knowledge and meaning of the information contained within this information resource.

9   The Relevance of Web Semantic within the Mental Health Domain

The World Health Organisation predicted that depression would be the world's leading cause of disability by 2020 [9]. The exact causes of many mental illnesses are unclear. Mental illness is a causal factor in many chronic conditions such as diabetes, hypertension, HIV/AIDS resulting in higher cost to the health system [10]. The recognition that mental illness is costly and many cases may not become chronic if treated early has lead to an increase in research in the last 20 years.

The complexity of mental illnesses adds further complications to research and makes control of the illness even more difficult. Mental illnesses do not follow Mendelian patterns but are caused by a number of genes usually interacting with various environmental factors [11]. There are many different types of severe mental illness [10], for example depression, bipolar disorder, schizophrenia. Genetic research has identified candidate loci on human chromosomes 4, 7, 8, 9, 10, 13, 14 and 17 [3]. There is some evidence that environmental factors such as stress, life-cycle events, social environment, economic conditions, climate etc. are important [10, 11].

Information regarding mental illness is dispersed over various resources and it is difficult to link this information, to share it and find specific information when needed. The information covers different mental illnesses with a huge range of results regarding different disease types, symptoms, treatments, causal factors (genetic and environmental) as well as candidate genes that could be responsible for the onset of these diseases. We need to take a systematic approach to making use of enormous amount of available information that has no value unless analysed and linked with other available information from the same mental health domain.

To overcome the currently complex and complicated situation, an intelligent and efficient information system needs to be designed that does not require researchers to sift through the same or similar results from different databases. This expert system needs to be able to intelligently retrieve information from the heterogeneous and disparate databases, and present it to the user in a meaningful way.

We propose a solution which includes the design of Generic Mental Illness Ontology (GMIO) and GMIO-based multi-agent system.

Generic Mental Illness Ontology (GMIO) can be developed to contain general mental health information. Four sub-ontologies can be designed as a part of the GMIO to represent knowledge about illness sub-groups (e.g., clinical depression, postnatal depression), illness causes (such as environmental and genetic), phenotypes (which describe illness symptoms) and possible treatments. The ontology and sub-ontologies will serve as template to generate Specific Mental Illness Ontologies (SHIO), the information specific to an illness in question (e.g., bipolar, depression, schizophrenia).

The GMIO needs to be effectively utilised within a multi-agent system. We need to define a multi-agent system architecture that will be based on GMIO and will enable the agents to collaborate effectively. A possible solution that includes different agent types (Interface, 'Manager', Information and 'Smart' agents) is represented in Figure 9. *Interface* agents assist the user in forming queries. Interface agents communicate user's request to the 'Manager' agents. *'Manager'* agents decompose the overall task into smaller tasks and assign these subtasks to the various Information agents. *Information* agents retrieve the requested information from a wide range of biomedical databases. Each information agent may have a set of databases assigned to it. The information agents send the retrieved information to the 'Smart' agents. *'Smart'* agents analyze this information, assemble it correctly and send to the Interface agent directing it back to the user as an answer to his/her query.

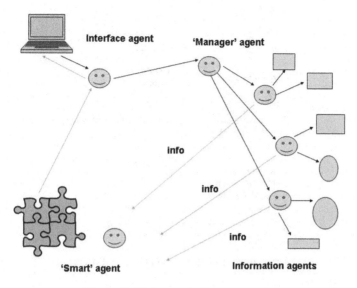

Fig. 9. GMIO-based Multi-agent System

Within the GMIO-based multi-agent system, ontology is used at the different levels to:

- *locate and retrieve* requested information. Information content within an information resource can be described using an ontology. Only then, an agent committed to this ontology is able to 'understand' the information contained within these resources and is able to exactly locate and retrieve the information requested by a user.
- enable cooperatively working agents to *communicate* with each other during the process of the information retrieval. Use of ontology permits coherent communication between the different agents and facilitates sharing of the information among different agents.
- *analyze and manipulate* the retrieved information. In this way, the redundant and/or inconsistent information is removed. Only relevant information is selected, assembled together and presented to the user.
- *present* the retrieved information to the user in a meaningful way. The information is presented to the user in a way that makes it easier for the researcher, physician or patient to have an overview of the requested knowledge regarding human disease of interest. Moreover, the inherited organisation of ontologies adds a taxonomical context to search results, making it easier for the users to spot conceptual relationships in data.

To understand the mechanism behind these different operations we refer the interested reader to [12, 15].

As a result of user's query, the overall problem to be solved is constructed as Specific Mental Illness Ontology (SMIO) template from GMIO by Interface agents. Retrieving and adding of relevant information upon this SMIO template results in Specific Mental Illness Ontology (SMIO). This is shown in Figure 10.

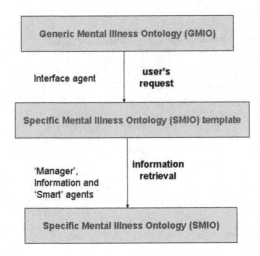

Fig. 10. GMIO, SMIO template and SMIO

Protégé [13] developed by Stanford University can be utilized for the modelling of the ontologies. JADE (Java Agent DEvelopment framework) [14] can be used to develop the multi-agent based system. JADE is a software framework that simplifies the implementation of agent applications in compliance with the FIPA (Foundation of Intelligent Physical Agents) specifications for interoperable intelligent multi-agent systems and it offers a general support for ontologies. The ontology can be stored as a computer readable description of knowledge and the agents can use this knowledge for intelligent actions.

The developed tool could be used internally in a specific organization such as a pharmaceutical company or consortium such as Universities Australia, or externally in a government-supported public information service e.g., PubMed.

10   Conclusion

In this paper, we talked about the issues associated with current use and management of the information on the web and discussed proposed solutions to these problems. We explained the concept of web semantics, introduced the Semantic Web project and discussed its vision. We briefly introduced two major technologies that may help realise this vision: ontology and agent-based technology, and illustrated their significance within the mental health domain.

The implementation of such a synergetic system will result in positive transformation of world-wide mental health research and management to a more effective and efficient regime. This system is highly significant and innovative, and represents a new generation of information-seeking tool. The beauty of this innovative tool is that the complexity is hidden from the client. Users will simply see targeted answers to their questions. The ontology and agents are working in concert to remove complexity where it should be, not on the human client's computer screen but within the tool.

Even though the concepts of Web Semantics and the Semantic Web project have been around for awhile, the big breakthrough has not happened yet. Either we need new advanced technologies or we will find missing pieces of the puzzle in the work ahead.

References

1. Goble, C.: The Grid Needs you. Enlist Now. In: Proceedings of the International Conference on Cooperative Object Oriented Information Systems, pp. 589–600 (2003)
2. Stevens, R., Baker, P., Bechhofer, S., Ng, G., Jacoby, A., Paton, N.W., Goble, C.A., Brass, A.: TAMBIS: Transparent Access to Multiple Bioinformatics Information Sources. Bioinformatics 16/2, 184–186 (2002)
3. Liu, J., Juo, S.H., Dewan, A., Grunn, A., Tong, X., Brito, M., Park, N., Loth, J.E., Kanyas, K., Lerer, B., Endicott, J., Penchaszadeh, G., Knowles, J.A., Ott, J., Gilliam, T.C., Baron, M.: Evidence for a putative bipolar disorder locus on 2p13-16 and other potential loci on 4q31, 7q34, 8q13, 9q31, 10q21-24, 13q32, 14q21 and 17q11-12. Molecular Psychiatry 8/3, 333–342 (2003)
4. Wooldridge, M.: An Introduction to Multiagent Systems. John Wiley and Sons, Chichester (2002)
5. Gómez-Pérez, A.: Towards a Framework to Verify Knowledge Sharing Technology. Expert Systems with Applications 11/4, 519–529 (1996)
6. Gómez-Pérez, A.: Knowledge Sharing and Reuse. In: The Handbook on Applied Expert Systems, pp. 1–36 (1998)
7. Berners-Lee, T., Hendler, J., Lassila, O.: The Semantic Web. Scientific American 284/5, 34–43 (2001)
8. Gruber, T.R.: A Translation Approach to Portable Ontology Specifications. Knowledge Acquisition 5/2, 199–220 (1993)
9. Lopez, A.D., Murray, C.C.J.L.: The global burden of disease, 1990-2020. Nature Medicine 4, 1241–1243 (1998)
10. Horvitz-Lennon, M., Kilbourne, A.M., Pincus, H.A.: From silos to bridges: meeting the general health care needs of adults with severe mental illnesses. Health Affairs 25/3, 659–669 (2006)
11. Smith, D.G., Ebrahim, S., Lewis, S., Hansell, A.L., Palmer, L.J., Burton, P.R.: Genetic epidemiology and public health: hope, hype, and future prospects. The Lancet 366/9495, 1484–1498 (2005)
12. Hadzic, M., Chang, E.: Ontology-based Multi-agent systems support human disease study and control. In: Czap, H., Unland, R., Branki, C., Tianfield, H. (eds.) Frontiers in Artificial Intelligence and Applications (special issues on Self-organization and Autonomic Informatics), vol. 135, pp. 129–141 (2005)
13. Protégé: an Ontology and knowledge-based editor, Stanford Medical Informatics, Stanford University, School of Medicine (2006), http://protege.stanford.edu/
14. JADE (Java Agent DEvelopment Framework), http://jade.tilab.com/
15. Hadzic, M., Chang, E., Wongthongtham, P., Dillon, T.: Ontology-based Multi-agent Systems. Springer, Heidelberg (to appear, 2008)

Ontologies for Production Automation

Jose L. Martinez Lastra and Ivan M. Delamer

jose.lastra@tut.fi, ivan.delamer@ut.fi

Abstract. The manufacturing sector is currently under pressures to swiftly accommodate new products by quickly setting up new factories or retrofitting existing ones. In order to achieve this goal, engineering tasks currently performed manually need to be automated. In this scenario, ontologies emerge as a candidate solution for representing knowledge about manufacturing processes, equipment, and products in a machine-interpretable way. This knowledge can then be used by automated problem-solving methods to (re)configure the control software that coordinates and supervises manufacturing systems. This chapter reviews current approaches to use ontologies in the manufacturing domain, which include use for vocabulary definition in multi-agent systems and use for describing processes using Semantic Web Services. In addition, current and emerging research trends are identified.

1 Introduction

As in many other domains, the pervasiveness of information and embedded computational power is causing a revolution in the way manufacturing systems are conceived. The drivers for such a revolution vary in different markets:

- In the high-tech and consumer product markets, the drivers are increased customization of product, higher new product introduction rates, and shorter time to market requirements.
- In other discrete product manufacturing markets, the drivers are reduction in lead time and supply chain responsiveness.
- In process and batch manufacturing markets, such as (petro)chemical, paper, mining, and so on, the drivers are increased quality and reduced factory downtime requirements.
- In developed markets, the drivers are improving competitiveness against economies of scale and emerging markets.
- In developing markets, the drivers are production in proximity to the target markets, and competition against established markets.

In order to respond to these requirements and remain competitive, manufacturing enterprises must be able to swiftly set up new factories, and more importantly to seamlessly re-configure existing factories in order to produce new product variants or new product families. However, these tasks require significant amounts of engineering efforts, which result in time delays in the order of months or even years, and major expenditures in highly-qualified engineers.

T.S. Dillon et al. (Eds.): Advances in Web Semantics I, LNCS 4891, pp. 276–289, 2008.
© IFIP International Federation for Information Processing 2008

One of the main reasons why engineering efforts are needed is that technical specifications for products, for the processes required to manufacture them, and for the devices and machines that are used are written in natural language, often aided by visual diagrams. These specifications are combined with the pre-existing know-how and experience of engineers to systematically develop the physical configuration of the factory and the automatic control programs that coordinate and supervise the processes. Whenever changes are needed to accommodate a new product or process, engineers are brought in to transform the specifications coming from the product design and the existing system into a modified manufacturing system.

In order to overcome most or all of the tasks that are currently performed manually by engineers on a case-by-case basis, the knowledge contained in technical specifications and engineering know-how must be given a machine-interpretable form, so that automated problem-solving methods can be used to (re-)configure systems. Ontologies have recently emerged as a strong candidate to represent the many different sources of knowledge used in production automation in a modular and reusable approach.

The rest of this chapter is an attempt to describe what types of knowledge need to be represented in order to facilitate the (re-)configuration of manufacturing systems, and what approaches are currently being followed in order to represent this knowledge using ontologies. The next subsection presents some background information about the manufacturing domain. Subsequently, the main domains of knowledge representation are introduced: processes, products, and equipment. The latter part of this chapter is then dedicated to explore how different researches have attacked the problem and what research directions are currently emerging.

2 Characteristics of Manufacturing Systems

Manufacturing systems can be coarsely defined to be composed by a series of processing station and by a material handling system that transfers and supplies products, parts, components and materials to the stations as required for each product variant being manufactured. Traditional manufacturing systems were composed of a linear process flow, called a *transfer line*, which transfers products through a fixed sequence of processing station, each of which perform a pre-defined process. Such transfer lines are usually highly efficient[1], but inflexible in that they are usually capable of producing only one type of product and require significant effort and down-time to be reconfigured for a new type of product. Thus, many newer manufacturing systems are composed of a flexible material handling system, which is able to provide multiple material routing paths and therefore many possible processing sequences, and of flexible and/or reconfigurable process cells.

Flexible production systems emerged during the 1980s, facilitated by an increase in availability of programmable robotized cells (used mostly for assembly and painting) and of programmable numerically-controlled machines (used for machining parts). Flexible process cells are able to perform a wide array of processes, therefore being

---

[1] Transfer lines are usually efficient while online, although they are less robust in case of breakdowns as the whole line needs to be stopped.

able to cope with a wide variety of products over its lifetime without need for altering or retrofitting the equipment. However, this flexibility implies that the equipment is typically underutilized, causing the excess capabilities to be an unwanted capital investment. As a result, reconfigurable process cells have gained attention as a paradigm that fosters the adoption of equipment with interchangeable parts, tools and mechatronic actuators that can be easily composed in different configuration according to the product requirements. Reconfigurable process cells meet without exceeding the process requirements for the products being manufactured, and can easily be changed or reconfigured when new products are introduced [1].

In both the cases where flexible or reconfigurable process cells are used, a number of mechatronic devices/modules/subsystems are coordinated to implement the process offered by the machine. These devices may include conveyors, fixturing actuators, articulated robot arms, tools such as screwdrivers, welding electrodes, drills, etc, and/or end effectors such as grippers, among others. Each device provides one or more physical operations, which can be coordinated with other devices' operations to implement complex processes, e.g. a conveyor, robot arm and gripper operations can be coordinated to implement a pick&place machine. Likewise, at a higher hierarchical level, the processes offered by the different stations can be sequenced in order to yield different products. Therefore, low-level physical operations can be coordinated to output different manufacturing processes, and these processes can then be sequenced to yield different types of products.

The currently available collection of industrial devices that are used to develop manufacturing equipment is extremely large. Likewise, the catalog of industrial equipment that can be used to generate different manufacturing systems is a large volume. However, the physical operations and manufacturing processes provided by the devices and equipment is a much more reduced set, which also changes slowly over time (even if frequent innovations cause the equipment to implement those processes in new ways). Therefore, production engineers are able to select appropriate devices/machines according to case-specific specifications based on their previous knowledge and experience on the required operations/processes.

This suggests that if existing knowledge on physical operations and manufacturing processes is represented in a machine-interpretable format, many of the (time- and resource-costly) design and configuration tasks that are currently performed by engineers could become more automatic. In this scenario, ontologies emerge as a candidate solution to represent knowledge about operations and processes, and about devices and equipment that offer those operations/processes. Such a representation would allow different reasoning mechanisms to be applied, such as subsumption for classifying new types of devices/equipment, or more complex problem-solving methods for composition and configuration purposes.

Furthermore, ontologies can be used to represent product knowledge. Such knowledge can be used during product design in order to create configurations or compositions of parts, sub-assemblies and components. However, the focus in this chapter shall be on the use of product knowledge in order to derive the process requirements that are used to develop the corresponding manufacturing system.

3 Process, Product and Equipment Ontologies

The first thing that is realized when developing an ontology for the manufacturing domain is that the common concept that is involved in all major portions of knowledge specification is the *process*. From a manufacturing equipment perspective, the main purpose or function is to offer a process. From a product perspective, the main requirement is a set of processes used to transform parts and raw materials into finished goods. The relationship between these major concepts is illustrated in Fig. 1.

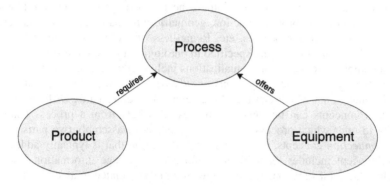

Fig. 1. Upper concepts in a manufacturing ontology

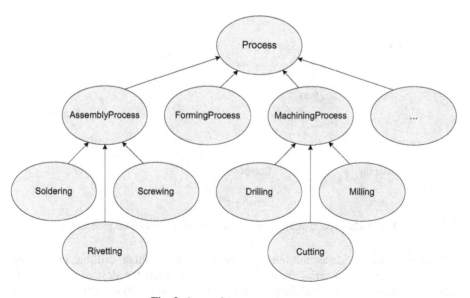

Fig. 2. A sample process taxonomy

3.1 Process Taxonomies

The three upper concepts, *Process*, *Equipment* and *Product*, typically serve as base concepts that are refined by sub-concepts representing more specific types of processes, equipment, and products. While the classification of equipment and products can be performed in numerous ways, processes are typically represented using a hierarchical taxonomy. This taxonomy can then be used to match required processes (from products) to offered processes (by equipment). A sample process taxonomy is illustrated in Fig. 2.

3.2 Product Ontologies

Conceptual specifications of products typically include both geometric information and non-geometric information. The hierarchical classification of concepts varies, and some ontologies may even place the same concepts in several parallel tree-like structures, e.g. according to product function, geometric information, product market type, required manufacturing processes, etc. Regardless of the adopted hierarchy model, certain product concepts that are specified in ontologies are predominant.

For assembled products, conceptualizations include *parts* and *connections* between parts. Parts may be atomic *components* or *subassemblies*, which are modular assemblies of components which can then be assembled to other parts. Different types of *connection* concepts can be linked to process concepts from a process taxonomy, allowing to infer which process is needed to connect or assembly two parts. In addition to *connection* concepts, non-geometric information that is typically added to the product concept includes manipulation information, storage information, and other lifecycle-related information. An overview of concepts related to assembled products is illustrated in Fig. 3.

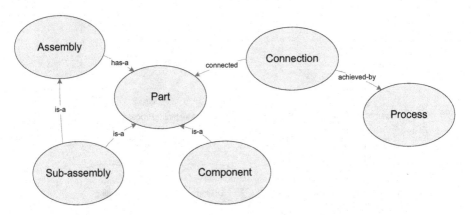

Fig. 3. Concepts typically found in ontologies for assembled products

For most metal products, machining processes such as cutting, drilling, and milling are used to remove material from a base cast part until the desired geometry is achieved. The specification of the product is usually achieved by composing primitive geometric shapes (such as cubes, cylinders, etc) which may be either filled (indicating the shape of the part) or blank (indicating where material is removed). In particular where material is removed, non-geometric information is included to indicate which types of processes can be used to achieve the geometry. For example, an empty cylinder may be associated to a drilling process, which is specified in a process taxonomy.

For virtually any type of product, traditional data structures for product representation have relied solely on 3D geometric data acquired using CAD software. The conceptualizations used in ontologies allow incorporating non-geometric information (in the form of connections, associated processes, etc) that represents the implicit knowledge that is traditionally held by engineers, but seldom conveyed with product specifications. This has often resulted in a knowledge gap between design engineers and manufacturing engineers, who need to figure out much of the semantic content that was left implicit during product design in order to develop a manufacturing system.

3.3 Equipment Ontologies

Traditionally, machines used in manufacturing were large, monolithic pieces of equipment. The drivers for flexibility and reconfiguration as well as time and cost pressures have led to developing machines from modular subsystems and mechatronic devices which provide different operations. An insight that is normally incorporated into modular equipment ontologies is that, like an assembled product, a machine is an assembly of devices. Therefore, similar structures as those used for representing products can be used to represent machines, utilizing geometric information and non-geometric information such as connections between modules and information about the physical operations facilitated by the device.

However, the unique feature of an equipment ontology is that an account must be given for the notion of *amplification of skills*: the composition of the atomic skills of devices to create more complex processes. As an example, the composition of vertical displacement and torque application results in a screw-driving process. Whenever amplification of skills occur, the resulting equipment is able to offer a process that is in addition to the atomic operations provided by the individual devices.

An equipment ontology will thus contain concepts for particular types of devices and hardware modules, and will also contain concepts corresponding to different types of machines and how these are composed. Therefore, by analyzing the properties of a particular configuration of devices and hardware modules, it is possible to classify a machine and determine what types of manufacturing processes it is able to offer.

3.4 Refined Process Ontologies

While a process taxonomy is sufficient to conceptualize the operations required to manufacture a product, and to analyze the configuration of a piece of equipment and infer what processes it may offer, a taxonomy is not sufficient to model the dynamic nature of processes. A taxonomy will suffice if all that is necessary is for software agents to model a manufacturing environment, infer the skills of a machine and determine the requirements of a product. However, it is not enough for software agents to dynamically evolve this model as processes occur. Therefore, process taxonomies are often either extended or complemented by an additional ontology in order to model the dynamic behavior observed before, during and after the execution of a process.

A refined process ontology will specify aspects such as:

- The states of the products and equipment in which the process can be executed.
- The coordination of operations needed to complete a process.
- The resulting state of the products and equipment after the process is completed.

Thus, agents are able to infer whether a required process can be achieved (or the associated capability to plan a set of actions that will lead to a valid state), can automatically coordinate the execution of operations and processes, and can evolve their internal representation of the physical environment to the resulting state.

3.5 Auxiliary Ontologies

In addition, some auxiliary ontologies are typically used in order to specify the concepts related to processes, equipment and products. These ontologies are typically developed so that they can be reused across domains, and are not specific to the manufacturing environment. The most commonly used auxiliary ontologies are used for conceptualizing:

- *Geometry*, such as shapes and volumes of parts, products, mechanical devices, and so on.
- *Physical units*, such as units of length, volume, temperature, force, pressure, velocity, acceleration, and so on.
- *Software interfaces*, used for invoking devices and machines.

4 Brief Survey and State of the Art

The use of ontologies in the manufacturing domain has evolved over the last 10 years starting since the inception of the Process Specification Language (PSL), developed by NIST [2]. Although it never gained a lasting acceptance, PSL would come around to influence several latter approaches, most notably those based on Semantic Web Services. Ontologies began to find its place in manufacturing automation architectures as a part of larger efforts focusing on multi-agent systems for manufacturing control. The use of ontologies has since been heavily influenced by developments outside the manufacturing domain, most notably the emergence of Web-based solutions and the Semantic Web.

This chapter section reviews some of the most prominent applications of ontologies in the manufacturing domain to date. The approaches can be classified into three groups:

- Multi-agent systems, which utilize ontologies as a shared vocabulary for inter-agent communications for production planning and scheduling.
- Design support systems, which utilize ontologies to guide the design process of manufacturing systems.
- Service-oriented systems, in which equipment offer processes as Web Services that are described by ontologies. Process-oriented Semantic Web Services can then be dynamically discovered, selected, composed and invoked.

4.1 Ontologies and Multi-agent Systems

During the late 1990's and early 2000's, the Holonic Manufacturing Systems (HMS) consortium merged a large portion of the research done on the application of multi-agent systems for manufacturing control. The initiative targeted the development of systems composed of modular and autonomous units, called *holons*, which would be able to self-organize and automate the control of the manufacturing system through the emergent behavior of a large group of holons. This target led to the adoption of multi-agent technologies developed in other domains, with particular emphasis on utilizing FIPA (Foundation for Intelligent Physical Agents) standards[2]. The body of work resulting from the HMS initiative has been very influential to the point that many recent developments continue to use the same technologies (e.g. KIF ontologies) in order to provide backwards compatibility.

Within the realm of multi-agent systems, both those developed under HMS and under other initiatives, ontologies have been mainly used as a means to provide a shared vocabulary that agents can use to construct messages and conversations [3] [4]. Following the original FIPA specifications, these ontologies were defined using the Knowledge Interchange Format (KIF). More recent developments have explored utilizing Web Ontology Language (OWL) ontologies, as well as utilizing ontology reconciliation techniques in order to enable interoperability among agents that use different ontologies [5].

An example ontology that is used in a multi-agent context is the Manufacturing Ontology in the ADACOR (ADAptive holonic COntrol aRchitecture for distributed manufacturing systems) Architecture [6], which is illustrated in Fig. 4. The ontology follows the Frame (F-Logic) formalism, as suggested by the FIPA standards. The main purpose of the architecture is to dynamically schedule and dispatch work orders, which result in material being sent to the appropriate process stations. The scope of the ontology is to support such scheduling functions.

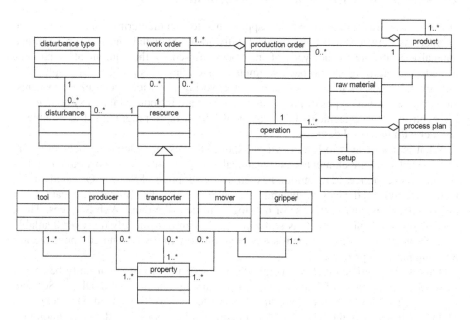

Fig. 4. Upper concepts of the manufacturing ontology in the ADACOR Architecture

---

[2] www.fipa.org

4.2 Ontologies for Design Support

As an intermediate solution towards automatic (re)configuration of manufacturing systems, the use of design tools that guide or suggest solutions through the design process can significantly reduce the required efforts. Inference mechanisms can then be used to rank and suggest alternative solutions on how to develop manufacturing equipment based on product requirements.

One such approach is proposed in [7], who developed a Web-based system that is used to develop assembly cells from mechatronic modules. The work reports an ontology framework for describing products, processes used for product assembly, and equipment that implement those assembly processes. The interactive Web-based system utilizes these ontologies in the background in order to guide the designer, starting from the product definition, in selecting devices/modules and their physical configuration that is required for the assembly the given product. The ontologies are specified using the Protégé frame-based language and following the design principles originally put forward in the CommonKADS methodology [8].

Although a big step forward in terms of facilitating the (re-)configuration of modular assembly systems, current design support solutions operate on functional and mechanical knowledge, i.e. the operations/processes provided by each module and how to mechanically attach them. However, the automatic control of the assembly cell is not considered, and must still be performed manually by control engineers once the modules are physically configured.

4.3 Semantic Web Services in Factory Automation

One of the shortcomings of existing approaches to manufacturing control is that the interactions between distributed control software components must be programmed manually according to the types of interfaces offered by the equipment to be controlled. The multi-agent approach attempts to overcome this problem by implementing several types of interactions aimed at covering the wide spectrum of systems. However, when new types of processes or devices are introduced into a system, existing control software elements (such as agents) must be re-programmed to be able to interact with the novel devices.

What is desirable is to have devices expose and describe their interfaces in a way that control software elements can dynamically acquire sufficient knowledge to interact with those devices. This notion has led to the adoption of Service-Oriented Architecture (SOA) for devices to dynamically publish and locate processes encapsulated as Web Services, and to the use of ontologies to describe those Web Services. Thus, control software entities such as software agents can dynamically discover available Web Services and infer sufficient knowledge on how to invoke and compose the underlying physical processes.

This is the approach that was originally reported in [9] and is currently being put forward in a number of European projects, among them SOCRADES[3] (Service-oriented Cross-layer Architecture for Distributed smart Embedded Devices), and SODA[4] (Service-oriented Device Architecture). The approach adopts technologies

---

[3] www.socrades.eu

[4] www.soda-itea.org

from the Web Services and Semantic Web domains in order to maximize interoperability. A Semantic Web Services ontology, such as OWL-S [10], is then used to describe the services offered by equipment in a way that they can be dynamically discovered, selected, composed and orchestrated.

In the work reported in [9], the process taxonomy is based on the DIN 8593 standard [11], which specifies and classifies joining processes, and which has been specified as an OWL ontology. In addition, it follows the previous work of Vos [12] who identified the basic physical operations that are needed in order to achieve certain assembly processes. Thus, the joining processes in the taxonomy (such as screwing, press-fitting, etc) are linked to atomic physical operations (such as forces, torques, displacements, etc). A sample view of the process taxonomy is given in Fig. 5.

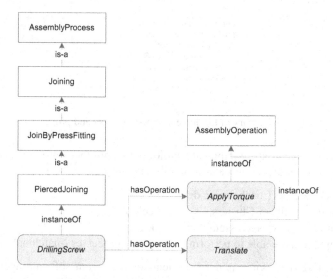

Fig. 5. Sample view of the process taxonomy

The product ontology is based heavily on the concept of assembly features, which is used to describe geometric and non-geometric (functional) information about how product parts are connected. Assembly features were originally proposed by Holland [13] and have been specified in an OWL ontology. Thus, a *Connection Feature* concept holds information about the geometry of the connection of the two parts, and the type of joint. The type of joint (e.g. a screw) can then be linked to a process (e.g. screwing) from the process taxonomy in order to dynamically define the required processes.

The equipment ontology is based on the Actor Based Assembly Systems (ABAS) reference architecture, reported by Martinez Lastra [14]. In this architecture, atomic mechatronic devices called *Actors* provide one of the atomic operations represented in the taxonomy. The Actors can then be grouped into *Clusters* in order to implement composite assembly processes, for example a vertical translation actor and a torque actor can be combined into a Screwing cluster. Clusters are modeled as an assembly of actors, using the same types of assembly features used for describing products. The properties of the upper Actor, Cluster and related concepts are illustrated in Fig. 6.

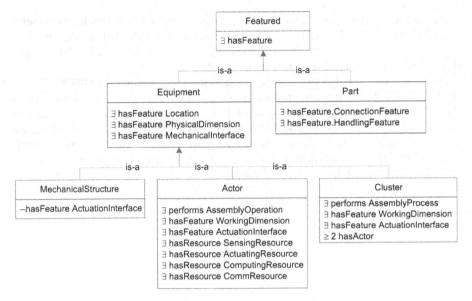

Fig. 6. Summary of Equipment upper concepts and properties

The process taxonomy, equipment ontology and product ontology offer a static model of the system that allows automatically inferring:

- The skills of Actors (mechatronic devices)
- The combined skills of Clusters (assembly cells)
- The required processes to assemble a product
- The clusters that are able to perform the required processes (matchmaking).

However, this static view is not sufficient for intelligent software agents to infer how the equipment should be used, i.e. how to invoke its processes in a changing environment. For this reason, a description is needed for:

- The software interface that allows invoking the process.
- The states of the environment in which the process can be invoked, i.e. the states of the product and of the equipment in which the process can be successfully executed.
- The resulting states of the environment after the process is executed, i.e. the resulting state of the product and of the equipment.

For this purpose, the OWL-S upper ontology for Web Services is used [10]. In particular, the Service Profile is used to model how operations are orchestrated to implement processes. Also, Preconditions and Effects are used to model the initial and consequent states of the world upon execution of the process. An example effect is the creation of a Connection Feature instance after a joining process is invoked (e.g. after a screwing process).

Compared to the inferences enabled by the static model offered by the process taxonomy, equipment ontology and product ontology, the addition of the dynamic model using a Semantic Web Service upper ontology allows the following tasks to be automated:

- The coordination or orchestration of Web Services that encapsulate operations in order to achieve higher-level processes.
- The generation of the necessary messages to invoke a Web Service (and its underlying physical process).
- The validation of preconditions necessary for the invocation of the process. Preconditions also provide a target state that can be used for planning in the case that the conditions are not immediately met.
- The evolution of the model of the product and of the equipment, according to the effects of the Web Service (and its underlying physical process).

As a result of the knowledge contained in all ontologies, intelligent agents are able to derive the specification of required processes for a product, to discover available equipment and select the appropriate devices according to the required processes, and to invoke those devices. The relationship of the different ontologies used is illustrated in Fig. 7.

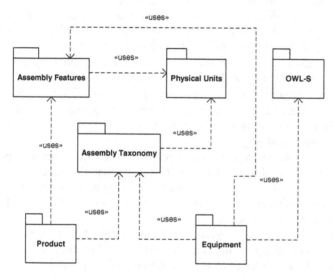

Fig. 7. Relationship between ontologies used for automation of assembly processes

4.4 State of the Art summary

Ontologies are currently the state-of-the-art technology in the manufacturing domain for the following purposes:

- Enabling interoperability of autonomous agents, specifically in the realm of dynamic process planning and scheduling where agent societies use ontologies to define a shared vocabulary in communications.
- For manufacturing equipment design support
- Dynamic system (re-)configuration using domain ontologies to derive required processes and Semantic Web Services to describe and encapsulate offered processes.

5 Emerging Research Directions

The experiences that have reported use of ontologies have generated interest into further developing the frameworks, infrastructures and architectures that have proposed their use. Some of the ongoing and emerging research directions targeting the use of ontologies in the manufacturing domain include:

- *Enhance process/task ontologies to include notions of time.* The modeling of the time taken to execute a process is currently missing or insufficient in most ontologies, which makes it more difficult to autonomously plan optimal process sequences.
- *Use First-Order Logic task ontologies.* Current approaches use Description Logics and struggle somewhat to describe dynamic aspects of the processes (such as preconditions and effects).
- *Optimized algorithms for dynamic A-Boxes.* Most inference engines are optimized for incremental additions of knowledge, and not for revision of knowledge (e.g. when the state of a product or equipment changes).
- *Application of ontology reconciliation algorithms.* Most current approaches operate only with shared ontologies and fail when different ontologies need to be used together (e.g. if devices from different vendors are used together).
- *Ontology mediation.* As an alternative to ontology reconciliation, mediators could be used to automatically translate between ontologies.
- *Consolidation of ontologies and development of ontologies for specific domains.*
- *Pilots and demonstrations.* Further demonstrators are needed to illustrate the application of ontologies, to show the feasibility of the approach, and to facilitate technology transfer to industry.
- *Applications to diagnostics.* Factories produce a constant stream of process data that reflects innumerable conditions in the factory, and that needs to be interpreted in real time. Ontologies may assist in modeling the relationships of data sources, modeling conditions and events, and facilitating reasoning that leads to diagnosis of caution and/or error conditions.
- *Applications to enterprise integration.* Ontologies have found extensive application to the domain of Enterprise Information Systems, and ongoing research is targeting at using ontologies to facilitate the integration of Enterprise and Factory Information Systems.

6 Conclusions and Outlook

Current market trends are driving manufacturers to set up and/or reconfigure their production systems faster and at lower costs. The engineering efforts that are currently invested need to be replaced by automation of (re)configuration tasks. In this scenario, ontologies emerge as the state-of-the-art mechanism to represent manufacturing domain knowledge in a modular and reusable approach – the representation of knowledge is a prerequisite for applying automated problem-solving methods.

Ontologies are being applied as part of large research initiatives involving prominent members of both academia and industry, with significant investments planned for the coming years. As ontologies continue to expand in other domains, the maturity of these related approaches will have a significant impact in the manufacturing domain, as is currently observed with the use of Semantic Web Services.

References

1. Mehrabi, M.G., Ulsoy, A.G., Koren, Y.: Reconfigurable manufacturing systems: Key to future manufacturing. Journal of Intelligent Manufacturing 11(4), 403–419 (2000)
2. Schlenoff, C., Grüninger, M., Tissot, F., Valois, J., Lubell, J., Lee, J.: The Process Specification Language (PSL): Overview and Version 1.0 Specification. NISTIR 6459, National Institute of Standards and Technology, Gaithersburg, MD (2000)
3. Foundation for Intelligent Physical Agents: FIPA Ontology Service Specification Geneva, Switzerland (2001), http://www.fipa.org/specs/fipa00086/XC00086D.html
4. Obitko, M., Marik, V.: Ontologies for Multi-Agent Systems in Manufacturing Domain. In: Hameurlain, A., Cicchetti, R., Traunmüller, R. (eds.) DEXA 2002. LNCS, vol. 2453, Springer, Heidelberg (2002)
5. Obitko, M., Mařík, V.: Integrating Transportation Ontologies. In: Mařík, V., William Brennan, R., Pěchouček, M. (eds.) HoloMAS 2005. LNCS (LNAI), vol. 3593, pp. 99–110. Springer, Heidelberg (2005)
6. Borgo, S., Leitao, P.: Foundations for a core Ontology of Manufacturing. In: Kishore, R., Ramesh, R., Sharman, R. (eds.) Ontologies: A Handbook of Principles, Concepts and Applications in Information Systems. Springer, Heidelberg
7. Lohse, N.: Towards an Ontology Framework for the Integrated Design of Modular Assembly Systems. Doctoral Thesis, University of Nottingham (2006)
8. Schreiber, G., Wielinga, B., de Hoog, R., Akkermans, H., Van de Velde, W.: Common-KADS: A Comprehensive Methodology for KBS Development. IEEE Expert, 28–37 (December 1994)
9. Delamer, I.M.: Event-based Middleware for Reconfigurable Manufacturing Systems: A Semantic Web Services Approach. Doctoral dissertation No. 631, Tampere University of Texhnology, Finland (2006) ISBN 952-15-1672-0
10. The OWL Services Coalition: OWL-S: Semantic Markup for Web Services. DAML-S Home Page (2007), http://www.daml.org/services/owl-s/
11. Deutsches Institute für Normung e.V., Manufacturing Processes Joining, DIN 8593 Parts 0-8 (2003) (in German)
12. Vos, J.A.W.M.: Module and System Design in Flexibly Automated Assembly. Ph.D. Thesis, Delft University of Technology, The Netherlands (2001)
13. Van Holland, W., Bronsvoort, W.F.: Assembly Features in modeling and planning. Robotics and Computer Integrated Manufacturing 16, 277–294 (2000)
14. Martinez Lastra, J.L.: Reference Mechatronic Architecture for Actor-based Assembly Systems. Doctoral dissertation No. 484, Tampere University of Technology, Finland (2004) ISBN 952-15-1210-5

Determining the Failure Level for Risk Analysis in an e-Commerce Interaction

Omar Hussain, Elizabeth Chang, Farookh Hussain, and Tharam S. Dillon

Digital Ecosystems and Business Intelligence Institute,
Curtin University of Technology, Perth, Australia
O.Hussain@cbs.curtin.edu.au

Abstract. Before initiating a financial e-commerce interaction over the World Wide Web, the initiating agent would like to analyze the possible Risk in interacting with an agent, to ascertain the level to which it will not achieve its desired outcomes in the interaction. By analyzing the possible risk, the initiating agent can make an informed decision of its future course of action with that agent. To determine the possible risk in an interaction, the initiating agent has to determine the probability of failure and the possible consequences of failure to its resources involved in the interaction. In this chapter as a step towards risk analysis, we propose a methodology by which the initiating agent can determine beforehand the probability of failure in interacting with an agent, to achieve its desired outcomes.

Keywords: Risk assessing agent, Risk assessed agent, FailureLevel and Failure scale.

1   Introduction

The development of the internet has provided its users with numerous mechanisms for conducting or facilitating e-commerce interactions. It has also provided its users with various functionalities which will facilitate the way e-commerce interactions are carried out. But along with the provision of the increased functionalities for facilitating e-commerce interactions, also comes the fear of loss or the fear of not achieving what is desired in an interaction. This fear of loss or not achieving what is desired is termed as 'Risk' in the interaction. The terms 'risk assessing agent' and 'risk assessed agent' defines the two agents participating in an interaction. The former refers to the one initiating the interaction, while the latter refers to the agent accepting the request. In other words, this is the agent with whom the risk assessing agent interacts with to achieve its desired outcomes. The significance of the risk assessing agent to analyze the possible risk before initiating an interaction with a risk assessed agent is substantial. The risk assessing agent, by analyzing the possible risk beforehand, could gain an idea of whether it will achieve its desired outcomes from the interaction or not. Based on this, it can safeguard its resources. Risk plays a central role in deciding whether to proceed with a transaction or not. It can broadly be defined as an attribute of decision making that reflects the variance of the possible outcomes of the interaction.

T.S. Dillon et al. (Eds.): Advances in Web Semantics I, LNCS 4891, pp. 290–323, 2008.

Risk & Trust complement what the risk assessing agent needs in order to make an informed decision of its future course of action with a risk assessed agent. But there is still confusion in the relationship between them. As Mayer et al [1] suggest 'it is unclear whether Risk is an antecedent or an outcome of Trust'. Different arguments can be given to this. It can be said that in an interaction risk creates an opportunity for trust, which leads to risk taking. In this case risk is an antecedent to trust. But it can also be said that when the interaction is done based on the level of trust, then there is a low amount of risk in it. In this case risk is an outcome of trust. Risk can also provide a moderating relationship between trust and the behaviour of the agent in an interaction. For example, the effect of trust on the behaviour is different when the level of risk is low and different when the risk is high. Similarly risk can have a mediating relationship on trust. For example, the existence of trust reduces the perception of risk which in turn improves the behaviour in the interaction and willingness to engage in the interaction. But it is important to understand that, although risk and trust are two terms that complement each other while making an informed decision, they express different concepts which cannot be replaced with each other. Further it is important to comprehend the difference between each concept while analyzing them. Risk analysis involves the risk assessing agent to determine beforehand the probability of failure and the subsequent possible consequences of failure to its resources in interacting with a risk assessed agent. On the other hand, trust analysis measures the belief that the risk assessing agent has in a risk assessed agent in attaining its desired outcomes, if it interacts with it. This analysis does not take into account the resources that the risk assessing agent is going to invest in the interaction. A lot of work has been done in the literature to determine and evaluate the trust in an interaction [6-14].

Risk analysis is important in the study of behaviour in e-commerce, because there is a whole body of literature based in rational economics that argues that the decision to buy is based on the risk-adjusted cost-benefit analysis [2]. Thus, it commands a central role in any discussion of e-commerce that is related to an interaction. The need to distinguish between the likelihood and magnitude of risk is important as they represent different concepts. Magnitude shows the severity of the level of risk, whereas the likelihood shows the probability of its occurrence. For example, the likelihood of selling an item on the web decreases as the cost of the product increases and vice versa. The likelihood of a negative outcome might be the same in both interactions, but the magnitude of loss will be greater in the higher cost interaction. Hence these two characteristics must be considered by the risk assessing agent while analyzing the possible risk in interacting with a risk assessed agent. Previous methods in the literature analyze risk by just considering the probability of failure of the interaction. However, in our approach apart from considering the probability of failure of the interaction, we also consider the possible consequences of failure while ascertaining the possible risk in an interaction. It should be noted that this is the first attempt in the literature to model and analyze risk by using the two aforesaid constituents in e-commerce interactions.

In this chapter, we propose a methodology to determine semantically one aspect of risk evaluation, namely determining the probability of failure of the interaction. We propose to determine the probability of failure in the interaction according to the magnitude or severity of failure, and the likelihood of its occurrence. The methodology is explained in the next sections.

2 Defining the Failure Scale

The risk assessing agent can determine the probability of failure in interacting with a risk assessed agent, by ascertaining its in-capability to complete the interaction according to the context and criteria of its future interaction with it. Context of the interaction defines the purpose or scenario for which the interaction is to be carried out [3], or it is a broad representation of the set of all coherently related functionalities, which the risk assessing agent is looking to achieve, or desires to achieve while interacting with a risk assessed agent. Subsequently in a context, there might be a number of different related functionalities which comes under it, and if a risk assessing agent wants to interact with a risk assessed agent in a particular context, then it is highly possible that it might want to achieve only certain functionalities, in the particular context and not all the available functionalities in it. So we term those desired functionalities that the risk assessing agent wants to achieve while interacting with a risk assessed agent in a particular context, as the 'assessment criteria' or 'criteria' or 'desired outcomes'. In other terms 'assessment criteria' represents the certain desired functionalities that the risk assessing agent wants to achieve specifically while interacting with a risk assessed agent, in the particular context. Hence it is logical to say that the risk assessing agent when ascertaining the possible risk in interacting with a risk assessed agent in a context, should determine it according to the specific criteria of its future interaction with it, which comes under that particular context.

We assume that before initiating the interaction, the risk assessing agent communicates with the risk assessed agent about the context, criteria or the desired outcomes that it wants to achieve in its interaction with it, and decide on the quantitatively expressed activities in the expected or mutually agreed behaviour [3]. These set of quantitatively expressed activities are termed as the 'expectations' of the risk assessing agent, which the risk assessed agent is expected to adhere to. So we propose that while determining the probability of failure in an interaction, the risk assessing agent should ascertain it according to the 'expectations' of its future interaction with a risk assessed agent.

In an interaction there might be various degrees of failure according to their severity. Subsequently, it would be more expressive and understandable if the levels of failure are expressed according to their severity, rather than being expressed by using just two superlatives or extremes, such as "High" or "Low". Hence, before determining the probability of failure in an interaction, it is first necessary to ascertain the different possible levels of failure possible in an interaction according to their severity, so that while determining the probability of failure of an interaction the risk assessing agent can determine the severity of failure and the probability of occurrence of that failure in interacting with a risk assessed agent according to its expectations for a given period of time.

To represent semantically the different levels of failure possible in an interaction according to their severity, we propose a 'Failure scale'. The Failure scale represents seven different varying degrees of failure according to their severity which could be possible in an interaction, while interacting with a risk assessed agent. We term each degree of failure on the Failure scale, which corresponds to a range of severity of failure as 'FailureLevel' (FL). We propose that the risk assessing agent while determining the probability of failure according to its severity and probability of occurrence, in interacting with a risk assessed agent, ascertains its FailureLevel on the Failure scale. FailureLevel quantifies the possible level of failure according to its severity on the failure scale, in interacting with the risk assessed agent. The risk assessing agent determines the FailureLevel in interacting with a risk assessed agent by ascertaining its in-capability to complete the interaction according to its expectations.

Semantics of Failure Level	Probability of Failure	FailureLevel
Unknown	-	-1
Total Failure	91-100 %	0
Extremely High	71-90 %	1
Largely High	51-70 %	2
High	26-50 %	3
Significantly Low	11-25 %	4
Extremely Low	0-10 %	5

Fig. 1. The Failure scale

To represent the varying degrees of failure according to their severity, we make use of seven different FailureLevel on the failure scale. The failure scale as shown in Figure 1 represents 7 different varying levels of failure according to their severity, which could be possible in an interaction. The failure scale is utilized by the risk assessing agent when it has to determine beforehand either direct interaction based probability of failure or reputation based probability of failure of an interaction. Each level on the failure scale represents a different degree or the magnitude of failure. The domain of the failure scale ranges from [-1, 5]. The domain on the failure scale is defined as the possible set of values from which a FailureLevel is assigned to the risk assessed agent, according to the severity of failure present in interacting with it. The reason for us to choose this domain for representing the FailureLevel of the risk assessed agent is that it is expressive, and the semantics of the values are not lost; as compared to the approach proposed by Wang and Lin [13]. The authors in that approach represent the possible risk in an interaction within a domain of [0, 1]. This domain for representation is not much expressive as either:

1. Any value which comes in between gets rounded off to its nearest major value. By doing so, the semantics and severity which the actual value represents is either lost or gets compromised, or;

2. If rounding off is not used then there might be number of values between this range, which gets difficult to interpret them semantically.

So in our method we use a domain which is more expressive and which can represent different levels of failure according to their severity, thus alleviating the above mentioned disadvantages. In our domain even when rounding is used, the representation of the severity of the level of failure does not get effected, as it gets rounded off to its nearest value which is of the same level of severity. Hence the features of the domain of the failure scale are:

- One level is used to represent the state of ignorance in the probability of failure (Level -1).
- Two levels to represent the high probability of failure in an interaction (FailureLevel 0 and 1). Out of those two levels, one represents the greater level of high probability of failure and the other represents the lesser level of high probability of failure in an interaction.
- Two levels to represent the medium probability of failure in an interaction (Level 2 and 3). From those levels, one represents the higher level of medium probability of failure and the other level represents the lower level of medium probability of failure in the interaction.
- Two levels to represent low probability of failure in an interaction (Level 4 and 5). One level represents the higher level of low probability of failure and the other level represents the lower level of low probability of failure in the interaction.

Hence the domain that we propose for the Failure scale ranges from [-1, 5], with -1 representing the level of failure as 'Unknown' and the levels from 0 to 5 representing decreasing severity of failure. In order to express each level of failure on the Failure scale semantically we have defined the semantics or meanings associated with each FailureLevel. We explain them below:

2.1 Defining the Semantics of the Failure Scale

- **Unknown**

The first level of the failure scale is termed as *Unknown Failure* and its corresponding FailureLevel is -1. This level suggests that the level of failure in interacting with the risk assessed agent is unknown.

Semantics: This level can only be assigned by the recommending agent to the risk assessed agent if it does not have any past interaction history with it, in the context and criteria in which it is communicating its recommendation. Hence we propose that, the recommending agent instead of recommending any random FailureLevel in the range of (0, 5) on the Failure scale, recommends the level -1 to the risk assessing agent soliciting for recommendations. An important point to note is that all new agents in a network begin with this value, and hence a FailureLevel of -1 is assigned to the risk assessed agent, when there are no precedents that can help to determine its FailureLevel.

- **Total Failure**

The second level of the failure scale is defined as *Total Failure* and its corresponding FailureLevel value is 0. A FailureLevel value of 0 suggests that the probability of failure in interacting with the risk assessed agent is between 91-100 %.

Semantics: This level on the failure scale suggests that at a given point of time and in the given criteria the risk assessed agent is totally or completely unreliable to complete the desired outcomes of the risk assessing agent. In other terms it will not complete the interaction according to the expectations at all and acts fraudulently in the interaction, thus resulting in total failure for the risk assessing agent in achieving its desired outcomes. The FailureLevel of 0 expresses the highest level of failure possible in an interaction.

- **Extremely High**

Extremely High is the third level on the failure scale with the corresponding FailureLevel value of 1. This level denotes that there is 71-90 % probability of failure in interacting with the risk assessed agent.

Semantics: This level on the failure scale depicts that at a given point of time and in the given criteria the risk assessed agent is unreliable most of the times to commit to the expectations of the risk assessing agent. In other terms it will deviate from the desired criteria most of the times, hence resulting in extremely high level of failure in the interaction accordingly.

- **Largely High**

The fourth level of the failure scale is termed as *Largely High* level of failure. The corresponding FailureLevel value of this level is 2. This level depicts that there is a 51-70 % probability of failure in interacting with the risk assessed agent.

Semantics: A FailureLevel of 2 on the failure scale indicates that there is significant high level of failure in the interaction, as the risk assessed agent at that given point of time will not commit to a greater extent to its expectations.

- **High**

The fifth level on the failure scale is termed as *High* level of failure and it is shown by a FailureLevel value of 3. This level outlines that there is 26-50 % probability of failure in the interaction.

Semantics: A FailureLevel value of 3 on the failure scale assigned to a risk assessed agent suggests that at that particular point of time, the risk assessed agent is unable to complete the interaction to a large extent according to its expectations, hence resulting in high level of failure in the interaction.

- **Significantly Low**

The sixth level on the failure scale is defined as *Significantly Low* level of failure with a corresponding FailureLevel value of 4. This level depicts that there is 11-25 % probability of failure in the interaction.

Semantics: This level on the failure scale suggest that at a given point of time the risk assessed agent can complete MOST but not ALL of the criterions of its expectations. A FailureLevel of 4 on the failure scale indicates that the risk assessed agent assigned with this value can be relied on to a greater extent in that time, to commit to the expectations of the interaction, thus resulting in significantly low failure level in the interaction.

- **Extremely Low**

Extremely Low is the seventh and the last level of the failure scale represented by the FailureLevel value of 5. This level shows that there is 0-10 % probability of failure in the interaction.

Semantics: This level on the failure scale implies that at a given point of time, the risk assessed agent can fully be relied upon to complete the interaction according to its expectations, hence minimizing the probability of failure in an interaction. The probability of failure in interacting with the risk assessed agent, if any will be minimal. A FailureLevel of 5 expresses the lowest level of failure possible in an interaction.

3   Determining the FailureLevel of an Interaction

As mentioned earlier, for risk analysis the risk assessing agent has to determine beforehand the FailureLevel and the possible consequences of failure in interacting with a risk assessed agent. The risk assessing agent can determine the FailureLevel in interacting with a risk assessed agent beforehand, by analyzing its in-capability to complete the interaction according to its expectations. The possible interaction of the risk assessing agent with the risk assessed agent is in the future state of time. Hence, for risk analysis, the risk assessing agent has to determine the FailureLevel in interacting with the risk assessed agent in that future state of time. In order to achieve that, we propose that the risk assessing agent analyze and determines the FailureLevel in interacting with a risk assessed agent in two stages. They are:

1.   Pre-interaction start time phase
2.   Post-interaction start time phase

Pre-Interaction start time phase refers to the period of time before the risk assessing agent starts its interaction with the risk assessed agent, whereas Post-Interaction start time phase is that period of time, after the risk assessing agent starts and interacts with the risk assessed agent. For risk analysis the risk assessing agent has to determine the FailureLevel in interacting with a risk assessed agent in this period of time, i.e. in the post-interaction start time phase. However, as this time phase is in the future state of time, the risk assessing agent can only determine it by using some prediction methods. So we propose that the risk assessing agent should first ascertain the FailureLevel of the risk assessed agent according to the specific context and criteria as that of its future interaction, in the pre-interaction start time phase. Based on those achieved levels, the risk assessing agent can determine its FailureLevel, in the post-interaction start time phase. The determined FailureLevel of the risk assessed agent in the post-interaction time phase depicts the probability of failure in interacting with it, in that time phase, according to the context and criteria of the risk assessing agent's future interaction with it.

3.1 Time Based FailureLevel Analysis

We define the perceived risk in the domain of financial e-commerce transaction 'as the likelihood that the risk assessed agent will not act as expected by the risk assessing agent resulting in the failure of the interaction and loss of resources involved in it' [4]. This 'likelihood' varies throughout the transaction depending on the behaviour of the risk assessed agent and, therefore, it is dynamic. As mentioned in the literature too, risk is dynamic and varies according to time. It is not possible for an agent to have the same impression of a risk assessed agent throughout, which it had at a particular time. Hence the risk assessing agent should take into account this dynamic nature of risk while doing risk analysis in its interaction with a risk assessed agent. In order to incorporate and consider this dynamic nature, we propose that the risk assessing agent should determine the FailureLevel in interacting with a risk assessed agent in regular intervals of time. By doing so, it ascertains the correct FailureLevel of the risk assessed agent, according to its incapability to complete criterions of its future interaction, in each particular interval of time, thus considering its dynamic nature while doing risk analysis. We will define some terms by which the total time can be divided into different separate intervals.

We quantify the level of failure on the failure scale in interacting with a risk assessed agent in a given context and at a given time 't' which can be either at the current, past or future time with the metric 'FailureLevel'. But for better understanding, we represent the FailureLevel of a risk assessed agent according to the time phase in which it is determined and hence corresponds to. For example, if the FailureLevel for a risk assessed agent is determined in the pre-interaction start time phase, then we represent it by the metric 'PFL' which stands for *'Previous FailureLevel'*. Similarly, if the FailureLevel for the risk assessed agent is determined in the post-interaction start time phase, then we represent it by 'FFL' which stands for *'Future FailureLevel'*. We define the total boundary of time which the risk assessing agent takes into consideration to determine the FailureLevel (previous or future) of a risk assessed peer as the *time space*. But, as mentioned earlier, risk varies according to time and if the time space is of a long duration, then the FailureLevel of the risk assessed agent might not be the same throughout. Hence we propose that the risk assessing agent divides the time space into different non-overlapping parts and it assess the FailureLevel of the risk assessed agent in each of those parts, according to its incapability to complete the criterions of its future interaction in that time slot, to reflect it correctly while doing risk analysis. These different non-overlapping parts are called as *time slots*. The time at which the risk assessing agent or any other agent giving recommendation deals with the risk assessed agent in the time space is called as *time spot*. The risk assessing agent should first decide about the total time space over which it is going to analyze the FailureLevel of a risk assessed agent. Within the time space, the risk assessing agent should determine the duration of each time slot. Once it knows the duration of each time slot, it can determine the number of time slots in the given time space, and subsequently analyze the FailureLevel of the risk assessed agent in each time slot, may it be either in past or future.

For explanation sake, let us suppose that a risk assessing agent wants to interact with a risk assessed agent for a period of 10 days from 01/02/2007 till 10/02/2007. This is the post-interaction start time phase. Before initiating the interaction, the risk assessing agent wants to determine the probability of failure of the interaction as a first step towards risk analysis. To achieve that, the risk assessing agent wants to determine the FailureLevel of the risk assessed agent according to the criteria of its future interaction with it, from a period of 30 days prior to starting an interaction with it, i.e. from 02/01/2007 till 31/01/2007. This is the pre-interaction start time phase. Hence, the total period of time which the risk assessing agent takes into consideration to determine the FailureLevel (PFL and FFL) of the risk assessed agent is of 40 days. This time space is a combination of pre and post interaction start time phase. Further, the risk assessing agent wants to analyze the FailureLevel of the risk assessed agent in a time interval basis of 5 days. The total time space is of 40 days and each time slot is of 5 days. The number of time slots in this time space will be 8 as shown in Figure 2.

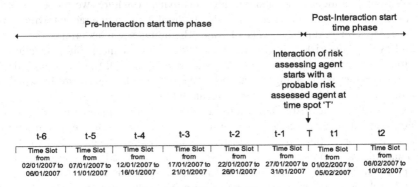

Fig. 2. Showing the division of the time space

Hence the risk assessing agent by determining the FailureLevel of the risk assessed agent in different time slots within the time space of its interaction is considering its accurate dynamic level of failure, according to its in-capability to complete the criterions in each of those time slots, thus reflecting it while doing risk analysis. The process for the risk assessing agent to ascertain the FailureLevel of the risk assessed agent in a time slot of its time space varies according to the time phase it comes in. We will briefly discuss the process by which the risk assessing agent can ascertain the FailureLevel of the risk assessed agent according to the expectations of its interaction with it, in each time slot of its time space depending upon the time phase it is in.

Scenario 1: The risk assessing agent determining the FailureLevel of the risk assessed agent in a time slot before the time spot of its interaction i.e. in the pre-interaction start time phase.

The risk assessed agent can determine the FailureLevel (PFL) of the risk assessed agent according to the expectations of its future interaction with it, in a time slot which is in the pre-interaction state time phase by considering either:

- its previous interaction history with it (if any) in the expectations of its future interaction, (direct past interaction-based probability of failure); or
- in the case of ignorance, then soliciting for recommendations from other agents and assimilating them according to the expectations of its future interaction, (reputation-based probability of failure).

A detailed explanation of how to determine the FailureLevel of the risk assessed agent in a time slot by using either direct past-interaction history or by soliciting recommendation from other agents is given in Section 4.

Scenario 2: The risk assessing agent determining the FailureLevel of the risk assessed agent in a time slot after the time spot of its interaction i.e. in the post-interaction start time phase.

Case 2.1: If the time spot and the duration of the interaction (post-interaction start time phase) is limited to the time slot in which the risk assessing agent is at present as shown in Figure 3, then it can determine the FailureLevel (FFL) of the risk assessed agent for the period of time in the post-interaction phase, by either considering its past-interaction history with the risk assessed agent (if any), or by soliciting recommendations from other agents.

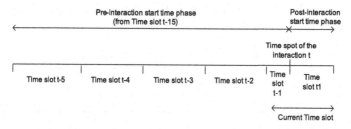

Fig. 3. The time spot and post-interaction phase of the interaction limited to the current period of time

The risk assessing agent can consider its past interaction history with the risk assessed agent only if it is in the same time slot, with the same expectations which had the same significance attached to each assessment criterion as for its future interaction with it. If this is the case, then the risk assessing agent can utilize the FailureLevel (AFL) that it had determined for the risk assessed agent in its past interaction as its FailureLevel (FFL) in the current interaction. This is based on the assumption made by Chang et al. [3] who state that the behavior of the risk assessed agent remains the same in a time slot, and subsequently the risk assessing agent can utilize the FailureLevel of the risk assessed agent from its past interaction if it is in the same expectations, significance and time slot of its future interaction as its FailureLevel (FFL) in that time slot. However, if the risk assessing agent does not have a past interaction history with the risk assessed agent in the expectations and in the time slot of its future interaction, or it has a past interaction history in the partial expectations in the time slot of its future interaction, then in such cases the risk assessing agent can solicit recommendations about the risk assessed agent from other agents for that particular time slot in the assessment criterion or criteria of its interest from its expectations, in which it does not have a past interaction history with it, and then assimilate them along with its past-interaction history (if any in the partial expectations) to determine the FailureLevel (FFL) of the risk assessing agent in the post-interaction start time phase. A detailed explanation of how to determine the FailureLevel of the risk assessed agent in a time slot by either using direct past-interaction history and/or by soliciting recommendation from other agents is given in Section 4.

It may be the case that the risk assessing agent may neither have any past interaction history nor obtains any recommendations from other agents for the risk assessed agent against all the assessment criteria of its expectations in the current time slot of its interaction. In such cases, the risk assessing agent should determine the FailureLevel (FFL) of the risk assessed agent in the current time slot by using the methodology proposed in case 2.2.

Case 2.2: If the time spot or duration of the interaction (post-interaction start time phase) begins or extends to a future point in time from the current time slot in which the risk assessing agent is at present as shown in Figure 4, then it should utilize the determined FailureLevel of the risk assessed agent from the beginning of the time space till the current time slot to predict and determine the future FailureLevel (FFL) of the risk assessed agent in each of the post-interaction start time slots. A detailed explanation of how to determine the FailureLevel of the risk assessed agent in future time slots is given in Section 5.

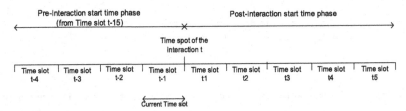

Fig. 4. The time spot and the post-interaction start time phase of the interaction extending to a future point in time

A point to be considered by the risk assessing agent while utilizing the FailureLevel of the risk assessed agent in the previous time slots to determine its FailureLevel during the time of its interaction, is that it should give more importance to the fresh status of the risk assessed agent (represented by its FailureLevel), which is in the time slots near or closest to the time spot of its interaction with it as compared with those which are in the less recent time slots from the time spot of its interaction. This takes into consideration the fact mentioned by Chang et al. [3] that 'recency is important' when utilizing the past values of an agent in order to determine its value/s in the future. They state that it is important for the risk assessing agent to weigh those values of the risk assessed agent obtained in the recent interactions or time slots more heavily among the values that it considers for it in the previous time slots, so as to avoid modeling its behavior in the future that may no longer be relevant according to the expectations of its future interaction. Hence, the prediction method should weigh the recent FailureLevel values of the risk assessed agent more heavily as compared to its FailureLevel values in the far recent time slots, progressively reducing the effect of the older FailureLevel values in order to take into consideration its fresh status while determining its FailureLevel value/s over a future period of time. We represent the weight to be given to the status of the risk assessed agent in a time slot before the time spot of the interaction by the variable 'w'. The weight (w) to be given to each time slot of the pre-interaction start time phase is represented in Figure 5 and is determined by:

$$
w = \begin{cases} 1 & \text{if } m \leq \Delta t \\ e^{\frac{-((n+\Delta t)-m)}{N}} & \text{if } m > \Delta t \end{cases} \tag{1}
$$

where, 'w' is the weight or the time delaying factor to be given to the status of the risk assessed agent,

'n' represents the current time slot,

'm' represents the time slot for which the weight of adjustment is determined,

'Δt' represents the time slots from the time spot of the interaction in which the risk assessing agent will give more importance to the fresh status of the risk assessed agent,

'N' is the term which characterizes the rate of decay.

We consider that the risk assessing agent among the 15 time slots of the pre-interaction start time phase, gives more importance to the FailureLevel of the risk assessed agent in the five time slots previous to the time spot of its interaction as compared to the other time slots, in order to consider the fresh status of the risk assessed agent while utilizing it to ascertain its FailureLevel in the future period of time. For the importance to be given to the status or FailureLevel of the risk assessed agent in the other time slots of the pre-interaction start time phase, the weight to be adjusted to it is a progressively declining value determined by using equation 1.

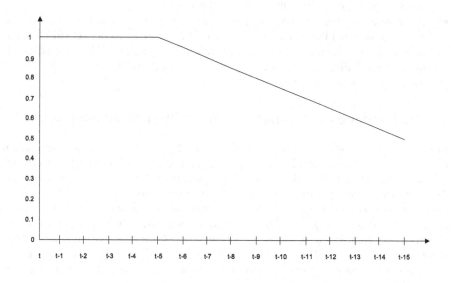

X-Axis represents the time slots in the pre-interaction start time phase
Y-Axis represents the weight to be given to each time slot

Fig. 5. The weight given to each time slot of the pre-interaction start time phase

To summarize the proposed methodology for the division of time in order to consider the dynamic nature of perceived risk while ascertaining the level of failure in an interaction:

- The risk assessing agent determines the 'time space' of its interaction over which it wants to analyze the FailureLevel of the risk assessed agent while ascertaining the performance risk in interacting with it.

- The time space is divided into different 'time slots' and then broadly divided into two phases, the pre-interaction start time phase and post-interaction start time phase according to the 'time spot' of the interaction.
- The risk assessing agent ascertains the FailureLevel of the risk assessed agent in each time slot of the pre-interaction time phase by either considering its past-interaction history with it or by soliciting recommendations from other agents.
- If the time spot and the post-interaction start time phase is limited to the current time slot at which the risk assessing agent is at present, then it determines the FailureLevel (FFL) of the risk assessed agent in the post-interaction start time phase by either considering its past-interaction history with it (if any) in the expectations and in the time slot of the interaction, or by soliciting recommendations from other agents, or by a combination of both.
- In the case of the risk assessing agent not being able to determine the FailureLevel of the risk assessed agent for each assessment criteria of its expectations in the current time slot, by using either its own past-interaction history or the recommendations from other agents, then it utilizes the approach mentioned in the next point to determine the FailureLevel (FFL) of the risk assessed agent in the post-interaction start time slot.
- If the time spot and the post-interaction start time phase extend to a point in time in the future, then the risk assessing agent utilizes the FailureLevel (PFL) that it determined for the risk assessed agent from the beginning of the time space till the preceding time slot, to determine its FailureLevel (FFL) in each time slot of the post-interaction time phase.

4   Determining the FailureLevel in the Pre-interaction phase

In this section, we will propose the methodology by which the risk assessing agent can ascertain the FailureLevel of the risk assessed agent according to the expectations of its future interaction with it, in the pre-interaction start phase time slots. As discussed earlier, the pre-interaction start time phase refers to that period of time in which the risk assessing agent considers the previous impression of the risk assessed agent, before determining its FailureLevel in the post-interaction start time phase of its interaction. Subsequently, this period of time ranges from the beginning of the time space to the time spot of the interaction. There are two methods by which the risk assessing agent can determine the FailureLevel of a risk assessed agent in the pre-interaction start time phase. They are:

a) Direct past Interaction-based Probability of Failure: by considering its past interaction history with the risk assessed agent in the expectations of its future interaction with it; and

b) Reputation-based Probability of Failure: by soliciting recommendations from other agents and then assimilating them to determine the inability of the risk assessed agent to complete the interaction according to the expectations of its future interaction with it.

In the next sub-sections we will explain in detail each method with which the risk assessing agent can determine the FailureLevel in interacting with the risk assessed agent in the each time slot of the pre-interaction start time phase.

4.1 Determining Direct Past Interaction-Based Probability of Failure in an Interaction

The direct past interaction-based probability of failure method refers to the risk assessing agent determining the probability of failure or FailureLevel in interacting with the risk assessed agent in a time slot, based on its past interaction history with it in that particular time slot. Further, the past interaction of the risk assessing agent with the risk assessed agent should be strictly according to the expectations and the same significance attached to each assessment criterion, as in its future interaction with it. This is necessary in order to take into consideration the property of dynamic nature of risk which varies according to the variation of the context and assessment criteria. Hence, if the risk assessing agent does have a past interaction history with the risk assessed agent in a pre-interaction start time slot, in the same context but in partial fulfillment of the assessment criteria of its expectations, then we propose that it cannot consider its past interaction history in order to determine the FailureLevel of the risk assessed agent in the total assessment criteria of its expectations in that time slot, due to the assessment criteria slightly varying from its past interaction as compared to the expectations of its future interaction. In such case, we propose that the risk assessing agent should determine the FailureLevel of the risk assessed agent in that time slot by using a combination of its direct past interaction history in the same assessment criteria from its past interaction as its expectations and the reputation of the risk assessed agent in the varying assessment criteria, to determine the FailureLevel of the risk assessed agent in that time slot. Three scenarios arise when the risk assessing agent determines the FailureLevel of the risk assessed agent in a pre-interaction start time slot by considering its past interaction history with it in that time slot. They are:

Scenario 3: The assessment criteria of the risk assessing agent's previous interaction and their significance are the same as those of its expectations of its future interaction.

If the context, assessment criteria and their significance of the risk assessing agent's previous interaction with the risk assessed agent in a time slot of the pre-interaction start time phase are exactly to the same as the expectations of its future interaction with it, then we propose that the risk assessing agent can utilize its risk relationship that it has formed with the risk assessed agent in that time slot, and consider the FailureLevel (AFL) that it had determined for the risk assessed agent in that interaction, as its FailureLevel (PFL) for that particular time slot. A detailed explanation of how the risk assessing agent ascertains the FailureLevel (AFL) of the risk assessed agent, after interacting with it is given in Hussain et al. [18].

In order to give more importance to the fresh status of the risk assessed agent which are in the time slots near or recent to the time spot of its interaction, the risk assessing agent should adjust the determined FailureLevel of the risk assessed agent in a pre-interaction start time slot 't-z' (PFL $_{Pt-z}$), according to the weight that it considers to give to that time slot depending on where it falls in the time space of its interaction. Hence the FailureLevel (PFL) of the risk assessed agent 'P' in a pre-interaction start time slot 't-z' based on the risk assessing agent's past interaction history with it in that time slot is represented by:

$$PFL_{Pt-z} = ROUND\ (w * AFL_{Pt-z}) \tag{2}$$

where, 'P' represents the risk assessed agent,

't' represents the time spot of the interaction,

'z' represents the number of time slots prior to the time spot of the risk assessing agent's interaction with the risk assessed agent,

'w' is the weight applied to the FailureLevel (AFL) of the risk assessed agent depending upon the time slot 't-z'.

The resultant value from equation 2 is rounded off to determine the crisp FailureLevel value for the risk assessed agent 'P' on the Failure Scale in the time slot 't-z' (PFL_{Pt-z}).

Scenario 4: The criteria of the risk assessing agent's previous interaction vary partially from the expectations of its future interaction, or the assessment criteria of the risk assessing agent's previous interaction are the same as those of its expectations, but the significance of these assessment criteria vary from those of the expectations of its future interaction.

Case 1: If the context of the previous interaction of the risk assessing agent with the risk assessed agent in a time slot of the pre-interaction start time phase is the same, but the assessment criteria differ partially as compared to the expectations of its future interaction, then we propose that the risk assessing agent from its previous interaction should consider only those partial criteria which are similar to the assessment criteria in the expectations of its future interaction and utilize them to determine the trustworthiness of the risk assessed agent in those, while considering the rest of the assessment criteria of its expectations by the reputation-based method, and then combine them to determine the FailureLevel (PFL) of the risk assessed agent in that time slot.

Case 2: If the assessment criteria of the risk assessing agent's previous interaction with the risk assessed agent in a time slot of the pre-interaction start time phase are identical to the expectations of its future interaction with it, but the significance of the criteria in its previous interaction vary from those of the assessment criteria of the expectations of its future interaction, then we propose that the risk assessing agent in such a case consider the criteria from its previous interaction and utilize them to determine the trustworthiness of the risk assessed agent in these.

In both the cases, the risk assessing agent cannot utilize the FailureLevel (AFL) that it had determined for the risk assessed agent in its previous interaction in a time slot of the pre-interaction start time phase as the FailureLevel (PFL) of the risk assessed agent in the pre-interaction start time slot of its current interaction, as was done in the previous scenario. This is because in the first case, the FailureLevel (AFL) of the risk assessed agent determined in the past interaction is not exactly according to the expectations of its future interaction; and in the second case, the FailureLevel (AFL) of the risk assessed agent determined in the past interaction is not according to the significance of the expectations of its future interaction. Therefore in such cases, we propose that the risk assessing agent take into consideration the relative 'assessment criteria' from its past interaction which are in the expectations of its future interaction, along with their corresponding 'Commitment Level' value that it had determined in its past interaction, and utilize them to determine the risk assessed agent's trustworthiness in those assessment criteria according to the weight to be given to its status in that time slot. 'Commitment Level' is a value which the risk assessing agent ascertains for each assessment criterion of its interaction with the risk assessed agent, when it determines its Actual FailureLevel (AFL) in the interaction. The Commitment Level value shows whether or not a particular assessment criterion was fulfilled by the risk assessed agent according to the expectations of the interaction, and is represented by a value of either 1 or 0. Further explanation of the way to determine the commitment level value for each assessment criterion of the interaction is given in the sub-section 4.2.1. Hence, the risk assessing agent by considering an assessment criterion along with its commitment level from its past interaction, which are in the expectations of its future interaction, should determine the trustworthiness of the risk assessed agent in those assessment criteria in a pre-interaction start time slot, according to the weight to be given to the status of the risk assessed agent in that time slot.

The risk assessing agent can determine the trustworthiness of the risk assessed agent 'P' in an assessment criterion (C_n) by considering its past interaction history with it a time slot 't-z' of the pre-interaction start time phase by:

$$T_{PCn\,t-z} = (w * \text{Commitment Level}_{Cn}) \tag{3}$$

where, 'P' represents the risk assessed agent,

'Cn' represents the assessment criterion, in which the trustworthiness of the risk assessed agent 'P' is being determined,

'Commitment Level $_{Cn}$' represents the level of commitment of the risk assessed agent in assessment criterion 'Cn',

'w' is the weight applied to the commitment level of the risk assessed agent to consider its status in the time slot 't-z'.

If there is more than one assessment criteria in the risk assessing agent's past interaction history with the risk assessed agent which matches the expectations of its future interaction with it, then the risk assessing agent by using equation 3 should determine the trustworthiness of the risk assessed agent for each of those assessment criteria. To consider the other assessment criteria of its expectations in which the risk assessing agent does not have any past interaction history with the risk assessed agent, we propose that it solicit recommendations from other agents and utilize them to determine the reputation of the risk assessed agent in those. It should then utilize the trustworthiness or reputation value of the risk assessed agent determined in each assessment criterion of its expectations to ascertain its FailureLevel for each of them. It should then combine the determined FailureLevel of each assessment criteria according to its significance in order to ascertain the FailureLevel (PFL) of the risk assessed agent in that time slot. The methodology for the risk assessing agent to ascertain the FailureLevel of the risk assessed agent in a time slot by utilizing its trustworthiness (determined by using its past interaction history) and/or its reputation (determined from the recommendations from other agents) in the assessment criteria of its expectations is mentioned in sub-section 4.2.4.

Scenario 5: The assessment criteria of the risk assessing agent's previous interaction are completely different from the expectations of its future interaction.

If the context of the risk assessing agent's previous interaction with the risk assessed agent in a time slot of the pre-interaction start time phase is the same, but the assessment criteria are completely different as compared to the expectations of the future interaction, then the risk assessing agent cannot utilize its past interaction history in determining the FailureLevel (PFL) of the risk assessed agent of that time slot. In such cases, we propose that the risk assessing agent determine the FailureLevel of the risk assessed agent by utilizing the reputation-based probability of failure method.

4.2 Determining Reputation-Based Probability of Failure in an Interaction

The reputation-based probability of failure method is utilized by the risk assessing agent in order to determine the probability of failure or FailureLevel in interacting with the risk assessed agent in a time slot of the pre-interaction start time phase, if it does not have any past interactions with it in that time slot, either in all or in the partial expectations of its future interaction with it. In such cases, we propose that the risk assessing agent rely on other agents by soliciting recommendations from those who have interacted in that time slot with the risk assessed agent in the assessment criteria of interest, and then utilize their recommendations to determine the reputation and then the FailureLevel in interacting with the risk assessed agent for those assessment criteria. The risk assessing agent, in order to determine the reputation of the risk assessed agent in the expectations or in partial expectations, issues a reputation query to solicit recommendations from other agents by specifying the risk assessed agent, the particular assessment criterion or criteria and the time in which it wants the recommendations to be. The agents who have had a previous interaction history with the risk assessed agent in the same time and assessment criterion or criteria, reply with their recommendations. The agents who reply with the recommendations are termed the 'Recommending Agents'. We consider that whenever an agent interacts with another agent, a risk relationship forms between them. This relationship is dependent on the time, context and assessment criteria of their interaction. We propose that when a risk assessing agent issues a reputation query soliciting recommendations for the risk assessed agent from other agents in a particular time and criteria related to a context, and if an agent has a previous interaction history with the risk assessed agent for those criteria and period of time for which its recommendation is being sought, then it communicates the risk relationship to the risk assessing agent that it had formed while interacting with the risk assessed agent in that time slot. Based on the risk relationships received from different agents, the risk assessing agent assimilates them and determines the reputation and then the FailureLevel of the risk assessed agent for the assessment criteria of interest for the particular time slot.

It is possible that the recommendations which the risk assessing agent receives for a risk assessed agent in a pre-interaction start time slot, might contain other criteria apart from the ones which are of interest to it in its interaction. Furthermore, it is possible that the risk assessing agent might receive more than one recommendation from different recommending agents for an assessment criterion of interest in a particular time slot. Subsequently, to utilize such recommendations, we propose that the risk assessing agent should classify all the recommendation that it receives from different recommending agents for its request according to the assessment criterion or criteria of its interest, and then utilize each of them in order to determine the reputation of the risk assessed agent in those assessment criterion or criteria. But it would be difficult for the risk assessing agent to comprehend and understand the risk relationship that it receives from each recommending agent and later assimilate them, if each agent when solicited gives its recommendation in its own format. So in order to alleviate this, we propose a standard format called the 'Risk Set' for the recommending agent to communicate its recommendation to the risk assessing agent. In the next sub-section we will propose the format for the risk set.

4.2.1 Defining the Format for Risk Set

Risk Set is defined as a standard format for the recommending agents to communicate their recommendations in an ordered way to the risk assessing agent. The risk assessing agent, by getting the recommendations from the recommending agents in an ordered way, can comprehend and classify them according to the criterion or criteria which are of interest to it in that time slot. The format of the Risk Set is:

{RA1, RA2, Context, AFL, (Assessment Criterion, Commitment level), Cost, Start time, End time}

where, *RA1* is the risk assessing agent in the interaction, which is also the recommending agent while giving recommendation.

RA2 is the risk assessed agent in the interaction.
Context represents the context of the interaction.

AFL represents the 'Actual FailureLevel' determined by the recommending agent after interacting with the risk assessed agent, by assessing the level of non-commitment in the risk assessed agent's actual behavior with respect to its expectations.

(Assessment Criterion, Commitment level) 'Assessment Criterion' represents the assessment criterion in the recommending agent's expectations of the interaction with the risk assessed agent. The combination of (Assessment Criterion, Commitment level) is represented for each assessment criterion in the expectations of the recommending agent's interaction with the risk assessed agent. This is the set of factors on which the recommending agent interacted with the risk assessed agent and later assigned it with the 'Actual FailureLevel' (AFL) in its interaction. These criteria are necessary to mention while giving recommendations so that a risk assessing agent who solicits recommendations knows the assessment criteria on which this particular risk assessed agent has been assigned the recommended FailureLevel (AFL). In this way, it can consider only those recommendations which are of interest to it according to the expectations of its future interaction. 'Commitment level' specifies whether or not the particular assessment criterion was fulfilled by the risk assessed agent according to the expectations of its interaction. A value of either 0 or 1 is assigned to it, based on the commitment of the risk assessed agent for that criterion. A value of 0 signifies that the assessment criterion was not fulfilled by the risk assessed agent

Table 1. The commitment level of each assessment criterion

Commitment Level	Semantics of the Value
0	The risk assessed agent did not commit to the assessment criterion as it was expected from it according to the expectations
1	The risk assessed agent committed to the assessment criterion exactly according to the expectations

according to the expectations, whereas a value of 1 signifies that the assessment criterion was fulfilled according to the expectations. Further explanation is given in Table 1.

Cost represents the total financial value of the recommending agent at stake in the interaction.

Start Time is the time at which the recommending agent started the interaction with the risk assessed agent.

End time is the time at which the interaction of the recommending agent ended with the risk assessed agent.

Once the risk assessing agent classifies all the recommendations that it receives from the recommending agents according to the criterion or criteria of interest to it in a particular time slot, it should then assimilate them in order to determine the reputation of the risk assessed agent according to those criterion or criteria in the particular pre-interaction start time slot. But before assimilating the recommendations from the recommending agents, the risk assessing agent should first classify them according to their credibility. We will discuss this in the next sub-section.

4.2.2 Credibility of the Recommendations

When the risk assessing agent issues a reputation query for a risk assessed agent, there is the possibility that some recommending agents will reply with recommendations which are incorrect. In order to omit and avoid such recommendations while determining the reputation and then the FailureLevel of the risk assessed agent in a pre-interaction start time slot, the risk assessing agent should first classify each recommendation of interest for a risk assessed agent according to its credibility, and then assimilate it accordingly. To achieve this, we adopt the methodology proposed by Chang et al. [3] of classifying the recommendations according to their credibility. In this methodology, the authors state that the risk assessing agent maintains the credibility value of all the recommending agents from which it took recommendations, which in turn denotes the correctness of the recommendations communicated by them to the risk assessing agent. We represent the credibility value of the recommending agents maintained by the risk assessing agent as the 'Recommending Agent's Credibility Value' (*RCV*). RCV of a recommending agent is context-based, and we consider that the risk assessing agent maintains the RCV for a recommending agent in each context for which it took its recommendation. This value is used to determine whether or not the particular recommending agent is credible while giving recommendations in the particular context.

An agent whose RCV is known to the risk assessing agent is termed as the 'Known' recommending agent, whereas an agent whose RCV is unknown to the risk assessing agent is termed as an 'Unknown' recommending agent. The known agents are further classified into two types, 'Trustworthy' agents and 'Untrustworthy' agents. Trustworthy agents are those agents whose RCV is within the specified range which is considered to be trustworthy by the risk assessing agent, whereas untrustworthy agents are those agents whose RCV is beyond the specified range which is considered as trustworthy. We consider that the credibility values of the recommending agents ranges from (-5, 5), and an agent whose RCV is within the range of (-1, 1) is considered as a trustworthy recommending agent by the risk assessing agent. Within that range, a value of 0 specifies that the recommendation communicated by the recommending agent for the risk assessed agent is exactly similar to what the risk assessing agent finds after its interaction with that agent. A positive value to the range of 1 specifies that the risk assessing agent finds that the recommending agent recommends a lesser value for the risk assessed agent, as compared to what it determines for the risk assessed agent after its interaction. A negative value to the range of -1 specifies that the risk assessing agent finds that the recommending agent recommends a higher value for the risk assessed agent, as compared to what it determines for the risk assessed agent after the interaction. The RCV of a recommending agent in a context is determined by the risk assessing agent based on its previous recommendation history with it, in that context. Further explanation of the way to determine the RCV of a recommending agent is given in Hussain et al. [17].

We consider that the risk assessing agent in the reputation-based probability of failure method to determine the FailureLevel of the risk assessed agent, considers only those recommendations which are from trustworthy and unknown recommending agents and omit the ones from untrustworthy recommending agents, in order to ascertain the correct reputation of the risk assessed agent. In other words, the risk assessing agent assimilates only those recommendations from agents whose credibility in communicating them in that context is trustworthy or unknown, and omits considering recommendations from those agents whose credibility is untrustworthy. Therefore to summarize, the risk assessing agent, while utilizing the recommendations of other agents to determine the reputation of the risk assessed agent in the assessment criteria of its expectations, should take into consideration:

- The credibility of the recommendations: The recommendations which the risk assessing agent should consider should be from trustworthy or unknown recommending agents, and not from untrustworthy recommending agents.
- The time slot of the recommendations: The time slot in which the risk assessing agent wants to determine the reputation of the risk assessed agent should match with the time of the recommendations that it considers.
- Expectations of its interaction: The recommendations considered by the risk assessing agent should be either in the exact or partial assessment criteria of its interest according to the expectations of its future interaction.

In the next section, we will propose a methodology by which the risk assessing agent can assimilate the recommendations after classifying them according to its credibility, time and criteria to ascertain the reputation of the risk assessed agent in the assessment criterion or criteria of its interest in a time slot of the pre-interaction start time phase.

4.2.3 Assimilating Recommendations for Ascertaining Reputation-Based FailureLevel of a Risk Assessed Agent

As mentioned earlier, it is possible that in a time slot the risk assessing agent might receive recommendations which contain other assessment criteria apart from those which are of interest to it in its present interaction. Further, it is possible that the risk assessing agent may receive more than one recommendation for an assessment criterion of interest in a time slot. Hence, in order to take into consideration all such types of recommendations, the risk assessing agent should determine the reputation of the risk assessed agent in each assessment criterion of interest from its expectations, by assimilating all the recommendations that it receives for the risk assessed agent for that particular criterion from the recommending agents. The risk assessing agent in such a case should consider the 'Commitment Level' value for the particular assessment criterion of interest, from all the recommendations which communicate in that criterion, and then adjust it according to the credibility of the recommendations (if it is from a trustworthy recommending agent) and the weight to be given to it according to the status of the risk assessed agent in that time slot, to ascertain the reputation of the risk assessed agent in that particular assessment criterion.

The reputation of a risk assessed agent 'P' in an assessment criterion 'Cn' (Rep $_{PCn}$) in a pre-interaction time slot 't-z' can be determined by assimilating the trustworthy and unknown recommendations that it receives from the recommending agents by using the following formulae:

$$\text{Rep }_{PCn\ t-z} = \quad (\alpha * (w * \frac{1}{K} (\sum_{i=1}^{K} RCV_i \oplus \text{Commitment Level } _{Cn}^{i}))) +$$

$$(\beta * (w * \frac{1}{J} (\sum_{o=1}^{J} \text{Commitment Level } _{Cn}^{o}))) \qquad (4)$$

where, 'RCV_i' is the credibility value of the trustworthy recommending agent 'i',

'Commitment level $_{Cn}$' is the level of commitment recommended by the recommending agent for assessment criterion 'Cn' for the risk assessed agent in the particular time slot 't-z',

'K' is the number of trustworthy recommendations that the risk assessing agent gets for the risk assessed agent in assessment criterion 'Cn' in time slot 't-z',

'J' is the number of unknown recommendations that the risk assessing agent gets for the risk assessed agent in assessment criterion 'Cn' in time slot 't-z',

'α and β' are the variables attached to the parts of the equation which will give more weight to the recommendation from the trustworthy known recommending agents as compared to those from the unknown recommending agents. In general $\alpha > \beta$ and $\alpha + \beta = 1$,

'w' is the weight applied to consider the status of the risk assessed agent in time slot 't-z'.

The reputation value of the risk assessed agent 'P' in an assessment criterion 'Cn' is determined in two parts as shown in equation 4. The first part of the equation calculates the reputation value of the risk assessed agent 'P' in the assessment criterion 'Cn' by taking the recommendations of the trustworthy recommending agents. The second part of the equation calculates the reputation value of the same risk assessed agent 'P' in the same assessment criterion 'Cn' by taking the recommendations of the unknown recommending agents. The recommendations from the untrustworthy recommending agents are left out and not considered. In order to give more importance to the recommendations from the trustworthy recommending agents as compared to ones from the unknown recommending agents, variables are attached to the two parts of the equation. These variables are represented by α and β respectively. It depends upon the risk assessing agent how much weight it wants to assign to each type of recommendation. Furthermore, each recommendation for the risk assessed agent in a time slot is adjusted according to the weight to be given to the status of the risk assessed agent in that time slot. The RCV of the trustworthy recommending agent is also considered while assimilating its recommendation. As shown in equation 4, the RCV of the trustworthy recommending agent is adjusted with the adjustment operator '$\oplus$' to its recommendation. This takes into consideration the accurate recommendation from the trustworthy recommending agent according to the credibility and accuracy by which it communicates its recommendations. The rules for the adjustment operator '$\oplus$' are:

$$a \oplus b = \begin{cases} a + b, & \text{if } 0 \leq (a + b) \leq 1 \\ 1, & \text{if } (a + b) > 1 \\ 0, & \text{if } (a + b) < 0 \end{cases}$$

It is possible that in a time slot 't-z', the risk assessing agent may not receive any recommendation for the risk assessed agent 'P' in an assessment criterion 'Cn' of its interest from its expectations, for which it does not have any past interaction history. In this case, we propose that the risk assessing agent should assume a value of '0' as the reputation of the risk assessed agent for that assessment criterion 'Cn' ($Rep_{PCn\ t\text{-}z}$) in that time slot. It is because the risk assessing agent assimilates the recommendations and determines the reputation of the risk assessed agent in an assessment criterion to ascertain its capability to complete that criterion. Hence, if there is no recommendation for the risk assessed agent in a time slot for a criterion, then in order to conduct a sensible risk analysis, we assume that the risk assessing agent considers that the risk assessed agent is incapable of completing the assessment criterion in that time slot. Hence, it should assign to it a value of '0' as its reputation for that assessment criterion.

The risk assessing agent should utilize equation 4 to determine the reputation of the risk assessed agent either in all or in partial assessment criteria of its expectations, in a pre-interaction start time slot for which it does not have any past interaction history. In the next section, we will propose an approach by which the risk assessing agent ascertains the FailureLevel of the risk assessed agent for each assessment criteria of its expectations, based on its determined trustworthiness in it (according to its past interaction history) or based on its determined reputation in it (according to the recommendations from other agents).

4.2.4 Ascertaining the FailureLevel (PFL) of the Risk Assessed Agent in a Pre-interaction Start Time Slot

Once the risk assessing agent ascertains the trustworthiness of the risk assessed agent in the partial assessment criteria of its expectations by using its past interaction history (discussed in scenario 4), and the reputation of the risk assessed agent by using recommendations from the other agents in the rest of the assessment criteria of its expectations (discussed in section 4.2.3), or the reputation of the risk assessed agent using the recommendations from other agents in all of the assessment criteria of its expectations, then it should combine them in order to determine the FailureLevel (PFL) of the risk assessed agent in the pre-interaction start time slot 't-z', according to the expectations of its future interaction. To achieve this, the risk assessing agent has to first ascertain the FailureLevel of the risk assessed agent for each assessment criterion of its expectations, from its determined trustworthiness or by its determined reputation.

The trustworthiness or the reputation of the risk assessed agent in against an assessment criterion shows its level of capability to meet the particular criterion. To determine the FailureLevel of the risk assessed agent for that criterion, the extent of its inability to complete the given assessment criterion has to be determined. To achieve this, we propose that the risk assessing agent should map the trustworthiness or the reputation of the risk assessed agent in each assessment criterion of its expectations in a pre-interaction start time slot 't-z', on the Failure Scale (FS). By doing so, the risk assessing agent knows the capability of the risk assessed agent to meet that assessment criterion on the Failure Scale, in the time slot 't-z'. It can then determine the probability of failure of the risk assessed agent in committing to that assessment criterion in that time slot according to its expectations, by ascertaining the difference between what it expects in that assessment criterion, and how far the risk assessed agent can fulfill it according to its trustworthiness or reputation for that criterion. The value achieved gives the probability of failure of that assessment criterion in that time slot. The FailureLevel of the assessment criterion in that time slot is then achieved by mapping the probability of failure of that assessment criterion to the Failure Scale.

As mentioned earlier, the levels on the Failure Scale between 0 and 5 represent varying degrees and magnitudes of failure. Hence, for ascertaining the FailureLevel of the risk assessed agent in an assessment criterion, its trustworthiness or reputation for that criterion should be mapped on the range of (0, 5) on the Failure Scale, as it is within these levels that its capability to complete the assessment criterion has to be ascertained on the Failure Scale. The trustworthiness or the reputation of the risk assessed agent in an assessment criterion can be represented on the Failure Scale (FS) by:

$$T_{PCn\,t-z\,FS} = ROUND\,(T_{PCn\,t-z} * 5) \qquad or,$$

$$Rep_{PCn\,t-z\,FS} = ROUND\,(Rep_{PCn\,t-z} * 5) \qquad (5)$$

where, '$T_{PCn\,t-z\,FS}$' represents the trustworthiness of the risk assessed agent in time slot 't-z' and in assessment criterion 'Cn' on the Failure Scale,

'$T_{PCn\,t-z}$' represents the trustworthiness of the risk assessed agent in assessment criterion 'Cn' and in time slot 't-z',

'$Rep_{PCn\,t-z\,FS}$' represents the reputation of the risk assessed agent in time slot 't-z' and in assessment criterion 'Cn' on the Failure Scale,

'$Rep_{PCn\,t-z}$' represents the reputation of the risk assessed agent in assessment criterion 'Cn' and in time slot 't-z'.

Once the risk assessing agent has determined the trustworthiness or the reputation of a risk assessed agent against an assessment criterion on the Failure Scale, it can then ascertain the probability of failure to achieve that particular assessment criterion in that time slot according to its expectations, by determining the difference between what it expects from the risk assessed agent in the assessment criterion and how far the risk assessed agent can fulfil it according to its trustworthiness or reputation in that. The risk assessing agent expects the risk assessed agent to complete the assessment criterion according to its expectations. This expectation of the risk assessing agent can be quantified with a value of 5 on the Failure Scale, as it represents the lowest probability of failure of the assessment criterion and expresses the maximum commitment by the risk assessed agent to its expectations. The probability of failure to achieve an assessment criterion 'Cn' according to the expectations in interacting with the risk assessed agent 'P' in a time slot 't-z', according to its trustworthiness or reputation in this can be determined by:

$$\text{Probability of Failure}_{PCn\,t\text{-}z} = (\frac{5 - T_{PCn\,t\,-\,zFS}}{5}) * 100 \quad \text{or,}$$

$$\text{Probability of Failure}_{PCn\,t\text{-}z} = (\frac{5 - \text{Rep}_{PCn\,t\,-\,zFS}}{5}) * 100 \qquad (6)$$

The determined probability of failure to achieve assessment criterion 'Cn' according to the expectations, in interacting with the risk assessed agent 'P' and in time slot 't-z' will be on a scale of 0-100 %. The risk assessing agent from this can determine the FailureLevel (PFL) of the risk assessed agent 'P' in assessment criterion 'Cn' and in time slot 't-z' on the Failure Scale (PFL $_{PCn\,t\text{-}z}$) by:

$$\text{PFL}_{PCn\,t\text{-}z} = \text{LEVEL (Probability of Failure}_{PCn\,t\text{-}z}) \qquad (7)$$

By using the above steps, the risk assessing agent should determine the FailureLevel of the risk assessed agent for each assessment criterion of its expectations in a pre-interaction start time slot. Once it does that, it can then determine the risk assessed agent's crisp FailureLevel in that time slot according to its expectations, by weighing the individual FailureLevel of each assessment criterion according to its significance. As discussed earlier, all assessment criteria in an interaction will not be of equal importance or significance. The significance of each assessment criterion might depend on the degree to which it influences the successful outcome of the interaction according to the risk assessing agent. The levels of significance for each assessment criterion (S_{Cn}) are shown in Table 2.

The crisp FailureLevel of the risk assessed agent 'P' in a pre-interaction start time slot 't-z' (PFL $_{Pt\text{-}z}$) is determined by weighing its FailureLevel to complete each

Table 2. The significance level of each assessment criterion

Significance level of the assessment criterion (S_{Cn})	Significance Rating and Semantics of the level
1	Minor Significance
2	Moderately Significant
3	Largely Significant
4	Major Significance
5	Highly or Extremely Significant

assessment criterion of the expectations in that time slot, with the significance of the assessment criteria. Hence:

$$\text{PFL}_{Pt\text{-}z} = \text{ROUND} \left(\frac{1}{\sum\limits_{n=1}^{y} SCn} \left(\sum\limits_{n=1}^{y} S_{Cn} * \text{PFL}_{PCn\,t\text{-}z} \right) \right) \tag{8}$$

where: 'S_{Cn}' is the significance of the assessment criterion 'Cn';

'PFL $_{PCn\ t\text{-}z}$' represents the FailureLevel of the risk assessed agent 'P' in assessment criterion 'C_n' in time slot 't-z'; and

'y' is the number of assessment criteria in the expectations.

By using the proposed methodology, the risk assessing agent should ascertain the FailureLevel of the risk assessed agent in each time slot of the pre-interaction start time phase according to the expectations of its future interaction, either by its past-interaction history or by the recommendations for the total assessment criteria of its expectations, or as a combination of its past interaction history in the partial assessment criteria of its expectations and the recommendations from the recommending agents for the other assessment criteria. Once the risk assessing agent has determined the FailureLevel (PFL) of the risk assessed agent in each of the pre-interaction start time slots according to the expectations of its future interaction, it can then utilize these to predict and ascertain the FailureLevel (FFL) of the risk assessed agent in the time slots of the post-interaction start time phase. As the FailureLevel of the risk assessed agent in the pre-interaction start time slots is according to the expectations of its future interaction, its determined FailureLevel in the time slots of the post-interaction start time phase will also be strictly according to the expectations of the risk assessing agent's future interaction with it.

5   Determining the FailureLevel in the Post-interaction Phase

In this section, we will propose the methodology by which the risk assessing agent can ascertain the FailureLevel of the risk assessed agent in the actual period of interaction and according to the expectations of its future interaction with it. As discussed earlier, the risk assessing agent's actual period of interaction with the risk assessed agent is represented by the post-interaction start time phase, and this period of time ranges from the time spot of the interaction to the end of the time space. Two scenarios arise when the risk assessing agent determines the FailureLevel of the risk assessed agent in the post-interaction start time phase. They are:

Scenario 6: The post-interaction start time phase of the risk assessing agent's interaction with the risk assessed agent is limited to the current time slot in which it is at present.

If the time spot and the duration of the risk assessing agent's interaction with the risk assessed agent is limited to the current time slot (as shown in Figure 3), then the risk assessing agent can determine the FailureLevel (FFL) of the risk assessed agent in the post-interaction start time slot by:

Case 6.1: Using its past interaction history with the risk assessed agent, if it is in the same expectations and time slot of its future interaction.

If the risk assessing agent has a past interaction history with the risk assessed agent in the time slot of its future interaction with it and in the same assessment criteria and significance, as the expectations of its future interaction with it, then it can consider the risk relationship of its previous interaction with the risk assessed agent and utilize the FailureLevel (AFL) which it had ascertained for the risk assessed agent in that previous interaction, as its FailureLevel (FFL) in the post-interaction start time slot. This is based on the assumption that the FailureLevel of the risk assessed agent in a time slot remains constant. Hence, the FailureLevel (FFL) of the risk assessed agent 'P' in a post-interaction start time slot 't_z', based on the risk assessing agent's past interaction history with it in that time slot and in the expectations of its future interaction is represented by:

$$FFL_{Ptz} = AFL_{Ptz} \qquad (9)$$

where, 'P' represents the risk assessed agent,

't_z' represents the time slot in which the risk assessing agent is determining the FailureLevel of the risk assessed agent,

Case 6.2: Using a combination of its past interaction history and the recommendations from other agents.

If the risk assessing agent has a past interaction history with the risk assessed agent in the same context and in the same time slot of its future interaction with it, but in the partial assessment criteria of its expectations, then it should utilize those partial assessment criteria and their corresponding 'Commitment Level' values to determine the trustworthiness of the risk assessed agent for those assessment criteria as discussed in scenario 4. It should then solicit recommendations from other agents for the remaining assessment criteria of its expectations by issuing a reputation query, and then assimilate them to ascertain the reputation of the risk assessed agent for those assessment criteria as discussed in section 4.2.3.

However, in each of the cases discussed previously (scenario 4 and section 4.2.3), the risk assessing agent adjusts the trustworthiness and/or the reputation of the risk assessed agent by the variable 'w' according to the weight to be given to the status of the risk assessed agent, depending upon the time slot in the pre-interaction start time phase. In the present case, where the risk assessing agent determines the trustworthiness of the risk assessed agent by using its past interaction history and/or the reputation of the risk assessed agent by soliciting recommendations from other agents, in the current time slot; the value for the variable 'w' should be 1. Hence, the risk assessing agent can determine the trustworthiness of the risk assessed agent 'P' in an assessment criterion 'Cn' by considering its past interaction history with it, in a post-interaction start time slot 't_z' by:

$$T_{PCn\,tz} = (Commitment\ Level_{Cn}) \qquad (10)$$

where, 'P' represents the risk assessed agent,

'Cn' represents the assessment criterion, in which the trustworthiness of the risk assessed agent 'P' is being determined,

'Commitment Level $_{Cn}$' represents the level of commitment of the risk assessed agent in assessment criterion 'Cn',

Similarly, the risk assessing agent can determine the reputation of the risk assessed agent 'P' in an assessment criterion 'Cn' by utilizing the recommendations of other agents in a post-interaction start time slot 't_z' by:

$$\text{Rep}_{PCn\,tz} = (\alpha * (\frac{1}{K} (\overset{K}{\underset{i=1}{\Sigma}} RCV_i \oplus \text{Commitment Level} \,_{Cn}^{i}))) +$$

$$(\beta * (\frac{1}{J} (\sum_{o=1}^{J} \text{Commitment Level} \,_{Cn}^{o}))) \tag{11}$$

where, 'RCV_i' is the credibility value of the trustworthy recommending agent 'i',

'Commitment level $_{Cn}$' is the level of commitment recommended by the recommending agent for assessment criterion 'Cn' for the risk assessed agent in the particular time slot 't_z',

'K' is the number of trustworthy recommendations that the risk assessing agent receives for the risk assessed agent in assessment criterion 'Cn' in time slot 't_z',

'J' is the number of unknown recommendations that the risk assessing agent receives for the risk assessed agent in assessment criterion 'Cn' in time slot 't_z',

'α and β' are the variables attached to the parts of the equation which will give more weight to the recommendation from the trustworthy recommending agents as compared to those from the unknown recommending agents. In general $\alpha > \beta$ and $\alpha + \beta = 1$.

The risk assessing agent should utilize equations 10 and 11 to ascertain the trustworthiness or the reputation of the risk assessed agent for each assessment criterion of its expectations, by using its past interaction history with it or by the recommendations from other agents respectively. Based on the determined trustworthiness or the reputation of the risk assessed agent for each assessment criterion of its expectations, the risk assessing agent can then determine the FailureLevel (FFL) of the risk assessed agent in the post-interaction start time slot by using the methodology proposed in Section 4.2.4.

It may be the case that the risk assessing agent does not have any past interaction history with the risk assessed agent in the time slot of its interaction, nor does it get recommendations from other agents for all the assessment criterion of its expectations in the time slot of its interaction. In this case, we propose that the risk assessing agent cannot utilize the above proposed methodology to determine the FailureLevel (FFL) of the risk assessed agent in the post-interaction start time phase of its interaction, and should utilize the methodology proposed in scenario 7 to determine the FailureLevel (FFL) of the risk assessed agent in that time phase.

Scenario 7: The post-interaction start time phase of the risk assessing agent's interaction with the risk assessed agent begins and extends till to a point in time in the future.

As discussed in the earlier sections, if the time spot or the duration of the risk assessing agent's interaction with the risk assessed agent extends to a point in time in the future (as shown in Figure 4), then the risk assessing agent has to determine the FailureLevel (FFL) of the risk assessed agent in those time slots by utilizing the prediction methods based on the previous impression that it has about the risk assessed agent. In other words, in order for the risk assessing agent to determine the FailureLevel (FFL) of a risk assessed agent in a post-interaction start time slot (if it is at a future point in time), it should know its FailureLevel according to the expectations of its future interaction from the beginning of the time space to the time slot preceding the one in which the FailureLevel (FFL) of the risk assessed agent has to be determined. The risk assessing agent should then utilize the determined FailureLevel of the risk assessed agent to that time slot, and predict its FailureLevel (FFL) in the time slots of the post-interaction start time phase. Hence, in our method we propose that the risk assessing agent, in order to determine the future FailureLevel of the risk assessed agent at time slot 't1' of the post-interaction phase in Figure 4, should consider all its FailureLevel values from the beginning of the time space to the time slot preceding it, i.e. to time slot 't-1'. Two cases arise when the risk assessing agent has to ascertain the FailureLevel (FFL) of the risk assessed agent in the future period of time of its interaction.

Case 7.1: The determined FailureLevel (PFL) of the risk assessed agent in the pre-interaction start time slots has features of either stochastic variation or trends in variation. In this case, we propose that the risk assessing agent, while determining the FailureLevel (FFL) of a risk assessed agent in a time slot of the post-interaction start time phase at a future period of time, should determine the magnitude of occurrence of each level of failure within the domain of (0, 5) on the Failure Scale in that time slot, rather than determining a crisp FailureLevel as it does in the Pre-Interaction start time slots. This is because determining the probability of failure of an interaction in the future period of time deals with uncertainty as it is being determined at a point in time in the future; and subsequently, when the FailureLevel series of the risk assessed agent has variability in it, the uncertainty of its behaviour over the future period of time should be captured, while ascertaining its FailureLevel during that time period. This uncertainty about the behaviour of the risk assessed agent is not totally captured when it is being represented by a crisp FailureLevel value. Hence, in order to take into consideration this uncertainty while ascertaining the FailureLevel of the risk assessed agent in a time slot at a future period of time, the risk assessing agent should ascertain the magnitude of the occurrence of each level of failure on the Failure Scale.

Our method for determining the FailureLevel (FFL) for a risk assessed agent at a future time slot 't1' (in Figure 4) by taking into consideration the uncertainness in its behaviour, is by taking its FailureLevel in each time slot from the beginning of the time space till time slot 't-1' and utilize the Gaussian Distribution to determine the probability of the future FailureLevel (FFL) in that time slot being any level on the Failure Scale (FS). As discussed earlier, the domain of the Failure Scale ranges from (-1, 5), with -1 denoting 'Unknown' level of failure. So the FailureLevel (FFL) of a risk assessed agent in the post-interaction start time slot should be determined in the domain of (0, 5) on the Failure Scale. Within this domain, there are six possible levels of failure. To determine the risk assessed agent's FailureLevel (FFL) at time slot 't1' within each of those levels, let us suppose that the risk assessing agent has determined the FailureLevel of the risk assessed agent in each time slot from the beginning of the time space till time slot 't-1'. These FailureLevel values of the risk assessed agent are represented as:

$$\{FL_{t-K}, \ldots \ldots FL_{t-3}, FL_{t-2}, FL_{t-1}\}$$

where, k is the number of time slots preceding the one in which the FFL is being determined.

The mean FailureLevel (μ_{FL}) is calculated as:

$$\mu_{FL} = \frac{1}{K} \sum_{i=1}^{K} FL_i \tag{12}$$

Accordingly, the unbiased Sample Variance (σ^2) is:

$$\sigma^2 = \frac{1}{K-1} \sum_{i=1}^{K} (FL_i - \mu_{FL})^2 \tag{13}$$

Since FFL $\sim (\mu, \sigma^2)$, then for any random variable FFL according to Gaussian distribution [15] the probability of FFL in a given range within the domain of (0, 5) on the Failure Scale can be determined according to equation 14.

$$P(a < FFL \leq b) = \frac{1}{\sqrt{2 \Pi} \sigma} \int_{\frac{a-\mu}{\sigma}}^{\frac{b-\mu}{\sigma}} e^{\frac{-t2}{2}} dt \tag{14}$$

By using equation 14, the risk assessing agent should ascertain the magnitude of the occurrence of each level of failure in the domain of (0, 5) on the Failure Scale, in a post-interaction start time slot. By doing so, the risk assessing agent would know the different levels of severity of failure and their level of occurrence in interacting with the risk assessed agent in a particular time slot of its interaction; and hence, the variability in the behaviour of the risk assessed agent over that particular future period of time of its interaction. The determined severities of failure are strictly according to the expectations of interaction between the risk assessing agent and the risk assessed agent. The risk assessing agent can also determine the FailureLevel (FFL) of the risk assessed agent in a time slot 't1' of the post-interaction start time phase, by utilizing the moments and cumulants of the obtained FailureLevel series of the risk assessed agent up to time slot 't-1'.

If there is more than one time slot in the post-interaction start time phase of the risk assessing agent's interaction with a risk assessed agent as shown in Figure 4, then the risk assessing agent has to determine the FailureLevel (FFL) of the risk assessed agent in each time slot of the post-interaction start time phase ('t1' till 't5' in Figure 4), to ascertain the performance risk in interacting with it. To ascertain the FailureLevel (FFL) of the risk assessed agent in the post-interaction start time slot 't2', the risk assessing agent, after determining the magnitude of occurrence of each level of failure in interacting with the risk assessed agent in the post-interaction start time slot 't1', should take the level with the highest probability of occurrence as the FailureLevel of the risk assessed agent in time slot 't1'. It should then consider the time slots from the beginning of the time space till time slot 't1' as shown in Figure 4, and utilize equations 12 - 14 to determine the magnitude of occurrence of different severities of failure in interacting with the risk assessed agent in post-interaction start time slot 't2'. By using the proposed methodology the risk assessing agent should determine the probability of occurrence of each FailureLevel on the Failure Scale in interacting with a risk assessed agent in each time slot of the post-interaction start time phase, according to its expectations.

Case 7.2: The determined FailureLevel (PFL) of the risk assessed agent in the pre-interaction start time phase has seasonal characteristics, and is the same in all the time slots of that time phase. In this case, the FailureLevel series of the risk assessed agent depicts a seasonality trend and the FailureLevel (FFL) of the risk assessed agent in the time slots of the post-interaction start time phase will be the same as that determined in the pre-interaction start time slots.

Scenario 8: The post-interaction start time phase of the risk assessing agent's interaction with the risk assessed agent extends till a point of time in the future, but the time spot is in the current period of time.

If the post-interaction start time phase of the risk assessing agent's interaction with the risk assessed agent extends to a point of time in the future, but the time spot of the interaction is in a time slot which has an overlap of the pre- and post- interaction start time phases as shown in Figure 6, then the risk assessing agent can ascertain the FailureLevel (FFL) of the risk assessed agent in time slot 't1' by using the methodology proposed in scenario 6, if it has either past interaction history with the risk assessed agent in that time slot or it gets recommendations from other agents in all the assessment criteria of its expectations in that time slot. In case the risk assessing agent does not have any of these, then it can utilize the methodology proposed in scenario 7 to ascertain the FailureLevel (FFL) of the risk assessed agent in all the time slots of the post-interaction start time phase.

Once the risk assessing agent ascertains the FailureLevel (FFL) of the risk assessed agent by using the methodology proposed either in scenario 6 or in scenario 7 in each time slot of the post-interaction start time phase, then it should ascertain the 'FailureLevel Curve' of the interaction in order to quantify the level of failure in interacting with the risk assessed agent.

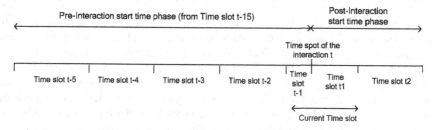

Fig. 6. The time spot and the post-interaction start time phase of the interaction

6 Determining the FailureLevel Curve of the Interaction

The 'FailureLevel Curve' (FLC) of the interaction quantifies and represents the performance risk in interacting with the risk assessed agent, based on its determined FailureLevel during the period of risk assessing agent's interaction with it. In other words, to the risk assessing agent the FailureLevel Curve represents the magnitude of the occurrence of different levels of severity of failure during the time period of its interaction with the risk assessed agent, according to its expectations. Hence, the FailureLevel Curve is such that the abscissa of the curve gives the level or severity of failure from the Failure Scale and the corresponding ordinate or impulse gives the probability of occurrence of that level. A point to be noted here is that the FailureLevel Curve of the interaction is determined by considering only the post-interaction start time phase of the time space. This is because the risk assessing agent wants to analyze the perceived risk in interacting with a risk assessed agent during the time in which it possibly interacts with it. This duration of time is represented by the post-interaction start time phase from its time space and subsequently the FailureLevel Curve of the interaction which represents the performance risk, should be ascertained by utilizing only the FailureLevel (FFL) of the risk assessed agent in each of the post-interaction start time slots. Two scenarios arise while ascertaining the FailureLevel Curve of the interaction. They are:

Scenario 9: The post-interaction start time phase of the risk assessing agent's interaction with the risk assessed agent is limited to the current time slot in which it is at present.

If the post-interaction start time phase of the interaction is limited to the current time slot as shown in Figure 3, and if the risk assessing agent ascertains the FailureLevel (FFL) of the risk assessed agent by utilizing either its past-interaction history or by soliciting recommendations from other agents, or a combination of both as discussed in scenario 6, then the determined FailureLevel (FFL) of the risk assessed agent in the post-interaction start time phase is a crisp value on the Failure Scale. In such cases, the FailureLevel Curve (FLC) of the interaction would represent just the determined FailureLevel (FFL) on the abscissa and its corresponding ordinate represents the probability of occurrence of that level, which in such cases is 1.

Scenario 10: The post-interaction start time phase of the risk assessing agent's interaction with the risk assessed agent extends till a point of time in the future.

If the time spot or the post-interaction start time phase extends to a future point in time as shown in Figure 4, and if there is a seasonal characteristics in the risk assessing agent's FailureLevel (PFL) in the pre-interaction start time phase as mentioned in case 7.2, then the FailureLevel (FFL) of the risk assessed agent in the post-interaction start time slots is the same as it is for the pre-interaction start time phase. In this case, the FailureLevel Curve (FLC) of the interaction would be determined as mentioned in scenario 9. On the contrary, if the FailureLevel (PFL) of the risk assessed agent variability in it (either stochastic variation or trends in variation) as mentioned in case 7.1, then the risk assessing agent ascertains the FailureLevel (FFL) of the risk assessed agent in each of the post-interaction start time slots as the probability of occurrence of each level of failure on the Failure Scale. In this case, the FailureLevel Curve (FLC) of the interaction is plotted by constructing the probability histogram of the sum of the probability of occurrence of each FailureLevel over the time slots of the post-interaction start time phase divided by the number of time slots within that time phase.

For example, consider an interaction scenario between risk assessing agent 'A' and the logistics company (termed as risk assessed agent 'B') in the context 'Transporting Goods' and in the assessment criteria 'C1-C4', if the risk assessing agent's 'A' interaction with the risk assessed agent 'B' is limited to the current period of time as shown in Figure 3 and if agent 'A' ascertain the FailureLevel (FFL) of agent 'B' in the time phase of its interaction as 2 on the Failure Scale, by either utilizing its past interaction history or recommendations from other agents as discussed in scenario 6, then the FailureLevel Curve (FLC) of the interaction in this case is shown in Figure 7. The FailureLevel Curve represents just one level of failure, as the FailureLevel (FFL) of the risk assessed agent is being determined over a period of time which is limited to the current time slot, by either using direct past interaction history or/and by using recommendations from other agents.

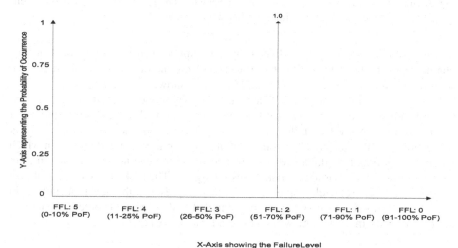

Fig. 7. The FailureLevel Curve when the interaction is limited to the current time slot and there is a single time slot in that time phase

If the risk assessing agent's 'A' interaction with agent 'B' extends to a point of time in the future as shown in Figure 4, then agent 'A' has to ascertain the FailureLevel (FFL) of the risk assessed agent 'B' in those time slots by using the methodology proposed in scenario 7. In this case, the FailureLevel Curve (FLC) of the interaction represents those levels of failure which occur in each of the post-interaction start time slots. The probability of the occurrence of each of these levels of failure is determined by the sum of the occurrence of a FailureLevel throughout the

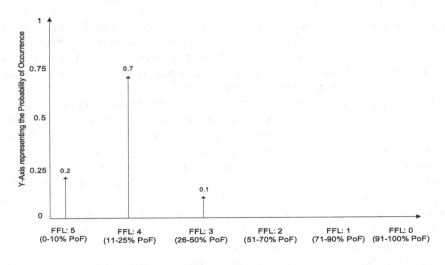

Fig. 8. The FailureLevel Curve of the interaction when the interaction extends to a point of time in the future and there are multiple time slots in that time phase

post-interaction start time slots, divided by the number of time slots within that time phase. An example of the FailureLevel Curve of the interaction determined in such scenario by considering the time slots of Figure 4 is shown in Figure 8.

The risk assessing agent 'A' by analyzing the magnitude of failure of a level and the probability of occurrence of that level in interacting with the risk assessing agent can determine the level of failure in achieving its desired outcomes in forming an interaction with that agent. This would help it to get an idea of the direction in which its interaction might head, and whether or not it will achieve its desired outcomes in interacting with the particular risk assessed agent. The risk assessing agent can consider the FailureLevel Curve (FLC) which represents the level of failure in interacting with a risk assessed agent, and utilize it to determine the other subcategory of perceived risk in interacting with it, i.e. the financial risk.

7 Conclusion

In an e-commerce interaction, it is possible that the risk assessing agent might have to decide and choose an agent to interact with from a set of risk assessed agents. It can ease its decision making process by analyzing the possible level of risk that could be present in interacting with each of them according to the demand of its interaction. Analyzing the possible level of risk gives the risk assessing agent an indication of the probability of failure of the interaction (FailureLevel) and the possible consequences of failure to its resources. In this chapter we proposed a methodology by which the risk assessing agent can determine the FailureLevel beforehand in interacting with a risk assessed agent. The determined FailureLevel is strictly according to the demand of the risk assessing agent's future interaction with the risk assessed agent.

References

1. Mayer, R.C., Davis, J.H., Schoorman, F.D.: An interactive model for organizational trust. Academy of Management Review 20(3), 709–734 (1995)
2. Greenland, S.: Bounding analysis as an inadequately specified methodology. Risk Analysis 24(5), 1085–1092 (2004)
3. Chang, E., Dillon, T., Hussain, F.K.: Trust and Reputation for Service-Oriented Environments: Technologies for Building Business Intelligence and Consumer Confidence, 1st edn. John Wiley and Sons Ltd., Chichester (2006)
4. Hussain, O.K., Chang, E., Hussain, F.K., Dillon, T.S.: Risk in Decentralized Communications. In: International Workshop on Privacy Data Management in Conjunction with 21st International Conference on Data Engineering, p. 1198 (2005)
5. Carter, J., Ghorbani, A.A.: Towards a formalization of Trust. Web Intelligence and Agent Systems 2(3), 167–183 (2004)
6. Wang, Y., Varadharajan, V.: A Time-based Peer Trust Evaluation in P2P E-commerce Environments. In: Zhou, X., Su, S., Papazoglou, M.P., Orlowska, M.E., Jeffery, K. (eds.) WISE 2004. LNCS, vol. 3306, pp. 730–735. Springer, Heidelberg (2004)
7. Wang, Y., Varadharajan, V.: Two-phase Peer Evaluation in P2P E-commerce Environments. In: Proceedings of the 2005 IEEE International Conference on e-Technology, e-Commerce and e-Service, Hong Kong, pp. 654–657 (2005)
8. Fan, J.C.: Trust and Electronic Commerce - A Test of an E-Bookstore. In: Proceedings of the IEEE International Conference on e-business Engineering, Shanghai, pp. 110–117 (2006)
9. Wojcik, M., Eloff, J.H.P., Venter, H.S.: Trust Model Architecture: Defining Prejudice by Learning. In: Fischer-Hübner, S., Furnell, S., Lambrinoudakis, C. (eds.) TrustBus 2006. LNCS, vol. 4083, pp. 182–191. Springer, Heidelberg (2006)
10. Su, C., Zhang, H., Bi, F.: A P2P-based Trust Model for E-Commerce. In: Proceedings of the IEEE International Conference on e-business Engineering, Shanghai, pp. 118–122 (2006)
11. Koutrouli, E., Tsalgatidou, A.: Reputation-Based Trust Systems for P2P Applications: Design Issues and Comparison Framework. In: Fischer-Hübner, S., Furnell, S., Lambrinoudakis, C. (eds.) TrustBus 2006. LNCS, vol. 4083, pp. 152–161. Springer, Heidelberg (2006)
12. Chien, H., Lin, R.: Identity-based Key Agreement Protocol for Mobile Ad-hoc Networks Using Bilinear Pairing. In: Proceedings of the IEEE International Conference on Sensor Networks, Ubiquitous, and Trustworthy Computing, Taichung, Taiwan, vol. 1, pp. 520–528 (2006)
13. Hussain, F.K., Chang, E., Dillon, T.: Trustworthiness and CCCI Metrics for Assigning Trustworthiness. International Journal of Computer Science Systems and Engineering 19(2), 173–189 (2004)
14. Cornelli, F., Damiani, E., Vimercati, S.C., Paraboschi, S., Samarati, P.: Choosing Reputable Servents in a P2P Network. In: Proceedings of the International WWW Conference, Honolulu, vol. (11), pp. 376–386 (2002)
15. Weiss, N.A.: A Course in Probability. Pearson Education, Inc., USA (2006)
16. Wang, Y., Lin, F.: Trust and Risk Evaluation of Transactions with Different Amounts in Peer-to-Peer E-commerce Environments. In: Proceedings of the IEEE International Conference on e-Business Engineering, Shanghai, China, pp. 102–109 (2006)
17. Hussain, O.K., Chang, E., Hussain, F.K., Dillon, T.: A Methodology for Determining the Creditability of the Recommending Agents. In: Gabrys, B., Howlett, R.J., Jain, L.C. (eds.) KES 2006. LNCS (LNAI), vol. 4253, pp. 1119–1127. Springer, Heidelberg (2006)
18. Hussain, O.K., Chang, E., Hussain, F.K., Dillon, T.S.: A methodology for risk measurement in e-transactions. International Journal of Computer Science Systems and Engineering 21(1), 17–31 (2006)

Process Mediation of OWL-S Web Services*

Katia Sycara and Roman Vaculín

The Robotics Institute, Carnegie Mellon University
{katia,rvaculin}@cs.cmu.edu

Abstract. The ability to deal with incompatibilities of service requesters and providers is a critical factor for achieving smooth interoperability in dynamic environments. Achieving interoperability of existing web services is a costly process including a lot of development and integration effort which is far from being automated. Semantic Web Services frameworks strive to facilitate flexible dynamic web services discovery, invocation and composition and to support automation of these processes. In this paper we focus on mediation and brokering mechanisms of OWL-S Web Services which we see as the main means to overcome various types of problems and incompatibilities. We describe the process mediation component and the hybrid broker component that present two different approaches to bridge incompatibilities between the requester's and provider's interaction protocols.

1 Introduction

The main goal of Web Services is to enable and facilitate smooth interoperation of diverse software components in dynamic environments. Due to the dynamic nature of the Internet and rapid, unpredictable changes of business needs, the ability to adapt to changing environments is becoming important. Existing service providers should be able to communicate with new clients that might use different data models and communication protocols. Furthermore, any changes in the provided services requires intensive modifications of existing interfaces and implementation to maintain interoperability. The possibility of achieving interoperability of existing components automatically without actually modifying their implementation is therefore desirable.

Process mediation services present a possible solution in situations where interoperability of components with fixed, incompatible communication protocols needs to be achieved. A process mediation component resolves all incompatibilities and generates appropriate mappings between different processes. Implementing the mediation component is complicated and costly, since it has to address many different types of incompatibilities. On the data level, components may be using different formats to encode elementary data or data can be represented in incompatible data structures. Furthermore, messages can be exchanged in different orderings, some pieces of information which are required by one process may be missing in the other one, or control flows can be encoded in very different ways.

---

* This research was supported in part by Darpa contract FA865006C7606 and in part by funding from France Telecom.

T.S. Dillon et al. (Eds.): Advances in Web Semantics I, LNCS 4891, pp. 324–345, 2008.

Current web services standards provide a good basis for achieving at least some level of mediation. WSDL [1] standard allows to declaratively describe operations, the format of messages, and the data structures that are used to communicate with a web service. BPEL4WS [2] adds the possibility to combine several web services within a formally defined process model, thereby allowing one to define the interaction protocol and possible control flows. However, neither of these two standards goes beyond the syntactic descriptions of web services. Newly emerging standards for semantic web services such as WSDL-S [3], OWL-S [4] and WSMO [5] strive to enrich syntactic specifications with rich semantic annotations to further facilitate flexible dynamic web services discovery and invocation. Tools for reasoning can be used for more sophisticated tasks such as, matchmaking and composition [6].

In this paper, we address the problem of automatic mediation of process models consisting of semantically annotated web services. Processes can act as service providers, service requesters or communicate in peer-to-peer fashion. We are focusing on the situation where the interoperability of two components, one acting as the requester and the other as the provider, needs to be achieved. We assume that both the requester and the provider behave according to specified process models that are fixed, incompatible and that are expressed explicitly. We use the OWL-S ontology for semantic annotations because it provides a good support for the description of individual services as well as explicit constructs with clear semantics for describing process models.

Depending on the nature of the environment in which the interoperability has to be achieved, different approaches must be considered. A relatively *closed corporate intranet environment* in which development of all components can be controlled by one authority allows high built-in interoperability of components. An agreement on the syntax and semantics of exchanged data and communication protocols is possible in advance. Typically, either both the client and the provider can be developed to cooperate together or the client is specifically developed to interact with some particular provider service. In a closed environment it is also much easier to develop ad-hoc mediation components for individual pairs of services, because limited number of components can interact with each other and all protocols and all incompatibilities are known.

However, the environment in corporations cannot always be considered as entirely closed because services of many diverse contractors and subcontractors are used as part of organizations' business processes. We call such an environment as *semi-open*. In semi-open environments, software components are *controlled and developed by independent authorities* which engenders both data and protocol incompatibilities. Semi-open environments are *dynamic with a controlled registration*, i.e., new components can be added, removed or replaced, there can be several interchangeable components (from several contractors) that solve the same problem. The system can define policies specifying how components are added and removed. Semi-open environments typically imply some level of *trust* among contractors. Contractors are motivated to publish descriptions of interaction protocols of their components to allow interoperability with other components.

The above mentioned characteristics of semi-open environments make the mediation a much harder problem. It is simply impossible to assume that various requesters and providers will interoperate smoothly without any mediation or that a one purpose mediation component can be developed for each new service provider or requester. However, since it is reasonable to assume that each component provides a description of its inter-action protocols and since mechanisms of registering components into the system can be controlled, it is possible (1) to analyze in advance if interoperability of some com-ponents is possible and if it is (2) to use mechanisms that utilize results of the analysis step to perform an automatic mediation. We will describe mechanisms for an automatic mediation in semi-open environments in Section 4.

Dynamic open environments add more levels of complexity to the mediation prob-lem. Since components can appear and disappear completely arbitrarily, it is necessary to incorporate appropriate discovery mechanisms. Also it is not possible to perform an analysis step in advance because requesters and providers are not known in ad-vance. Therefore the automatic mediation process must rely only on *run-time* medi-ation. In Section 5 we describe a standalone Runtime Process Mediation Service that can be combined with suitable discovery mechanisms. We also show how the Runtime Process Mediation Service can be integrated into a Broker component that performs both, the discovery of suitable service providers and the runtime mediation.

The rest of the paper is structured as follows. In Section 2, we present an overview of OWL-S and componets for discovery and invocation of OWL-S web services. In Sec-tion 3, we will discuss categories of problems and mismatches that must be resolved during the mediation process, we specify assumptions of our approach and we also in-troduce an example that demonstrates several mismatches between the requester's and the provider's process models. In Section 4, we describe an algorithm for process me-diation in the semi open world[1]. In Section 5, we discuss the process mediation in the open world. Namely, we first introduce a Runtime Mediation Service that can be used as a standalone component, and next we describe how this component can be embedded into the Broker that combines the discovery with the process mediation. In Section 6, we give an overview of the related work and we conclude in Section 7.

2   Overview of Relevant Concepts and Components of OWL-S

OWL-S [4] is a Semantic Web Services description language, expressed in OWL [7]. OWL-S covers three areas: web services capability-based search and discovery, specifi-cation of service requester and service provider interactions, and service execution. The Service Profile describes what the service does in terms of its capabilities and it is used for discovering suitable providers, and selecting among them. The Process Model spec-ifies ways of how clients can interact with the service by defining the requester-provider interaction protocol. The Grounding links the Process Model to the specific execution infrastructure (e.g., maps processes to WSDL [1] operations and allows for sending messages in SOAP [8]). Corresponding Profiles, Process Model and Groundings are connected together by an instance of the Service class that is supposed to represent the whole service.

The elementary unit of the Process Model is an atomic process, which represents one indivisible operation that the client can perform by sending a particular message to the service and receiving a corresponding response. Processes are specified by means of their inputs, outputs, preconditions, and effects (IOPEs). Types of inputs and outputs are

---

[1] In the rest of the paper we will use words *environment* and *world* interchangeably.

usually defined as concepts in some ontology or as simple XSD data-types. Processes can be combined into composite processes by using the following control constructs: sequence, any-order, choice, if-then-else, split, split-join, repeat-until and repeat-while. Besides control-flow, the Process Model also specifies a data-flow between processes.

From the perspective of necessary tools, there are two main areas that need to be covered: *service search and discovery* and *service invocation*.

In order to make its capabilities known to service requesters, a service provider advertises its capabilities with infrastructure registries, or more precisely middle agents [9], that record which agents are present in the system. UDDI registries [10] are an example of a middle agent, with the limitation that it can make limited use of the information provided by the OWL-S Profile. The OWL-S/UDDI Matchmaker [11,12] is another example, which combines UDDI and OWL-S Service Profile descriptions. The OWL-S/UDDI matchmaker supports flexible semantic matching between advertisements and requests on the basis of ontologies available to the services and the matchmaking engine. After a requester has found the contact details of a provider through matchmaking, then the requester and the provider interact directly with one another. Since in an open environment they could have been developed by different developers, incompatibilities during interoperation may happen.

Brokers present another mechanism of discovery and synchronization [9,13]. Brokers have been widely used in many agents applications such as integration of heterogeneous information sources and Data Bases [14], e-commerce [15], pervasive computing [16] and more recently in coordinating between Web services in the IRS-II framework [17]. A brokering component used as a middle agent between a requester and a provider addresses several problems: a broker can perform discovery and selection of providers incorporating a decision procedure assessing compatibility issues, it can perform mediation and so to allow interactions of otherwise incompatible partners, it can be used to maintain anonymity of a provider and a requester by acting as a proxy and effectively hiding their identities, brokers can perform a range of coordination activities such as load balance between several provider's, etc. In this paper we focus specifically on the broker ability to combine discovery and process mediation functionalities.

A tool for execution of OWL-S web services must be able to interpret the Process Model of the service according to its semantics and provide a generic mechanism for invocation of web services represented as atomic processes in the Process Model. The OWL-S Virtual Machine (OVM) [18] is a generic OWL-S processor that allows Web services and clients to interact on the basis of the OWL-S description of the Web service and OWL ontologies. Specifically, the OWL-S Virtual Machine (OVM) executes the Process Model of a given service by going through the Process Model while respecting the OWL-S operational semantics [19] and invoking individual services represented by atomic processes. During the execution, the OVM processes inputs provided by the requester and outputs returned by the provider's services, realizes the control and data flow of the composite Process Model, and uses the Grounding to invoke WSDL based web services when needed. The OVM is a generic execution engine which can be used to develop applications that need to interact with OWL-S web services.

The architecture of the OVM and its relation with the rest of the Web service is described in Figure 1. On the left side the provider is displayed together with its OWL-S

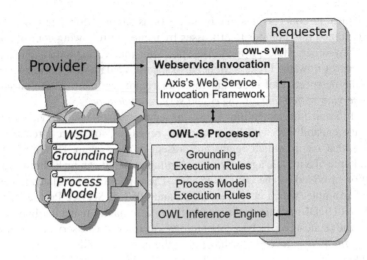

Fig. 1. The OWL-S Virtual Machine Architecture

Process Model, Grounding and WSDL description that together define how clients can interact with this service. The OVM is displayed in the center of the picture. It is logically divided in two modules: the first one is the OWL-S Processor which uses the OWL-S Inference Engine and a set of rules implementing the operational semantics of the OWL-S Process Model and Grounding to manage the interaction with the provider. The second component is the Web service Invocation module that is responsible for the information transfer with the provider. Finally, the OVM is shown as a part of the requester which can use it to interact with the provider.

3   Mediation Problem

In this section, we analyze the various types of incompatibilities that must be addressed during the mediation, describe how these mismatches can be generally handled in the context of OWL-S, introduce the problem setting and its assumptions and provide an example that demonstrates several mismatches.

Interoperability of a requester and a provider might be complicated by diverse types of incompatibilities. In the context of process mediation the following types of mismatches can be identified:

1. *Data level mismatches*:
    (a) *Syntactic / lexical mismatches*: data are represented as different lexical elements (numbers, dates format, local specifics, naming conflicts, etc.).
    (b) *Ontology mismatches*: the same information is represented as different concepts
        i. in the same ontology (subclass, superclass, siblings, no direct relationship)
        ii. or in different ontologies, e.g., (Customer vs. Buyer)
2. *Service level mismatches*:
    (a) a requester's service call is realized by several providers' services or a sequence of requester's calls is realized by one provider's call

(b) requester's request can be realized in different ways which may or may not be equivalent (e.g., different services can be used to to satisfy requester's requirements)

(c) reuse of information: information provided by the requester is used in different place in the provider's process model (similar to message reordering)

(d) missing information: some information required by the provider is not provided by the requester

(e) redundant information: information provided by one party is not needed by the other one

3. *Protocol / structural level mismatches*: control flow in the requester's process model can be realized in very different ways in the provider's model (e.g., sequence can be realized as an unordered list of steps, etc.)

In OWL-S, syntactic and lexical level mismatches (category 1a) are handled by the service Grounding which defines transformations between syntactic representation of web service messages and data structures and the semantic level of the process model. The Grounding provides mechanisms (e.g., XSLT transformations) to map various syntactic and lexical representations into the shared semantic representation.

Mediation components (or mediators) can be used to resolve other types of incompatibilities. In our work we distinguish two types of mediation: *data mediation* and *process mediation*. We assume, that *data mediators* are responsible for resolving data level mismatches (category 1) while *process mediators* are responsible for resolving service level and protocol level mismatches (categories 2 and 3).

Typically, when trying to achieve interoperability, process mediators and data mediators are closely related. A natural way is to use data mediators within the process mediation component to resolve "lower" level mismatches that were identified during the process mediation. As opposed to WSMO methodology [5], OWL-S does not introduce mediators as first class objects. [20] shows that in the OWL-S framework mediators can be naturally represented as web services and described in the same way as any other web service. This is particularly the case for data mediators, which typically work as transformation functions from one domain into the other and hence can be easily described as atomic processes by specifying appropriate input and output types. We comply with this approach since we believe that the view of mediators as web services naturally fits into the Semantic Web Services Architecture [21] and allows us, for example, to use the same discovery and invocation mechanisms for mediators as for any other services.

3.1 Process Mediation

When requesters and providers use fixed, incompatible communication protocols interoperation can be achieved by applying a *process mediation component* which resolves all incompatibilities, generates appropriate mappings between different processes and translates messages exchanged during run-time. Figure 2 shows this problem setting.

Implementing the mediation component is complicated and costly, since it has to address all different types of incompatibilities described in the previous section. We describe mechanisms how the process mediation can be solved automatically. In particular, we show how the workflow and dataflow mismatches can be resolved.

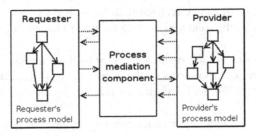

Fig. 2. Mediation of process models

We assume that both the requester and the provider behave according to specified process models and that both process models are expressed explicitly using OWL-S ontologies[2]. Because the problem of process mediation is complex and extensive, we address the data mediation (mismatches of type 1) only in a very limited way. For details on the data mediation see, e.g., [22,23,24]. We assume that data mediators are given to the process mediation component as an input. In our system data mediators can have a form of a converter that is built-in to the system or of an external web service [20]. We support basic type conversions as up-casting and down-casting based on reasoning about types of inputs and outputs. By up-casting or down-casting we mean a conversion of an instance of some ontology class to a more generic or more specific class respectively.

We assume that both process models describe services that belong to the same domain. By this we mean that inputs, outputs, preconditions and effects are defined in the same ontology and that both partners target conceptually the same problem. We want to avoid the situation when, for example, the provider is a book selling service and the requester needs a library service. Both process models could be using the same ontology but the mediation would not make much sense in this case. This requirement can be easily achieved either by appropriate service discovery mechanisms [25] or simply be a consequence of the real situation when only applications from within the same domain need to be integrated.

3.2 Motivating Example

Figure 4 depicts a fragment of the process model of a hypothetical provider from the flights booking domain. The requester's model, presented in Figure 3, represents a straightforward process of purchasing a ticket from some airlines booking web service, while the provider's process model represents a more elaborate scenario that allows the requester, besides booking the flight, to also rent a car or to book a hotel. Boxes in figures represent atomic processes with their inputs, outputs, preconditions, and effects, while ovals stand for control constructs. The control flow proceeds in the top-down and left to right direction. Inputs and outputs types used in process models refer to a very

---

[2] OWL-S process model pertains mainly to describing service providers. However, its constructs can be used to describe the requester in the same way as if describing the provider. The only conceptual difference in using the OWL-S process model to describe the behavior of a requester is that it describes the behavior the requester expects a provider to have.

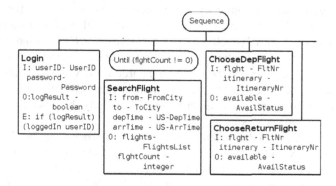

Fig. 3. Requester's process model

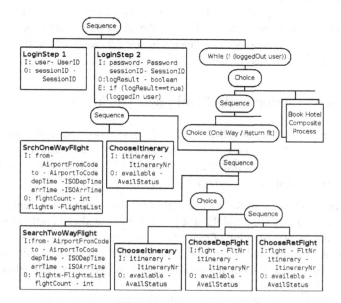

Fig. 4. Provider's process model

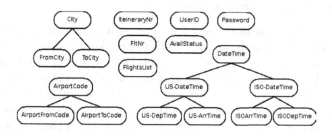

Fig. 5. Simple flights domain ontology

simple ontology showed in Figure 5 (ovals represent classes and lines represent subsumption relations). The requester's process model starts with the *Login* atomic process that has two inputs — *userId* which is an instance of the *UserID* class and *password* of *Password* type — one output *logResult* of *boolean* type and the conditional effect saying that the predicate *(loggedIn userID)* will become true if the value of *logResultOutput* equals to true. In the next step, the *SearchFlight* is executed within the repeat-until loop which is repeated until some flight is found. Similarly the process continues by executing other atomic processes.

This example demonstrates several types of inconsistencies that we have to deal with. The *Login* step in the requester's model is represented by two separate atomic processes in the provider's model (mismatches 2a, 2e, see Section 3). Types of inputs and outputs do not always match exactly, e.g., *AirportFromCode* and *FromCity* are not directly related in the ontology (mismatch 1b), *US-DepTime* and *ISODepTime* based on ISO 8601 are subclasses of the common superclass *DateTime* (mismatch 1b). *SearchFlight* in the requester's process model can be mapped either to *SrchOneWayFlight* or to *SrchTwoWayFlight* (mismatch 2b). Finally, the structure of both processes is quite different and it is not obvious at first glance whether the requester can be mapped into the provider's process model (mismatch 3).

4   Process Mediation in the Semi-open World

An environment in corporations that need to dynamically interact with services provided by other trusted partners, e.g., their subcontractors can be usually characterized as a semi-open environment. From the point of view of process mediation several characteristics of semi-open environments are important:

1. Components are not controlled by one authority which implies a presence of incompatibilities. Also it is not possible or it is too complicated and costly to modify the interfaces and implementation of individual components.
2. Components can be added to the system, removed or replaced dynamically.
3. There can be several interchangeable components (from several contractors) that solve the same problem.
4. The registration of a new component to the system can be controlled.
5. Typically some level of trust among contractors is necessary which allows us to assume that all components can publish descriptions of their interaction protocols to allow interoperability with other components.

These characteristics allow us to perform the process mediation of the given requester and the provider in two steps: (1) during the registration process of the component to the system an analysis of the process models of the requester and the provider is performed to find possible mappings between provider's and requester's process models, or to identify incompatibilities that cannot be reconciled with given set of available data mediators and external services. Results of the analysis for the given requester/provider pair are stored. (2) During run-time, when some requester and provider need to interact, saved mediation mappings for this pair are used in the mediator runtime component.

4.1 Execution Paths Analysis Approach

The problem of process mediation can be seen as finding an appropriate mapping between requester's and provider's process models. The mapping can be constructed by combining simpler transformations representing different ways of bridging described mismatches. We need to decide if *structural differences* between process models can be resolved. Assuming that the requester starts to execute its process model, we want to show that for each step of the requester the provider (with some possible help of intermediate translations represented by *data mediators* or *built-in conversions*) can satisfy the requester's requirements (i.e., providing required outputs and effects) while respecting its own process model. This can be achieved by exploring possible sequences of steps (execution path) that the requester can execute.

Requester's execution path is any sequence of atomic processes which can be called by the requester in accordance with its process model, starting from the process model first atomic process and ending in one of the last atomic processes of the process model. An atomic process is last in the process model if there is no next atomic process that can be executed after it (respecting the control constructs, as e.g. loops).

Since any of all possible requester's execution paths can be chosen by the requester, we need to show that each requester's execution path can be mapped into the provider's process model (assuming some data translation facility). If there exists a possible requester's execution path which could not be mapped to any part of the provider's process model, we would know that if this path were chosen, the mediation would fail. Thus the existence of a mapping for each possible requester's execution path is a necessary precondition of successful mediation. Indeed, it is only a necessary condition of successful process mediation for the following reason. Since the possible mappings are being searched before actual execution, some of them can turn out not to work during execution (e.g., because of failing preconditions of some steps). Still, by analyzing requester's execution paths and trying to find mappings for them, we can partially answer the question of mediation feasibility.

Finding possible mappings means to explore the search space generated by combining allowed execution paths in the provider's process model with available translations (data mediators in our case). We explore the search space by simulating the execution of the provider's process model with possible backtracking if some step of the requester's path cannot be mapped or if more mappings are possible. During the simulation, data mediators are used to reconcile possible mismatches.

Finally, during the execution we need to decide, what actions should be performed in each given state. Generated mappings are used to decide, if and what services of the provider's process model should be executed, or if a translation (or a chain of translations) is necessary after the requester executes each step.

4.2 Mediator Algorithm Overview

The following procedure provides a top-level view of the whole process mediation:
1. Generate requester's paths: based on the process model of the requester, possible requester's paths are generated (see Section 4.3)
2. Filter out those requester's paths that need not be explored: as the result we get the *minimal set of requester's paths*. (see Section 4.3)

3. Find all appropriate mappings to the provider's process model for each requester's path from the minimal set of paths and store them in the *mappings repository*: if for a path no mapping is found, user is notified with pointing out the part of the path for which the mapping was not possible[3]. (see Section 4.4)

During the mediation process following steps are performed to choose the best available action and to execute it:

1. Retrieve possible actions from the *mappings repository* that are available in this context
2. Remove inconsistent actions: actions that are not consistent with actual variables bindings (e.g., preconditions fail)
3. When no suitable mediation action is available, fail
4. When more actions are available, choose the best: Having execution paths and mappings precomputed, we can easily figure out, if the suggested mediation action, if chosen, allows to finish all paths that can be taken by the requester from this state of execution. If there is no such an action, we choose the one that allows to finish the most paths.
5. Execute selected action: depending on the type of an action either the *OWL-S Virtual Machine* [18] is called to execute the external service or the provider's atomic process, or the built-in converter is called, or a response to the requester is generated by the mediation component.
6. Update state of the mediator: *mappings repository* and variables values and valid expressions are updated.

4.3   Generating the Minimal Set of Requester's Execution Paths

When generating requester's execution paths we potentially have to deal with combinatorial explosion caused by chains of branching in the requester's process model. We want to find out what reconciliation actions are available or necessary in given state of execution which depends on possible combinations of available variables and valid expressions in this state. Because the current state depends on actions performed preceding this state, we might be in principle interested in every possible requester's path. In [26] we describe some heuristics for pruning those paths that provide no additional information. The path pruning also reduces the number of requester's execution paths for which appropriate mappings to the provider's process model need to be found.

4.4   Finding Mappings for the Requester's Path

In order to find all the mappings for a given requester's path we simulate the execution of the provider's process model and try to map each step of the requester's path to some part of the provider's model (atomic process or several atomic processes) with help of *data mediators*. If some step of the requester's path cannot be mapped to the provider's process model, the simulation backtracks to the last branching (e.g., *choice* or *any-order*). The mapping is constructed during the simulation and is represented as

---

[3] At this point service discovery could be used to find a service capable of resolving the mismatch.

a sequence of actions that the mediation component should execute during the runtime mediation (see Figure 6 for an example of a mapping).

The *reconciliation algorithm* for the given requester's path works as follows:

Input: requester's path *requesterStepsSequence*
1. Initialize the simulator state by adding *requesterStepsSequence* to it
2. Call *executeNextRequesterCalls* method
3. Simulate the execution of the provider's process model until no requester's steps need to be reconciled, or the provider finishes, or reconciliations fails
   – when the atomic process P is reached during simulation, call the *reconciliation method for P*

The *executeNextRequesterCalls* is a simple method that removes the first call from the requester's path *requesterStepsSequence* and adds inputs of this call to the simulator's state and for each output and effect that are expected to be produced creates an appropriate goal.

The *reconciliation method for a provider's atomic process P* first tests if all inputs of process P are available in the simulator's state and all preconditions are satisfied. In such a case it simulates the execution of the atomic process P. If some inputs or preconditions of P are missing, a backward chaining algorithm is used to find a *combination of data mediators* which can provide missing inputs or preconditions. If some of the outputs or effects required by the requester are missing after P is executed, the same backward chaining algorithm is used to find a *combination of data mediators* which can translate generated outputs or generate missing effects. After the process P is reconciled, the *executeNextRequesterCalls* is called to simulate next requester's call. Details of the reconciliation algorithm can be found in [26].

The requester's path:
Login, SearchFlight, ChooseDepFlight, ChooseRetFlight, ...
A possible mapping for first two steps:
requester-Login s1-userID s1-password
provider-LoginStep1 s1-user sessionID
provider-LoginStep2 s1-password sessionID logResult
mediator-prepare-to-send logResult
mediator-send
requester-SearchFlight s2-from s2-to s2-depTime s2-arrTime
external-AirportCityToCode s2-from apt-code-gener1
mediator-explicit-down-casting apt-code-gener1 AirportToCode
external-AirportCityToCode s2-to apt-code-gener2
mediator-explicit-down-casting apt-code-gener2 AirportToCode
external-USTimeToISO s2-depTime iso-time-gener1
mediator-explicit-down-casting iso-time-gener1 ISODepTime
external-USTimeToISO s2-arrTime iso-time-gener2
mediator-explicit-down-casting iso-time-gener2 ISODepTime
provider-SearchReturnFlight apt-code-gener1 apt-code-gener2 iso-time-gener1 iso-time-gener2 flights flght-Count
mediator-prepare-to-send flights
mediator-prepare-to-send flghtCount
mediator-send

Fig. 6. Example solution for a requester's path

4.5 Example Mapping

Figure 6 shows part of one mapping generated for a requester's execution path that can be executed by a requester as defined in Figure 3 in Section 3.2. The mapping was generated for a provider's process model defined in Figure 4. This example assumes that we have provided the system with the *AirportCityToCode* external web service for translating instances of *City* to instances of *AirportCode*, and the service *USTimeToISO* for translating between US and ISO time formats. Each step name is prefixed by *requester*, *provider* and *external* to indicate to which component it is related. Requester's steps show names of inputs parameters, while for the provider, translators and external services also output variables are included. This example also illustrates implicit up-casting of types and explicit down-casting which is enforced by the fact, that *AirportCityToCode* and *USTimeToISO* are defined to work with more generic types than those provided by requester and requested by the provider. Due to the requirement for explicit down-casting, the user is prompted whether the chosen casting is allowed or not. In this example all the castings are allowed. See [22] for details on analyzing casting operations for ontology classes.

5 Process Mediation in the Open World

Dynamic open world of the Internet imposes some restrictions on the mediation process. Since components can appear and disappear completely arbitrarily, it is necessary to incorporate appropriate discovery mechanisms. It is not possible to perform an analysis step in advance because requesters and providers are not known in advance. Therefore the automatic mediation process must rely only on the run-time mediation. Also reasoning possibilities are quite restricted during runtime, since the mediation component must respond instantly and it cannot afford to delay the execution of the requester.

In Section 5.1 we first describe a standalone Runtime Process Mediation Service (RPMS). This component works as an ordinary web service. Therefore, it can for example, be registered with discovery registries and discovered later. In this scenario we assume that the requester first contacts a discovery registry (e.g., the OWL-S/UDDI matchmaker [11]) to find an appropriate provider and then it uses the the RPMS (which also can be discovered) to mediate interactions with the discovered provider.

Next, we show how the Process Mediation Service can be integrated into a Broker component that performs both, the discovery of suitable service providers and the runtime mediation (Section 5.2).

5.1 Runtime Process Mediation Service

Similar to the approach in the semi-open world, the Runtime Process Mediation Service needs to select an appropriate mediation action after it receives a request from the requester. In the run-time scenario, however, the mediation service cannot use any analysis of process models. The only things that the mediation service can see are the current request message received from the requester, the provider's process model, the state of the execution of the provider and a set of available data mediators. Available mediation actions include (1) executing an appropriate service (atomic process) of the provider's process model, (2) applying an appropriate translation (i.e., executing a *data mediator* or a *built-in conversion*) to some inputs provided by the requester or outputs returned by the provider, and (3) preparing required results and sending them to the requester.

Figure 7 shows an architecture of the *Runtime Process Mediator Service*. The *server port* is used for interaction with the requester and the *client port* for interaction with the provider. The *client port* uses OWL-S Virtual Machine to interact with the provider. Another instance of the OVM is used to execute external data mediation services if it is necessary. The *Execution Monitor* is the central part of the RPMS. It executes the mediation algorithm and links all the other components together. It uses the *Knowledge Base (KB)* to store information about received inputs and produced outputs.

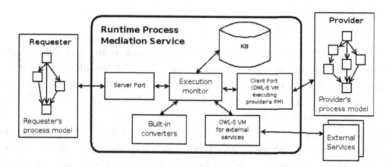

Fig. 7. Runtime Process Mediation Service architecture

The following high-level steps represent a mediation procedure executed by the *Execution Monitor* in the *Runtime Process Mediation Service*:

for each requester's request
    do *processRequest*
until requester or provider finishes successfully or the execution fails

Procedure *processRequest:*
1. Receive a requester's atomic process call *requesterCall* via the *server port*
2. Store inputs of *requesterCall* and required results in the *KB*
3. **While** required results (outputs and effects) of *requesterCall* are not available
    3.1 Choose the best available mediation action A
        if some provider's atomic process that can be executed and that matches inputs
            provided by the requester and produces required results[4] **then**
            select the atomic process as the mediation action A
        else find a *data mediator* or a *combination of available data mediators* that are
            able to translate between the *requesterCall* atomic process and some of the
            possible provider's atomic processes

---

[4] In [27] we described extensions of the OVM that support monitoring and execution introspection which allows an easy tracking of the current execution context. It is for example possible to use the OVM during the execution to get a list of atomic process that are allowed by the process model in the given execution context.

        if some combination is found **then** use it as the mediation action A
        else fail
3.2 Execute action A
    – **if** A is an atomic process of the provider's process model **then**
      use the OVM of the *client port* to execute it
    – **elseif** A is a data mediator represented by an external service **then**
      use OVM for external services to execute it
    – **elseif** A is a built-in conversion **then**
      use built-in converters
    – **elseif** A is return results to requester **then**
      use Requester's *server port*
3.3 Update the *KB*
4. Return required results of *requesterCall* to the requester

The most obvious shortcoming of the run-time approach is the fact that it uses only current state information without analyzing requester's and provider's process models. While this local reasoning is efficient, in cases when more mediation actions are available it can lead to choosing a wrong mediation action that eventually causes the failure of the whole mediation process.

Consider the situation of the requester from Figure 3 in which it executes the *Search-Flight* atomic process. Let us assume that this step can be mapped either to the *Search-OneWayFlight* process or the *SearchReturnFlight* process of the provider. If the mediator used only the current state information (as, e.g., in [28]), these two options would appear as indistinguishable since there is no difference in their IOPEs. Therefore the mediator could choose the *SearchOneWayFlight* which would be wrong since no mapping exists for following two steps (*ChooseDepFlight* and *ChooseRetFlight*) in this context, while in case of selecting the *SearchTwoWayFlight* the mapping exists.

In the previous example the runtime mediation failed because *SearchOneWayFlight* process and the *SearchReturnFlight* process have exactly the same set of inputs and outputs with the same types, and are therefore indistinguishable for the runtime mediation service. Even though such a situation is possible in real-life process models, we believe that in well formed process models it is rather an anomaly than a common situation.

5.2 Broker Hybrid Discovery and Mediation

Although the Runtime Process Mediation Service can be used as a standalone web service, it can be convenient to embed it into the Broker component. The Broker combines the mediation with the discovery process and thus simplifies the interaction process of the provider because it does not have to contact a discovery and mediation service separately. Also the Broker can maintain anonymity of the provider and the requester.

In this paper we build on the broker system described in [29] and extend it by incorporating the process mediation component. We adopt the definition of the Broker protocol based on [13], as graphically summarized in Figure 8. Any transaction involving Broker requires three parties. The first party is a *requester* that initiates the transaction by requesting information or a service to the Broker. The second party is a *provider* which is selected among a pool of provider as the best suited to resolve the problem of the requester. The last party is the *Broker* itself.

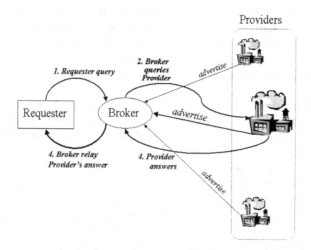

Fig. 8. The broker protocol

The protocol in Figure 8 can be divided in two parts: the *advertisement protocol*, and the *mediation protocol*. In the advertisement protocol, the Broker first collects the advertisements of Web service providers that are available to provide their services. These advertisements, shown in Figure 8 by straight thin lines, are used by the Broker to select the best provider during the interaction with the requester. The mediation protocol, shown in Figure 8 using thick curve lines, requires (1) the requester to query the Broker and wait for a reply while the Broker uses its discovery capabilities to locate a provider that can answer the query. Once the provider is discovered, (2) the Broker reformulates the query for that provider, and finally queries it. Upon receiving the query, (3) the provider computes and sends the reply to the Broker and finally (4) the Broker replies to the requester.

In general, the execution of the protocol may be repeated multiple times. For example, the requester may have asked the Broker to book a flight from Pittsburgh to New York. Since there are multiple flights between the two cities, the provider may ask the Broker, and in turn the requester, to select the preferred flight. These interactions are resolved with multiple loops through the protocol. For example, the Broker translates the list of flights retrieved by the provider for the requester, through steps (3) and (4) of the protocol, and then translates the message with the selected flight from the requester to the provider, via steps (1) and (2) of the protocol. The only exception is that step (1) does not require any discovery since the provider is already known.

The protocol described above shows that the Broker needs to perform a number of complex reasoning tasks for both the discovery and mediation part of its interaction. The discovery task requires the Broker to use the query to describe the capabilities of the desired providers that can answer that query, and then match those capabilities with the capabilities advertised by the providers. During the mediation process, the Broker needs to interpret the messages that it receives from the requester and the provider to decide how to translate them, and whether it has enough information to answer directly the provider without further interaction with the requester. If the requester and provider use fixed, incompatible protocols, the process mediation component as described in the previous section can be used in the broker to mediate between the requester and the provider. In the next two sections, we will analyze these reasoning tasks in more detail.

Discovery of Providers. The task of discovery is to select the provider that is best suited to reply to the query of the requester. Following the protocol, providers advertise their capabilities using a formal specification of the set of capabilities they posses, i.e. the set of functions that they compute. These capability specifications implicitly specify the type of queries that the provider can answer.

The discovery process requires two different reasoning tasks. The first one is to abstract from the query of the requester to the capabilities that are required to answer that query. The second process is to compare the capabilities required to answer the query with the capabilities of the providers to find the best provider for the particular problem.

The first problem, the abstraction from the query to capabilities, is a particularly difficult one. Capabilities specify what a Web service or an agent does, or, in the case of information providing Web services, what set of queries it can answer. For example, the capability of a Web service may be to provide weather forecasting, or sell books, or register the car with the local department of transportation. Queries instead are requests for a very specific piece of information. For example, a query to a weather forecasting agent may be to provide the weather in Pittsburgh, while a query to a book-selling agent may be to buy a particular book. The task of the Broker therefore is to abstract from the particular query, to its semantics, i.e. what is really asked. Finally, the Broker must identify and describe in a formal way the capabilities that are needed to answer that query. The abstraction mechanism is described in detail in [29].

As an alternative to abstracting from the query, the requester could specify its request directly in terms of required provider's capabilities before sending the specific initial query. This would allow to skip the abstraction step in the broker. Such a solution is appropriate, for example, in a situation when the initial broker query does not provide enough information to select an appropriate provider.

The second task of the discovery process is to match the capabilities required to answer the query with the advertisements of all the known providers. Since it is unlikely that the Broker will find a provider whose advertisement is exactly equivalent to the request, the matching process can be very complicated, because the Broker has to decide to what extent the provider can solve the problems of the requester. The matching of the service request against the advertised capabilities was implemented using the OWL-S matching engine reported in [11] and [12].

Management of Mediation. Once the Broker has selected a particular provider, the second reasoning task that the Broker has to accomplish is to transform the query of the requester into the query to send to the provider. This process of mediation has two aspects. The first one is the efficient use of the information provided by the requester to the Broker; the second one is the mapping from the messages of the requester to messages to the provider and vice versa.

Since the requester does not a priori know when it issues the initial query which is the relevant provider, the (initial) query it sends to the Broker and the query input that the (selected) provider may need in order to provide the service may not correspond exactly. The requester may have appended to the query information that is of no relevance to the provider, while the provider may expect information that the requester never provided to the Broker. In the example above, we considered the example of a requester that asks to book the cheapest flight from Pittsburgh to New York. However, besides the trip origin and destination, the selected provider may expect date and time of departure. In the example, the requester never provided the departure time, and the provider has no use for the "cheapest" qualifier. It is the task of the Broker to reconcile the difference between the information that the requester provided and the information that the provider expects, by (1) recognizing that the departure time was not provided, and therefore it should be asked for, and (2) finding a way to select the cheapest flight among the ones that the provider can find.

The approach to mediation proposed in [29] assumes that the requester is flexible enough to be able to provide missing pieces of information that are required by the provider. In our example, the requester must be able to provide the departure time. The problem occurs, when the requester is not flexible because it uses a fixed protocol. It could be perfectly possible that the requester would not be able to provide the departure time immediately after the first query was processed because its own protocol dictates the provider to do something else, as e.g. providing the arrival time first and the departure time afterwards. In such a case the broker must take also the requester's process model into considerations to accomplish mediation.

In general, during mediation, the broker has to deal with exactly the same problems that we identified in Section 3 on process mediation and therefore incorporating a process mediation component into the broker is a natural extension of the original broker implementation.

The Broker Architecture. The overall architecture of the Broker incorporating the mediation component is shown in Figure 9. It is based on the broker architecture described in [29]. To interact with the provider and the requester the Broker instantiates two ports: a *server port* for interaction with the requester (since the Broker acts as a provider vis a vis the requester) and a *client port* for interaction with the provider (since the Broker acts as a client vis a vis the provider).

The *client port* uses the *OWL-S Virtual Machine* (OVM) [18] to interact with the provider. The *server port* which is used to communicate with the requester exposes the process model that is equivalent with the requester's process model. Specifically, the *server port* receives messages sent by the requester and replies to them in the format that corresponds to definitions of the provider's Process Model.

The reasoning of the Broker happens in the *Execution Monitor / Query Processor* that is responsible for translation of messages between the two parties and for the implementation of the mediation algorithms. Specifically, the *Execution Monitor / Query Processor* stores information received from the requester in a *Knowledge Base* that is instantiated with the information provided by the requester. Furthermore, the *Execution Monitor / Query Processor* interacts with the *Discovery Engine,* which provides the

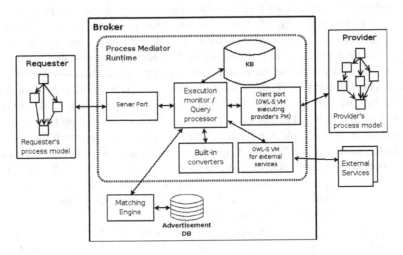

Fig. 9. The hybrid broker architecture

storage and matching of capabilities, when it receives a capability advertisement and when it needs to find a provider that can answer the query of the requester.

6   Related Work

The work in [30] provides a conceptual underpinning for automatic mediation. The work closest to ours is in [28]. Mediation between two WSMO based process is performed strictly during runtime. Besides structural transformations (e.g., change of message order) also data-mediators can be plugged into the mediation process. Aberg et. al [31] describes an agent called sButler for mediation between organizations' workflows and semantic web services. The mediation is more similar to the brokering, i.e., having a query or requirement specification, the sButler tries to discover services that can satisfy it. The requester's process model is not taken into considerations. OWL-S broker [29] also assumes that the requester formulates its request as query which is used to find appropriate providers and to translate between the requester and providers. The work in [32] and [33] describe the IRS-III broker system based on the WSMO methodology. IRS-III requesters formulate their requests as goal instances and the broker mediates only with providers given their choreographies (explicit mediation services are used for mediation). Authors in [34] apply a model-driven approach based on WebML language. Mediator is designed in the high-level modeling language which supports semi-automatic elicitation of semantic descriptions in WSMO. In [22], data transformation rules together with inference mechanisms based on inference queues are used to derive possible reshapings of message tree structures. An interesting approach to translation of data structures based on solving higher-order functional equations is presented in [23] while [24] argues for published ontology mapping to facilitate automatic translations.

7   Conclusions and Further Work

In this paper we described mediation mechanisms of two OWL-S process models that can be applied in different environments. Specifically, we considered a semi-open environment that frequently occurs for inter and intra corporation business process. We described an algorithm based on the analysis of provider's and requester's process models for deciding if the mediation is possible and for performing the runtime mediation in the semi-open environment. Next, we described a Runtime Process Mediation Service that can be used in an open environment of the Internet. Finally we demonstrated how discovery can be combined with the process mediation within the Broker component. Embedding discovery and process mediation into the Broker facilitates interactions between requesters and providers because requesters do not have to contact discovery and mediation services separately and Broker further allows to maintain anonymity of the requester and the provider.

In future work, we will address some of the current limitations such as a better support for data mediation and improve efficiency. We want to explore a user assisted mediation and the top-down analysis of process models to allow more selective exploration of requester's paths and so to better deal with possible combinatorial explosion. Also we want to explore the possibility of discovering data mediators during runtime.

References

1. Christensen, E., Curbera, F., Weerawarana, G.M.S.: Web Services Description Language (2001)
2. Andrews, T., Curbera, F., Dholakia, H., Goland, Y., Klein, J., Leymann, F., Liu, K., Roller, D., Smith, D., Thatte, S., et al.: Business Process Execution Language for Web Services, Version 1.1 (2003)
3. Akkiraju, R., Farrell, J., Miller, J., Nagarajan, M., Schmidt, M.T., Sheth, A., Verma, K.: Web Service Semantics - WSDL-S (2005),
   http://www.w3.org/Submission/WSDL-S/
4. The OWL Services Coalition: Semantic Markup for Web Services (OWL-S),
   http://www.daml.org/services/owl-s/1.1/
5. Roman, D., Keller, U., Lausen, H., de Bruijn, J., Lara, R., Stollberg, M., Polleres, A., Feier, C., Bussler, C., Fensel, D.: Web Service Modeling Ontology. Applied Ontology 1(1), 77–106 (2005)
6. Sycara, K., Paolucci, M., Ankolekar, A., Srinivasan, N.: Automated discovery, interaction and composition of semantic web services. Journal of Web Semantics 1 (1), 27–46 (2004)
7. Bechhofer, S., van Harmelen, F., Hendler, J., Horrocks, I., McGuinness, D.L., Patel-Schneider, P.F., Stein, L.A.: OWL web ontology language reference, W3C Recommendation (February 10, 2004), http://www.w3.org/TR/owl-ref/
8. SOAP: Simple object access protocol (SOAP 1.1), http://www.w3.org/TR/SOAP
9. Wong, H.C., Sycara, K.P.: A taxonomy of middle-agents for the internet. In: ICMAS, pp. 465–466. IEEE Computer Society, Los Alamitos (2000)
10. OASIS: Universal description, discovery and integration v3.0.2 (2005),
    http://www.oasis-open.org/committees/uddi-spec/doc/spec/v3/
    uddi-v3.0.2-20041019.htm
11. Paolucci, M., Kawmura, T., Payne, T., Sycara, K.: Semantic matching of web services capabilities. In: First Int. Semantic Web Conf. (2002)

12. Paolucci, M., Sycara, K.P., Kawamura, T.: Delivering semantic web services. In: WWW (Alternate Paper Tracks) (2003)
13. Decker, K., Sycara, K., Williamson, M.: Matchmaking and brokering. In: Proceedings of the Second International Conference on Multi-Agent Systems. The AAAI Press, Menlo Park (1996)
14. Lu, J., Mylopoulos, J.: Extensible information brokers. International Journal on Artificial Intelligence Tools 11(1), 95–115 (2002)
15. Jennings, N.R., Norman, T.J., Faratin, P., O'Brien, P., Odgers, B.: Autonomous agents for business process management. Applied Artificial Intelligence 14(2), 145–189 (2000)
16. Chen, H., Finin, T.W., Joshi, A.: Semantic web in the context broker architecture. In: Proceedings of the IEEE Conference on Pervasive Computing and Communications (PerCom), pp. 277–286. IEEE Computer Society, Los Alamitos (2004)
17. Motta, E., Domingue, J., Cabral, L., Gaspari, M.: IRS-II: A Framework and Infrastructure for Semantic Web Services. In: Fensel, D., Sycara, K.P., Mylopoulos, J. (eds.) ISWC 2003. LNCS, vol. 2870, pp. 306–318. Springer, Heidelberg (2003)
18. Paolucci, M., Ankolekar, A., Srinivasan, N., Sycara, K.P.: The DAML-S virtual machine. In: Fensel, D., Sycara, K.P., Mylopoulos, J. (eds.) ISWC 2003. LNCS, vol. 2870, pp. 290–305. Springer, Heidelberg (2003)
19. Ankolekar, A., Huch, F., Sycara, K.P.: Concurrent semantics for the web services specification language DAML-S. In: Arbab, F., Talcott, C.L. (eds.) COORDINATION 2002. LNCS, vol. 2315, pp. 14–21. Springer, Heidelberg (2002)
20. Paolucci, M., Srinivasan, N., Sycara, K.: Expressing wsmo mediators in owl-s. In: International Semantic Web Conference (2004)
21. Burstein, M., Bussler, C., Finin, T., Huhns, M., Paolucci, M., Sheth, A., Williams, S., Zaremba, M.: A semantic web services architecture. Internet Computing. IEEE 9(5), 72–81 (2005)
22. Spencer, B., Liu, S.: Inferring data transformation rules to integrate semantic web services. In: International Semantic Web Conference, pp. 456–470 (2004)
23. Burstein, M., McDermott, D., Smith, D.R., Westfold, S.J.: Derivation of glue code for agent interoperation. Derivation of glue code for agent interoperation V6(3), 265–286 (2003)
24. Burstein, M.H., McDermott, D.V.: Ontology translation for interoperability among semantic web services. The AI Magazine 26(1), 71–82 (2005)
25. Sycara, K.P., Klusch, M., Widoff, S., Lu, J.: Dynamic service matchmaking among agents in open information environments. SIGMOD Record 28(1), 47–53 (1999)
26. Vaculín, R., Sycara, K.: Towards automatic mediation of OWL-S process models. In: 2007 IEEE International Conference on Web Services, pp. 1032–1039. IEEE Computer Society, Los Alamitos (2007)
27. Vaculín, R., Sycara, K.: Monitoring execution of OWL-S web services. In: European Semantic Web Conference, OWL-S: Experiences and Directions Workshop, June 3-7 (2007)
28. Cimpian, E., Mocan, A.: WSMX process mediation based on choreographies. In: Business Process Management Workshops, pp. 130–143 (2005)
29. Paolucci, M., Soudry, J., Srinivasan, N., Sycara, K.: A broker for owl-s web services. In: Cavedon, M., Martin, B. (eds.) Extending Web Services Technologies: the use of Multi-Agent Approaches. Kluwer, Dordrecht (2005)
30. Wiederhold, G., Genesereth, M.R.: The conceptual basis for mediation services. IEEE Expert 12(5), 38–47 (1997)
31. Aberg, C., Lambrix, P., Takkinen, J., Shahmehri, N.: sButler: A Mediator between Organizations Workflows and the Semantic Web. In: World Wide Web Conference workshop on Web Service Semantics: Towards Dynamic Business Integration (2005)
32. Cabral, L., Domingue, J., Galizia, S., Gugliotta, A., Tanasescu, V., Pedrinaci, C., Norton, B.: IRS-III: A Broker for Semantic Web Services Based Applications (2006)

33. Domingue, J., Galizia, S., Cabral, L.: Choreography in irs-iii - coping with heterogeneous interaction patterns in web services. In: Proc. 4th Intl. Semantic Web Conference (2005)
34. Brambilla, M., Celino, I., Ceri, S., Cerizza, D., Valle, E.D., Facca, F.M.: A software engineering approach to design and development of semantic web service applications. In: Cruz, I.F., Decker, S., Allemang, D., Preist, C., Schwabe, D., Mika, P., Uschold, M., Aroyo, L. (eds.) ISWC 2006. LNCS, vol. 4273, pp. 172–186. Springer, Heidelberg (2006)

Latent Semantic Analysis – The Dynamics of Semantics Web Services Discovery

Chen Wu, Vidyasagar Potdar, and Elizabeth Chang

Digital Ecosystems and Business Intelligence Institute
Curtin University of Technology, Perth 6845, WA, Australia
{Chen.Wu,Vidyasagar.Potdar,Elizabeth.Chang}@cbs.curtin.edu.au
http://debii.curtin.edu.au

Abstract. Semantic Web Services (SWS) have currently drawn much momentum in both academia and industry. Most of the solutions and specifications for SWS rely on ontology building, a task needs much human (e.g. domain experts) involvement, and hence cannot scale very well in face of vast amount of web information and myriad of services providers. The recent proliferation of SOA applications exacerbates this issue by allowing loosely-coupled services to dynamically collaborate with each other, each of which might maintain a different set of ontology. This chapter presents the fundamental mechanism of Latent Semantic Analysis (LSA), an extended vector space model for Information Retrieval (IR), and its application in semantic web services discovery, selection, and aggregation for digital ecosystems. First, we explore the nature of current semantic web services within the principle of ubiquity and simplicity. This is followed by a succinct literature overview of current approaches for semantic services/software component (e.g. ontology-based OWL-s) discovery and the motivation for introducing LSA into the user-driven scenarios for service discovery and aggregation. We then direct the readers to the mathematical foundation of LSA – SVD of data matrices for calculating statistics distribution and thus capturing the 'hidden' semantics of web services concepts. Some existing applications of LSA in various research fields are briefly presented, which gives rise to the analysis of the uniqueness (i.e. strength, limitations, parameter settings) of LSA application in semantic web services. We provide a conceptual level solution with a proof-of-concept prototype to address such uniqueness. Finally we propose an LSA-enabled semantic web services architecture fostering service discovery, selection, and aggregation in a digital ecosystem.

1 Introduction

Semantics play an important role in the complete lifecycle of Web services as it is able to help service development, improve service reuse and discovery, significantly facilitate composition of Web services and enable integration of legacy applications as part of automatic business process integration. Unfortunately, current Web Service Description Language (WSDL) standard operates at the syntactic level and lacks the semantic expressivity needed to represent the requirements and capabilities of Web

T.S. Dillon et al. (Eds.): Advances in Web Semantics I, LNCS 4891, pp. 346–373, 2008.

Services. This gap has motivated a lot existing research effort towards the Semantic Web Services (SWS). The fundamental idea underlying current SWS community is that in order to achieve machine-to-machine integration, a markup language must be descriptive enough that a computer can automatically determine its meaning. Following this principle, many semantic annotation markup languages have thus come into view, among them are OWL-S (formerly known as DAML-S) and WSDL-S that have gained great momentum in recent years. The main goal of both OWL-S and WSDL-S is to establish a framework within which service descriptions are made and shared.

The premise of such an ontology-based markup language approach is that every SWS user (be it normal website or end customer) is able to employ a standard ontology, consisting of a set of basic classes and properties, for declaring and describing services. One concern about this descriptive annotation-driven approach is its feasibility: since it would be much more time-consuming to create and publish ontology-annotated (WSDL) content as they would need to be done by domain human experts and powerful editing tools for common users. Other problems might occur when different groups of users and communities want to manage the shared ontology. With this being the case, it would be much less likely for industry companies to adopt these practices as it would only slow down their progress.

In this chapter we carry out SWS research, in particular the service discovery, from another empirical perspective. We believe that one thing distinguishes Web (services) semantics from other forms of semantics is its 'user-centred' commitment towards ubiquity and simplicity, the two most renowned factors leading to the great success of today's Web. By ubiquity, we mean that the underlying technology (such as HTTP and TCP/IP) has to be very robust, lightweight, and non-human intervened to serve for various applications and users. By simplicity, we mean millions of end users can easily access to and personally use the technology without too much expertise in both domain and IT areas. This idea drives us to come up with a novel method to approach the SWS using Latent Semantic Analysis (LSA) technique – the main theme of this chapter. Nevertheless, proposing alternate approach does not mean we completely go against ontology-based approach. On the contrary, we acknowledge that adding semantics in the form of ontology to represent the requirements and capabilities of Web services is essential for achieving unambiguity and machine-interpretability for web services, and hence becomes our long term research objectives. For example, we are currently seeking effective ways that can convert some of our research result – the higher-order association (the very initial stage semantic space) in this chapter – into lightweight ontology in a semi-automatic manner.

The chapter is organised as follows. Section 2 presents OWL-S and WSDL-S, the two widely accepted SWS specifications and characteristics. Section 3 provides in-depth mathematics technique and working mechanism on LSA. Applications of LSA in IR, cognitive, and psychology are introduced in Section 4. This is followed by the rationale to apply LSA in the area of SWS, which has some issues to address. Section 5 proposes the conceptual model of LSA-based SWS. Section 6 then presents semantic search engine prototype based on our conceptual model. Experiment results are reviewed in Section 7, where three LSA and one WSDL parameters are manipulated to gain further understanding of our approach. The chapter concludes in Section 8.

2 Related Work

In this chapter we provide a succinct survey on existing SWS approaches. In particular, the OWL-s and WSDL-s specifications are examined from an empirical perspective.

Semantic Web Services (SWS) attempts to address the problem associated with automatic service discovery, dynamic service selection, composition and aggregation, and other relevant tasks involved in implementing web services. In this section we provide a succinct survey on existing SWS approaches. In particular, the OWL based Web service ontology (OWL-S), Web services modelling ontology (WSMO), and WSDL-S specifications are examined from an empirical perspective. Of these three approaches OWL-S and WSMO can be categorized together because they propose a model for developing WS Ontologies, on the other hand WSDL-S utilizes the existing web service description and attempts to add a layer of semantics to it [28, 29].

From the first category i.e. OWL-S and WSMO, OWL-S is more successful compared to WSMO although both of these approaches define their own set of detailed semantic models, which need to be used when expressing the semantic meaning to web services. On the other hand WSDL-S incorporates the existing WSDL file and adds semantics to it, thus reducing a reasonable amount of time to reconstruct the whole WSDL using a model similar to OWL-S or WSMO. Thus it takes more of a bottom up approach when adding semantics to web services. We now explain each of these approaches in more detail [28].

2.1 OWL Based Web Service Ontology

The W3C standard OWL is based on RDF(S) and supports the modelling of knowledge/Ontologies. Within the semantic web services framework, OWL-S provides a set of markup language constructs that can be used to unambiguously describe the properties and functionalities of web service, which in turn adds machine understanding capabilities to web services [29]. This machine understandability can facilitate dynamic service discovery, selection, composition and aggregation. OWL-S describes each and every instance of web service using an OWL ontology which comprises of service profile, service model and service grounding. *Service profile* describes what the service does, *service model* explains how the service works or how it can be used and finally *service grounding* details how to interact with this service. These three components within OWL-S provide the backbone for adding machine intelligence to web services. There are several editors which can be used to create OWL-S. Protégé provides an OWL-S extension, which can be used for creating semantic service descriptions [http://owlseditor.semwebcentral.org/documents/tutorial.pdf].

2.2 Web Services Modelling Ontology

WSMO has four top-level elements, which needs to be described in order to express Semantic Web services [28, 30]. These include:

- Ontologies, which provide the terminology to capture the relevant aspects of the domain.
- Web services, which describe the computational entity providing access to services.

- Goals, which represent user's desires, which can be fulfilled by discovery and executing a Web service and finally
- Mediators, which describe elements that overcome interoperability problems between different WSMO elements. WSMO handled mediation at three levels – data level, protocol level and process level, which is essential for interoperability.

The WSMO elements referred above includes Concepts, Relations, Functions, Instances and Axioms. *Concepts* refer to the subsumption hierarchy and their attributes, including range specification e.g. "person" with attributes like "name, family, DoB etc". *Relations* describe interdependencies between a set of parameters. *Functions* are a special type of relations that have an unary range beside the set of parameters. Instances define the concepts explicitly by specifying concrete values for attributes and *Axioms* are specified as logic expressions and help to formalize domain specific knowledge.

2.3 Web Service Description Language - Semantics

WSDL describes web services so that they can be interoperable however it lacks the semantic information which is required to facilitate machine understanding. OWL as described earlier supports developing of ontologies in a powerful way, but it lacks the technical details required to express web services. WSDL-S incorporates these advantages by adopting semantic annotation to web services using OWL. In other words WSDL-S connects WSDL and OWL in a very practical manner [27, 28]. It is much better than OWL-S and WSMO because of the following advantages:

- It is compatible to WSDL as semantic information is added to the WSDL file itself by relating it to an external domain model.
- There is a option to choose any modeling technology like OWL or UML
- It is very simple to apply and is based on a stabile standard, which is important for practical use

WSDL-S is the latest submission to W3C. It extends the functionality of WSDL by defining new elements and annotations for already existing elements. This annotation provides a mechanism for adding semantics by linking it to semantic concepts defined in some external domain model or ontology. Semantics can be associated with inputs, outputs, operations, preconditions etc. The main advantage with this approach is that the semantic models are not necessarily being tied with OWL; any existing semantic models can be reused as long as they can be referenced from within the WSDL. This is a huge advantage because there are already a lot of semantic model developed over time and this approach can adopt the same rather than replicating from scratch. The following extensibility elements and attributes can be used in WSDL-S to add semantics to web services. These include

- modelReference – Element: Input and Output Message Types:
  - This extending attribute supports the connections between a WSDL document and an ontology
- schemaMapping - Element: Input and Output Message Types
  - This extending attribute solves differences in schemas by XSL transformation

- modelReference - Element: Operation
    - o Captures the semantics of the functional capabilities of an opera-
      tion, which can be used for dynamic web service composition or
      aggregation.
- pre-conditions – Parent Element: Operation
    - o This extending attribute defines the pre-conditions that need to be
      satisfied before the operation can be invoked.
- effects – Parent Element: Operation
    - o This extending attribute defines the set of semantic statements that
      must be true after the operation is executed.
- category - Parent Element: Operation
    - o This element is used in the classification of the service interface. It
      is basically used for semantic lookup in a service registry.

The flexibility offered by WSDL-S in incorporating existing semantic domain models shows that this approach is going to be very successful as well as practical. Creating semantic markup of web services is a realization of true semantic web and can only
be possible by a mix of these technologies. This concludes the section on semantic web
services and we now discuss focus on web services discovery using LSA.

To our best knowledge, [1] is the only work to date that has attempted to leverage
LSA in facilitating web services discovery. However, both implementation and the
experiment result is not very thoroughly stated and analysed.

3 Background

3.1 LSA and VSM

Latent Semantic Analysis (LSA), also known as Latent Semantic Indexing, is a technique for document retrieval based on text vector representation. It evolves from the
Vector Space Model (VSM), a widely used IR model. In VSM, each document is
encoded as a vector, where each vector component reflects the importance (e.g. frequency or $tf * idf$) of a particular term in representing the meaning of that document.
The vectors for all documents in a collection are stored as the 'columns' of a single
matrix – the term-by-document matrix A, in which each document is represented as a
vector with a set of dimensions, each of which measures the importance of a keyword
in that particular document. On the other hand, each term is considered as a vector,
each component of this vector measures the weight of this term in the document. A
user query is then converted to a 'pseudo-document' that can be compared with all
existing documents through basic vector operations, such as the cosine value of two
vectors (vector dot product divided by the norm product). The cosine value is referred
to as the 'similarity' between user query and each document, and is used for ranking
the document relevance against the user query.

The main problem of VSM is its reliance on keyword-based matching that has several well-known limitations such as low recall, issues caused by synonymy and
polysemy. This is due to the assumption in VSM that keyword can precisely represent
the concepts inferred in the human natural langue. As stated in [2], "users want to
retrieve on the basis of conceptual content, and individual words provide unreliable

evidence about the conceptual topic or meaning of a document. There are usually many ways to express a given concept, so the literal terms in a user's query may not match those of a relevant document. In addition, most words have multiple meanings, so terms in a user's query will literally match terms in documents that are not of interest to the user." To solve these limitations, LSA is firstly proposed in [2] that supports a different idea – "a particular concept is stated not just through a unique word, but with a distribution of terms focused around that concept". This different approach – indexing and retrieving based on matching concept distribution rather than a particular term/keyword – should firstly reduce the consequence of synonyms as the searching is not based on a single term that may have many synonyms, but a distribution of terms around that concept. It should also reduce the influence of polysemy as the distribution of terms will differ for each unique concept, on which the indexing and searching are based. Evidently, the main challenge in LSA is to convert keyword-based term space into a concept-centred semantic space. While a thorough discussion on mathematics detail of LSA is beyond the scope of this chapter, a brief introduction of mathematics aspects, especially the matrices techniques, is presented as follows.

3.2 Mathematical Foundation

LSA is a variant of the VSM in which the original VSM matrix is replaced by a low-rank approximation matrix. Therefore, the original vector space is reduced to a 'subspace' as close as possible to the original matrix in order to: (1) Remove extraneous information or noise from the original space, (2) Extract and infer more important and reliable relations of expected contextual usage of terms. As a result, LSA overcomes traditional VSM problems by building 'hidden' meaningful concepts from existing keywords. This is achieved by finding the higher-order association between different keywords using pure mathematical model: a two-mode factor analysis, i.e. the Singular Value Decomposition (SVD).

Formally, given a rectangular m-by-n matrix A (m $\geq$ n), and rank $(A) = r$ the singular value decomposition of A, denoted by $SVD(A)$, is any factorisation of the form:

$$A = T_0 \, S_0 \, D_0^t,$$

where T_0 is an m-by-m orthogonal matrix $(T_0^t \, T_0 = I_n)$ having the left singular vectors of A as its columns, D_0 is an n-by-n orthogonal matrix $(D_0^t \, D_0 = I_n)$ having the right singular vectors of A as its columns, and S_0 is an m-by-n diagonal matrix $S_0 = \mathrm{diag}(\sigma_1, \sigma_2, \cdots, \sigma_n)$, $\sigma_i > 0$ for $1 \leq i \leq r$, $\sigma_j = 0$ for $j \geq r + 1$, and $\sigma_1 \geq \sigma_2 \geq \cdots \geq \sigma_r > \sigma_{r+1} = \cdots = \sigma_n = 0$ of S_0 in decreasing order along its diagonal. The white area in S_0 stands for zeroes. That is to say the rank r of the matrix A is equal to the number of non-zero singular values. This factorisation exists for any matrices. There exists dedicated research on calculating SVD in the literature such as [3]. Here we focus on how and why SVD can be applied in the application of Information Retrieval (IR). Figure 1 depicts the application of LSA in IR, where the element a_{ij} in term-document matrix A denotes the frequency in which term i occurs in document j. In practice, this frequency value is replaced with local and global weights to adjust the importance of terms within or amongst documents. Based on the two-mode factor analysis theory, the columns in T_0 represent term vectors (i.e. the 'entity' set) and columns in D_0 (or rows in D_0^t) represent document vectors (i.e. the 'attribute' set).

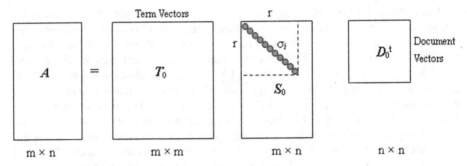

Fig. 1. SVD before dimension reduction

A primary concern of factor analysis with such a matrix is to determine "whether the table may be simplified in some way" [4]. Hence, it aims to reduce the rank of A and obtain the structural information of matrix. The decreasing order of σ_i values gives rise to a simple strategy for such an optimal approximation – reducing the rank of original A by keeping the first k largest singular values in S_0 and set others small σ_i ($k < i \le r$) to zero. In fact, rigorous mathematical proofs [5] have demonstrated that an reduced matrix A_k constructed from the k-largest singular triplets of A, is the best approximation to A. It can reveal important information about the structure of a matrix. The representation of S_0 can then be simplified by deleting the zero rows and columns to obtain a new k-by-k diagonal matrix S. Likewise, the corresponding columns of T_0 and D_0 are removed to derive a new left singular vector T and a new right singular vector S. The resultant matrix is a reduced model:

$$A \approx A_k = T\,S\,D^t,$$

Where T is the m-by-k matrix whose columns are the first k columns of T_0, D is the n-by-k matrix whose columns are the first k columns of D_0. The new model contains k factors that form the basis of the semantic vector space A_k that is closest to the original vector space A in the sense of least-square-fit. This is illustrated in Figure 2, where shaded rectangles represent the reduced matrices.

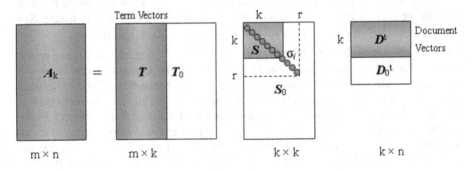

Fig. 2. SVD after reduction to k dimension

It is this reduced shaded A_k that states the hidden relations – i.e. the semantics – between terms, terms and documents, and documents. For example, the term-by-term comparison can be conducted using inner product of smaller factorised matrices as follows: $AA^t \approx A_k A_k^{\ t} = TSD^t (TSD^t)^t = TSD^t DS^t T^t$, since D is orthogonal, so $D^t D = I$, hence $TSD^t DS^t T^t = TSS^t T^t = TS (TS)^t$. That is to say the term-by-term similarity is the dot product of two rows in the smaller matrix TS. Similarly, the document-by-document comparison can be formulated as $A^t A \approx A_k^{\ t} A_k = DS (DS)^t$, i.e. the dot product between two rows in the smaller matrix DS. Moreover, the similarity between a term and a document is literally the value of the cell in the reduced matrix A_k, thus $A_k = TSD^t = TS^{1/2} (DS^{1/2})^t$, i.e. the dot product between two rows in smaller matrix $TS^{1/2}$ and $DS^{1/2}$ respectively.

For information retrieval, a generic user query can be converted into a pseudo-document $D_q = A_q^t TS^{-1}$, where A_q is the query's term vector in the original vector space A. TS^{-1} means to scale the original query term vector using the inverse of a corresponding singular values on the T space. Once such a conversion is achieved, any user queries can be compared to existing documents and terms in the reduced semantic space A_k using a score function. While inner product, as illustrated earlier, has been proven the most natural score function in the reduced space [6], most LSA applications have used the cosine between two vectors in the classical VSM model as shown in Section 3.1. Research in [7] has summarised four variants of LSA cosine score functions used by existing LSA research and applications, but the author was also unable to find a theoretical basis for opting for one score function over another. For applications other than IR, the original documents are often used for the purpose of training – i.e. creating the *semantic space*. We will discuss the semantic space in following sections.

3.3 Rationale

The hidden association derived by LSA are not just simple proximity frequencies, co-occurrence counts, or correlations in usage, but are based on SVD, which captures not only the surface adjacent pair-wise co-occurrences of terms (keywords) but the thorough patterns of occurrences of a great number of terms over very large numbers of text corpus. The word constraint information provided by higher-order associations is extracted by LSA by analysing tens of thousands of different episodes of past sentences and paragraphs. This extracted information enables computers to generate meaning similarity and distance judgments.

The central idea of SVD is to derive matrix A_k not by recreating the original term document matrix A exactly. This, in one sense, captures most of the important underlying structure in the association of terms and documents, yet at the same time removes the noise or variability in word usage that affect keyword-based retrieval methods. For example, sine the number of dimension k is much smaller than the number of documents n and number of original terms m, some seemly different terms will be 'squeezed' into the vicinity locations on this newly created the semantic space only because they occur in similar documents rather than in the same document.

Why SVD, and in particular, the dimension reduction, is so helpful in finding the true relations (e.g. the similarity)? While this is still an open topic that is lack of

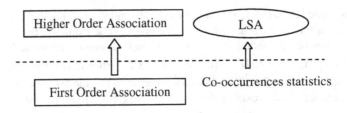

Fig. 3. Higher-order association

rigorous mathematical proof for the potential answers (e.g. removing the noise data, finding the statistical distribution), authors in [8] hypothesise that one reason could be that the original data are generated from a source of the same dimensionality and general structure as the reduction. This assumes that there exists a true semantic space, from which the original documents corpora are generated. This true semantic space has the same dimensions and structure of the reduced SVD matrix. In other words, the reduced matrix reflects the truth of the semantics, by which the data source is created. Furnas et al. in the first paper [9] on SVD applied in IR state that "the problem, of course, is that we do not have access to the true matrix; we must settle for only an estimate". Here the true matrix is the true semantic space, the estimate could be the original (what they call 'observed') data source matrix, or could be the reduced estimated low-dimensional matrix. But the reduced matrix has made a 'better' estimation due to it attempts to capture the "structural model of associations/patterns among terms". One reason could be that the cells of the observed matrix are used to estimate parameters of the underlying model. Since there are fewer parameters than data cells, this model can be more statistically reliable than the raw data, and hence can be used to reconstruct an improved estimate of the cells of the true matrix (the semantic space).

It has been found that most of the hidden association that LSA has inferred among words is in documents where particular words did not occur. As described in [8], relationships inferred by LSA are "relations only of similarity that has mutual entailments of the same general nature, and also give rise to fuzzy indirect inferences that may be weak or strong and logically right or wrong". Such a higher-order hidden structure of the association of terms and documents is revealed with the use of linear algebra (SVD). This is different from "readily appreciated" [10] first-order associations (as shown in Figure 3), which do not contain the only type of information that exists amongst words; nor do such associations necessarily hold the most important information. Higher-order association complies with the basic way how human brain relates similar words: humans have an intuitive grasp of which meanings are similar and which are not, and no philosophical or other analysis to can even fully explain how or why people understand intuitively which words can be swapped in certain contexts.

4 Applications

Most applications of LSA come from two major research fields: Information Retrieval and Psychology-related disciplines such as psycholinguistic, cognitive science, etc. The abundance application of LSA in IR field is not surprising as mathematical foundation of LSA is built on top of the VSM, the most popular IR model to date. On the other hand, the nature and motivation of LSA (SVD) is the two-mode factor analysis. As well known, it is in the field of psychology that factor analysis has its start and where undoubtedly the greatest amount of theoretical and empirical work has been done. Hence, it is very natural for LSA flourishing in this field during the past two decades.

4.1 Application in Information Retrieval (IR)

LSA has been extensively applied in the area of Information Retrieval. In IR field, LSA is more often referred to as Latent Semantic Indexing (LSI) as automatic indexing is the central issue dealt with in most IR research, and LSA adds an important step to this document indexing process [2]. One prominent advantage of LSA over existing VSM is that LSI does not require an exact match to return useful results under the circumstances that two documents may be semantically very close even if they do not share a particular keyword. Where a plain keyword search will fail if there is no exact match, LSI will often return relevant documents that don't contain the keyword at all as will be shown in our prototype experiment, but have semantically overlapping (context or topic) vectors with the used keyword. LSA will perform better is more keywords are used, thus providing context information for the search and mapping more document vectors. On the whole, using LSA for indexing can significantly improve three important characteristics of a search engine [11]: recall (find every document relevant to the query), precision (no irrelevant documents in the result set) and ranking (most relevant results come first). We are also particularly interested in LSA in component retrieval area as the concept of Web service evolves from the concept of component. Investigating component retrieval can provide insightful information for service discovery. Authors in [12] has built a Java reuse repository using LSA for component retrieval. Similarly, research in [13] proposed an active component repository systems that support "reuse-within-development" using real-time LSA component retrieval.

4.2 Applications in Human Knowledge Acquisition

LSA has also been successfully applied in modelling human conceptual knowledge. The capability of LSA's reflection of human knowledge has been established in a variety of ways. For example, its scores overlap those of humans on standard vocabulary and subject matter tests; it mimics human word sorting and category judgements; it simulates word-word and passage-word lexical priming data; and it can even accurately estimates passage coherence, learnability of passages by individual students, and the quality and quantity of knowledge contained in an essay. LSA acquires words at a similar pace to human children, sometimes exceeding the number of words to which it is exposed [14]. LSA's knowledge of synonyms is as good as that of second-language English speakers as evidenced by scores on the Test of English as a Foreign Language (TOEFL [14]). LSA can tell students what they should read next to help them learn (see Wolfe, 1998). Some research reported that LSA can even interpret metaphors like, "My lawyer is a shark" [15].

5 Issues

In this section, we discuss several well-known issues associated with LSA that have been constantly reported from the literature and applications. Further more, we identify three major issues needed to be addressed when applying LSA in semantic Web services, in particular, service discovery.

5.1 Issues in LSA

The first issue for LSA is that it makes no use of word order, thus of syntactic relations or logic, or of morphology. As shown in [16], LSA does not do very well on single sentences as it does pretty satisfactory on single words or non-syntactic texts such as a long passages of words. However, several approaches have been proposed to address this well-known LSA weakness. One method [17] seems promising: it uses surface parsing of sentences so that the components can be compared separately with LSA and String Edit Theory is employed to infer structural relations.

The first dimension of the k-dimensions is problematic [18]. Since all elements in the original matrix A are non-negative, the entries in first dimension of T always have the same sign. And the mean of their values is much larger than that of other dimensions. As a result, the cosine value between any two documents, if the term method (i.e. the text vector can be computed as a function of the term vectors) is employed, is inevitably related to the number of terms in the document. This property makes it very hard to reveal the real similarity without considering the sizes of the documents. However, this issue is not very prominent when retrieving documents rather than comparing documents is desired. In a nutshell, when using LSA cosine value for the measure of similarity between documents, the size of the texts has to be thoroughly considered.

Following the first dimension issue, a more fundamental question is: "what do all these dimensions really mean?" So far, there is no clear definition or interpretation on these latent (hidden) factors, which are used in a rather abstract mathematical manner. First dimension phenomena is the only observed findings regarding the pattern of these dimensions (factors), no further research has done beyond that. Such a blur way of conducting LSA causes many confusing, if not totally wrong, solutions in face of various applications. For example, the appropriate number of dimensions (factors) is a well-known question plaguing LSA researchers. Currently, only practical experiments can perhaps provide some hints, which is however only limited to those experiments settings.

There is also considerable debate as to what extent LSA captures first order co-occurrence or higher order co-occurrence. Recent evidence from [19] shows that although second order effects do occur, large changes in the similarity measure between two words can be seen when a document is added to a training corpus in which both words occur (first order co-occurrence). In other words, first order association seems also receive adequate support from LSA.

Last, performance is also an issue that needs to be addressed since large matrix reduction such as SVD does take tremendous resources in terms of CPU and RAM. We will discuss the LSA complexity and scalability in Section 7.6 and 7.7.

5.2 Issues in SWS

Introducing LSA to semantic Web service is a novel yet challenging task. It is true that LSA can extract complex patterns of association contained in word usage from many tens of thousand of sentences. However, the uniqueness of Web services, in particular WSDL files, has raised a number of issues. First of all, WSDL is not a human natural language. It is an XML-based interface language meant to be processed automatically by machines. Therefore, it is far more structural and compact than general natural language. For example, an online foreign exchange Web service can have an operation with signature "decimal AUD2USD(decimal quantity)". Intuitively, it would be difficult for LSA to build the association between this operation and the concept of 'Currency Conversion' that a service consumer is looking for. This is because WSDL is designed for machine-to-machine interaction at the syntactic level. Therefore, word meanings that LSA can infer accurately might not apply WSDL. One promising approach appears to conduct rigorous analysis on WSDL corpus with sublanguage [20] patterns.

One may wonder that WSDL has provided a flexible extension mechanism to support annotations and comments written in natural language. To our best knowledge, most existing WSDL files are automatically generated from binary interfaces written in advanced programming languages such as Java, C#, or Delphi, etc. To avoid the error-prone XML syntax and seemingly daunting schema definitions, software vendors and open sources have provided powerful tools that can transparently convert object models into WSDL (e.g. Java2WSDL from Axis[1]). As a result, most developers provide Web services without dealing with actual WSDL files. For example, this can be done by just pressing the "import from WSDL" or "export to WSDL/Endpoint" buttons in many IDEs. We believe this has its own benefits for wider and quicker adoption of Web services. However, it does bring some problems. For example, it is not feasible for Web services users to annotate thousands of WSDL overnight. Nor can we have them include well-built ontology definitions in WSDL files at this stage. This calls for a solution that can suffice most Web users at the scale of Web in terms of time, space, and human resources. This chapter is our initial work towards this end.

Third, unlike traditional Information Retrieval (IR), service consumers cannot choose relevant Web services by just browsing the WSDL. While a document (e.g. a Web page) has static ('as is') information for a person to read, a Web service provides some level of skills, ingenuity, and experience that is consumed by various software applications. Such fuzzy and abstract characteristics of a service made service discovery and selection far more difficult than what traditional search engines have provided. Moreover, interacting with any web service always necessitates a transaction between two distantly located software programs that have never 'talked' with each other before. Such uncertainty and loose-coupling also bring about many technical issues that have not been addressed by existing search engines and IR techniques. We leave this part as our future work. In this chapter, we provide our preliminary experiment results of LSA application in semantic Web services.

---

[1] http://ws.apache.org/axis

6 Experiment

We have conducted several experiments, where LSA has been utilised in finding web services based on semantic concepts rather than simple keywords. This section discusses the experiment settings, i.e. the environment where Web service discovery can be conducted using the LSA model. In order to provide an overall picture to readers, we first introduce the experiment conceptual model that converts the service discovery problem into an IR problem that can be solved by LSA. Based on the conceptual model, we then implement a proof-of-concept software prototype that has realised the LSA techniques and that is able to convert Web services into regular documents for LSA process. Last, we provide parameter setting information that will be manipulated during the experiment to verify the performance of LSA applied in SWS discovery.

6.1 Conceptual Model

Central to the conceptual model is our main idea, which aims to convert the semantic Web service discovery problem into IR problems that can be processed by the LSA model. To that end, Web service discovery can be partially considered as WSDL discovery. This is because WSDL file contains the textual information describing the capability of a Web service in a standardised format. This way, each Web service can be seen as an individual document containing important service advertisement information. These documents can then be used by IR models such as LSA or VSM. It is true that some important information (e.g. reputation, trustworthiness, quality of services) is missing in the current WSDL specification. However, in this Chapter, we focus solely on the functional capability of a Web service as we believe this is the fundamental discovery problem need to be solved. We leave for our future work factors (e.g. QoS) determining other aspects of the problem.

With this basic idea, we then present the conceptual model. As shown in Figure 5, the conceptual model consists of two parts: run time and design time. Design time provides techniques (e.g. LSA) that can semantically examine and analyse Web services from system perspective. Based on what we have discussed earlier, Web services in this model have been instantiated into WSDL files corpus. The initial corpus is then processed using traditional IR methods (e.g. VSM), which produce WS indices, the compact form of WS corpus with metadata stored in a search-efficient way. With the help of semantic techniques such as LSA, a semantic space is thus constructed to help users to find the most appropriate services using intelligent techniques such as matchmaking, aggregating, and clustering available at run time. Run time provides infrastructure (WS-discovery UI and intelligent algorithms) that enables 'social' activities from a user-centred perspective. We believe it is essential to incorporate adequate user participation such as recommending, blogging, and tagging to facilitate the semantics services discovery. In the rest part of this chapter, we will however focus solely on dealing with the semantic techniques provided in design time, i.e. using LSA to capture the 'latent' semantics hidden in WS indices and WSDL corpus. We believe this will pave the way for building a self-organised complete semantic space in the long run to foster increasingly growing user activities.

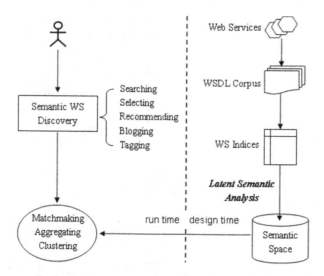

Fig. 4. Conceptual Model for Semantic Service Discovery Experiment

6.2 Software Prototype

To examine the effect of LSA indexing and retrieval accuracy for semantic Web services discovery, we have developed a semantic service search engine prototype based on the conceptual model (Figure 4). The main hindrance of implementing LSA lies in several practical engineering problems in matrix processing. We had attempted to use the JAMA (a Java Matrix Package[2]) to process the SVD. Unexpectedly, it constantly receives the "out of memory" exception and takes unacceptable long time (maximum RAM heap is set to 1G bytes) to achieve the SVD for a 24230 (terms) × 3577 (WSDL) matrix. This is because the SVD method used in JAMA applies orthogonal transformations directly to the sparse matrix A. Previous studies have shown that [3] this normally costs tremendous amount of memory consumption. Moreover, the matrix representation in JAMA is simply implemented as an in-memory two dimensional array. When the dimension of a matrix reaches the magnitude level of ten thousands, the actual memory consumption will reach the magnitude level of a hundred million that cannot be met by the physical RAM.

We eventually used software tool General Text Parser (GTP[3]) for LSA indexing and querying. This works gracefully in terms of time and space complexity due to several reasons. First of all, during the process of SVD, it generates binary interim matrix on the disk, and only partial matrix information is read into memory at a time when necessary to participate the calculation. Moreover, Harwell-Boeing [21] compressed matrix is used to reduce the memory consumption. Most importantly, the LAS2, a single-vector method suitable for solving large, sparse, symmetric eigenproblems is used for handling the very large-scaled sparse SVD. Readers can refer to [3] for the detailed algorithm of LAS2.

---

[2] http://math.nist.gov/javanumerics/jama/
[3] http://www.cs.utk.edu/~lsi

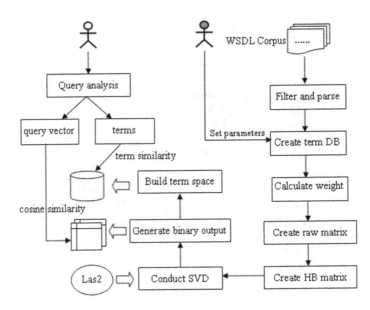

Fig. 5. Workflow of LSA in our prototype

We have also used VSM approach to generate WS indices in order to compare the performance of these two methods in semantic Web service discovery. In particular, for the same user queries, we provide a quantitative comparison between LSA method and VSM method using two IR measurement metrics – **Precision** and **Recall** (Section 7.2). A full description of this software prototype is beyond the scope of this chapter. In this chapter, we focus on the LSA experiment settings and results analysis. As shown in Figure 5, the basic LSA experiment environment consists of a number of steps that constitute a 'U' shape workflow. It is our intention to ensure the workflow is consistent with the conceptual model defined in Figure 4.

The right part of 'U' illustrated detailed steps for LSA indexing, and the left part of 'U' depicts how a user query is processed using the LSA index and the semantic term space. The workflow starts from the top right, where a collection of WSDL files is provided as the input of the software prototype. The source of this WSDL collection can be obtained through various mechanisms such as UDDI business registry, public Web services registries (e.g. *XMethod*), or data sources from previous studies (such as [22]). The WSDL corpus is then sent to the 'Filter and parse, which is fairly crucial for the success of LSA. This is because we have found that only 30% of WSDL files that we obtained from public registries are valid. They thus might not represent the capability of a Web service precisely. Without losing generality, we screen out all valid WSDL files and discarding all invalid ones. More importantly, WSDL files are XML documents with predefine XML tags. These tags (WSDL elements) do not contribute to representing the capability of a Web services. Therefore, a WSDL parser is essential to remove all XML tags and produce XML-striped plain texts ready for IR models such as VSM and LSA. Moreover, the filter can easily select desired part of the WSDL file to manipulate the dataset as we will show in Section 7.5.

When all preparation work is ready, the software administrator (shown as the grey human in Figure 5) needs to set the parameters required by the LSA experiment. Detailed parameter information will be discussed in 58633056. When parameters setting are completed, the system starts to work in a batch mode. The analysis work starts from generating the term DB, also known as the 'vocabulary' widely used in VSM to create the inverted index data structure. As the central task in this stage is to construct the term-by-document matrix, the term weights, which fill in each entry of the matrix, is of great concern. After all term weights are obtained using the scheme specified in the parameter setting, a raw matrix A can thus be built. The raw matrix is then compressed into the HB matrix, and is written to the disk. The SVD module reads the HB matrix from the disk, and loads the LAS2 algorithm to conduct the SVD, which generates the binary output consisting of a number of triples "<singular values, left vector, and right vector>".

The SVD result serves for two purposes. First, it can be directly used by service consumer (shown as the white human in Figure 5) query for general service retrieval, where each WSDL vector is compared with the query vector in order to rank WSDL files based on the cosine value. Second, the SVD result can be used for building the term space, a preliminary version of semantic space. The term space records semantic distance between any two terms collected from the term DB. The distance is calculated based on the dot product between two k dimensional term vectors. Each component in the term vector represents a factor value. The term space enables service consumer to enhance the query results, discover related Web services, and even provide more implications such as the semantic service composition.

We have also fixed up some problems in the original GTP HB matrix file representation that might cause infinite SVD calculation loop. The cosine score function discussed in [23] is used to calculate the similarity measure between a query vector and document vectors. Query vectors are generated by (1) summing the term vectors of terms in the query; (2) scaling up each term vector dimension by the inverse of a corresponding singular value. The scaling is done to emphasise the more dominant LSA factors. We have experimented in our prototype the effect of such a scaling as shown in Section 7.3. The whole prototype is developed on Java2 (JDK1.4.1) platform with JSP web user interface running on Tomcat 5.0.

6.3 Parameter Settings

As mentioned earlier, the prototype administrator needs to set several important parameter values before the experiment goes to the batch mode. In this section, we discuss these essential parameters listed in Table 1. The default values will be used if the administrator does not explicitly change them. While most parameters can be manipulated, some parameters have constant values, and thus cannot be changed. In what follows, the meaning of each parameter is introduced.

"# docs" represents the number of valid WSDL files in the corpus. By default, the size of the corpus is 2,817. However, the administrator can reduce this value by excluding certain number of WSDL files. This is sometimes necessary in order to examine the scalability of the LSA discussed in Section 7.7. Note that excluding WSDL files will affect the matchmaking performance. Therefore, this value stays at 2,817 except for the scalability test which only focuses on time and memory consumption.

Table 1. Default parameter settings for the LSA experiment

Parameter	Description	Default Value
# docs	Number of valid WSDL docs	3577
# terms	Number of terms parsed from the docs	24230
# factors	Number of reduced dimensions	105
corpus type	Different parts of WSDL	All
scale to singular value?	Whether query vector is scaled	Yes
normalise doc vector?	Whether docs vector is normalised	No
local weight	local term weight scheme	log
global weight	global term weight scheme	entropy

"# terms" records the number of terms parsed from the WSDL corpus. As mentioned earlier, these terms are all XML-stripped words in the plain texts. This value also indicates the size of the term DB, where each term has an individual entry. "# factors" represents the number of factors retained in the reduced matrix. If not specified, GTP software will find the 'optimised' value. However, as discussed in the issues of LSA, there are rigorous proofs exist supporting the optimised number of this value. Hence, it is useful to test this parameter to examine its effectiveness as will be discussed in Section 7.3.

"Corpus type" refers to the WSDL elements that will be indexed and searched during the LSA-based service discovery. Currently, it has enumeration values such as "All", "PortType", "Operation", and "Messages". Section 7.5 will discuss the different matchmaking performance based on different values for this parameter. "Scale to singular value?" refers to an option as to whether the query vector is scaled by the singular values before calculating the cosine similarity. Such a scaling can be used to emphasise the more dominant LSA factors. "Normalise doc vector?" is another option that determines whether or not the length of the document vector should be taken into account. By default, it is switched off. This means that the length of the document vector is considered as a factor affecting the similarity score of each WSDL file. Section 7.3 will demonstrate the effect of these two parameters. "Local weight" refers to how to calculate the importance of a term in one WSDL file, whereas "global weight" refers to how to work out the importance of a term in the whole WSDL corpus. As discussed previously, these two parameters determine the term weight, i.e. the original entry value for matrix A. Test results related to these two will be discussed in Section 7.4.

7   Evaluations

In this section, we provide the experiments result collected from our prototype system. We start by two qualitative experiments which demonstrate the semantics that our LSA-enabled prototype can bring to standard web services – the improved *recall* measure and potential term ontology learning. Given that improved *recall* might accompany the poorer *precision*, we then present experimental *precision* measure by varying three eminent LSA parameters and one web services specific parameter. For each manipulation we conduct predefined 10 queries, which contain the most frequently entered query terms that had been captured by prototype's logging system.

For the moment we only test the single term query, future work will be carried out on complex terms queries. The mean retrieval precision is then collected for the first 20 returned services under each manipulation. Experiment performance is provided in the end to show the empirical applicability of our prototype system.

7.1 Higher-Order Association

The higher-order association is one of the most appealing and unique characteristics of LSA. It enables the semantic (topic)-based discovery rather than keyword-based search. In this section, we demonstrate the terms-based higher order association, which is constructed through LSA service indexing techniques.

Figure 6 shows a screenshot of our prototype system GUI – a typical search engine web page that displays a list of Web services on the topic "calculator", which has been input in the search box. Our LSA-based search engine has returned twenty Web services that can do certain kind of 'calculation'. Perusing the first service in the ranking list, (http://ausweb.scu.edu.au/aw02/papers/refereed/kelly/MathService.wsdl), we are unable to find any occurrences of string "calculator" in its WSDL file, or in its URL. However, this service is ranked at the first place when a service consumer needs a service such as a 'calculator'. Hence, LSA has automatically built a hidden semantic association between 'calculator' and 'maths' even though they do not co-occur in any WSDL files. Such a higher-order association cannot be captured by the VSM model with the original term-by-document matrix or any keyword-based service searching mechanisms.

Table 2 presents a partial term space based on higher-order associations. For each term (bold face) in the left column, we provide ten semantically-close terms shown in the right column. These ten predefined query terms ("maths", "book", "weather", "search", "credit card", "fax", "stock", "quote", "bank", and "SMS") are chosen because they are frequently submitted by service consumers according to the search engine

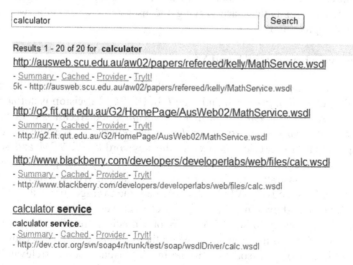

Fig. 6. Higher-order association between "Calculator" and "Math"

Table 2. Higher-order association between terms

maths	parameter - 0.91 number - 0.7 result - 0.31 add - 0.23 complex - 0.16 text - 0.16 address - 0.14 document - 0.12
book	bible - 0.97 word - 0.93 order - 0.92 key - 0.83 title - 0.82 app - 0.24 set - 0.23 address - 0.18 real - 0.16 music - 0.13
weather	city - 0.99 zip - 0.87 global - 0.72 arg - 0.4 search - 0.39 info - 0.33 service - 0.32 job - 0.23 set - 0.17 add - 0.1
search	google - 0.73 query - 0.72 simple - 0.46 page - 0.45 key - 0.39 cached - 0.37 shopping - 0.29 email - 0.16 find - 0.15 set - 0.14
credit card	process - 0.67 send - 0.41 product - 0.37 expiration - 0.36 password - 0.29 address - 0.27 sale - 0.26 submit - 0.17 detail - 0.16 cc - 0.16
fax	send - 0.96 address - 0.58 message - 0.48 image - 0.43 document - 0.35 code - 0.19 incoming - 0.18 order - 0.18 outgoing - 0.16 elements - 0.09
stock	quote - 0.96 index - 0.91 daily - 0.67 headline - 0.65 symbol - 0.41 earning - 0.31 schema - 0.27 license - 0.21 key - 0.17 split - 0.15
quote	stock - 0.92 symbol - 0.64 document - 0.3 text – 0.28 top - 0.27 gainer - 0.2 loser - 0.2 ticker - 0.19 currency - 0.18 translate - 0.15
bank	number - 0.95 detail - 0.91 sale - 0.77 card - 0.48 voice - 0.46 credit - 0.38 key - 0.34 arg - 0.28 check - 0.19
SMS	send - 0.98 email - 0.63 status - 0.52 address - 0.2 process - 0.19 inp - 0.16 credit - 0.14 currency - 0.12 oid - 0.1

query log. The number right after each term indicates the similarity between this term and the term in the left column. For example, the similarity between 'maths' and 'add' is "0.23".

The higher-order associations between terms have perhaps provided a very promising approach to build a light-weight ontology or taxonomy in a semi-automatic manner. An interesting research direction is thus to integrate these higher-order associations with end user activities such as feedback, blogging, and tagging to build and maintain a generic semantic space serving the user-centred semantic web services discovery and aggregation.

7.2 Comparison between LSA and VSM

In order to objectively compare these two methods, we use two metrics from the Information Retrieval (IR): *precision* and *recall*. In IR [24], **Precision** is defined as "*the fraction of the retrieved documents which is relevant*"; and **Recall** is defined as "*the faction of the relevant documents which has been retrieved*". We thus conduct two sets of experiments for service discovery: the keyword-based VSM and semantic-based LSA. In each experiment setting, we carry out the same 10 queries as we did in Section 7.1 for demonstrating the higher order association. We then measure the Precision and Recall respectively. We finally calculate the average precision values [24] at each recall level as shown in the Precision-Recall curves (Figure 7).

In Figure 7, the concept-based LSA has poorer precision result (74% compared to 100% for keyword-based matchmaking) when the recall is low. This is caused by the inaccuracy of higher-order association found in LSA. Hence, some irrelevant services are included and ranked very high in the discovery result list. For example, when we

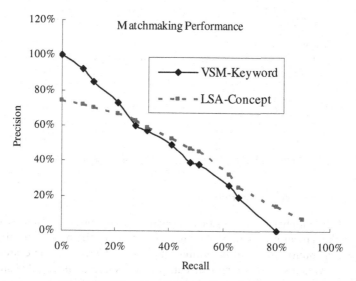

Fig. 7. Precision-Recall Curves for VSM and LSA

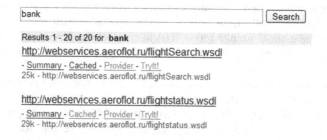

Fig. 8. LSA-based service discovery shows poorer precision

query "bank", the LSA presents the first two services as shown in Figure 8. However, neither of them ("flightSearch" and "flightstatus") is relevant to Web services that are related to banking services.

On the other hand, LSA has generally shown better recall result especially when the precision becomes lower. In the keyword-based VSM (Figure 7), however, precision at levels of recall higher than 80% drops to 0. This is because keyword-based discovery approach fails to find the remaining 20% relevant Web services due to the limitation of literal keyword matching. Take another example, when we query "credit card", the LSA-based service discovery can find two extra Web services that do not contain the keyword "credit card" as illustrated in Figure 9. The "address finder" service is chosen because we have found most online credit card Web services provide services to validate the address of card holders. To some degree, such a semantic-based discovery has implications for semantic-based service composition. Similarly, the "GeoFinder" service is returned as it is an essential part of the "address validation". The hidden "has-a" relationship between two concepts "credit card" and "address" is implicitly captured by LSA.

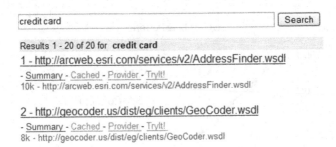

Fig. 9. LSA-based service discovery shows better recall

7.3 K-Factor and Score Function

Selection of the number of dimensions (k) is a challenging task. There is so far no rigorous mathematical proofs to show which values have the best performance, neither is there a intuitive way to determine the best k. Currently, there are repeated empirical experiments evaluations that can provide some hints for choosing the k, leaving it an open issue for LSA. Author in [25] have found that the precision of her results set peaked at around 100 factors and then slowly dropped as the factors increased. She also notes that "the number of dimensions needed to adequately capture the structure in other collections will probably depend on their breadth". In our experiment, we have chosen range (50 – 300) to conduct the service discovery and the result is shown in Figure 10, which shows the average precision is at a maximum for 50 to 300 factors.

The score function computed in the reduced dimensional space is normally the cosine between vectors. Empirically, this measure tends to work well, and there are some weak theoretical grounds for favouring it [7, 23]. In order to preserve cosine similarities in the original space, one can length-normalise the documents before SVD. However, some research has shown that the additional use of the length of LSA vectors to be useful. This is because the length reflects how much was said about a concept rather than how central the discourse was to the concept. Therefore, we have made it a parameter in our experiment – document normalisation, i.e. whether or not

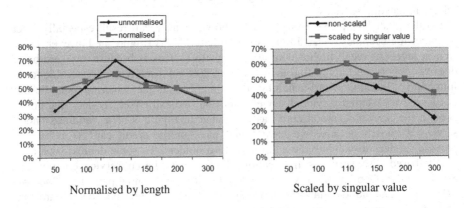

Fig. 10.

to take into account the length of the document vector. The second variable that we have manipulated is to control whether the query vector is scaled by the singular values before calculating the cosine similarity.

7.4   Term Weights

The settings and calculation of term weights (global or local) are crucial for IR models such as VSM and LSA. In order to obtain these two weights, one needs to process the entire corpus to obtain some important measures such as the frequency of occurrences. In our prototype, we have considered the following different types of weighting schemes in Table 3. Readers can refer to [25] for a complete understanding on these schemes.

Table 3. Term weight

Local	Application
tf	The corpus spans general topics
log	$1\big/\sqrt{\log(1+tf)^2}$
binary	Only the presence as denoted by "1", or absence by "0" of a term is included in the vector

Global	Application
idf	Only if the corpus is relatively static
idf2	$\log(ndocs)\big/\log 2 - \log(df+1)\big/\log 2$
entropy	Consider to take out noises

We have thus carried out 10 predefined queries for each one of these nine (3×3) combinations of local-global term weight schemes. The precision is measured to examine the effectiveness of these weightings. The overall average precision under each weigh scheme is given in Figure 11, whereas the average precision for all weight schemes is shown in Figure 12. Figure 11 indicates that in our experiment "log-entropy'

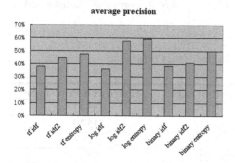

Fig. 11. Average precision for each term weight

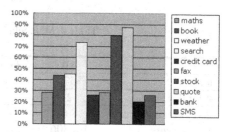

Fig. 12. Average precision for each query term

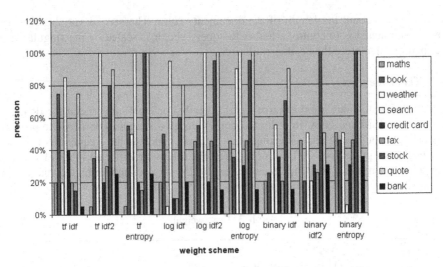

Fig. 13. Individual query precision under different term weights

term weight has obtained the best precision result. Figure 13 shows the individual query result under different term weights. It would be very interesting to explore the hidden pattern or reason that why some query terms have much better precision result than others, which might provide some useful information on the nature of the WSDL files as well as domain-specific data.

7.5 WSDL Corpus

Literature [8] has shown that the size of document corpus can affect the LSA performance. Unfortunately, there is little hard evidence on what the "ideal" size of LSA corpus might be. Intuitively, adding additional texts is unlikely to reduce performance, so a basic best practice will be "the more the better". In this experiment, the corpus contains 3577 WSDL documents with 24230 individual terms. In order to test the effect of size on the LSA performance, we have conducted three sets of experiments. In each experiment, we build LSA index on one type of data elements of WSDL files. This way, the number of terms in the WSDL corpus is manipulated as shown in Table 4.

Table 4. Manipulate the size of the Corpus

	All	Operation only	Service and Port Type
# docs	3577	3577	3577
# total terms	24230	3657	1520
Avg. #terms/docs	8.6	1.3	0.53
% non-zero	102806	22926	9847

Note: The value of Avg. #terms/docs could be less than 1 because some too frequent words are filtered out before the LSA.

The test result is shown in Figure 14and Figure 15. In the "Operation only" corpus, we found that the search performance is very different from the "All" corpus. We notice that the low precision performance for query "stock" in the "Operation only" corpus, but the high precision performance for the query "quote". Moreover, it is surprising to observe that "Operation only" corpus has the worst precision. Presumably, operation names appear to capture the nature of a Web services, and hence can semantically represent the capability of a Web service. However, our experiment does not support this assumption.

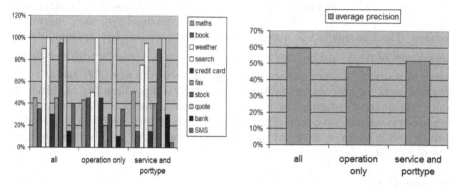

Fig. 14. Average precision for each corpus   **Fig. 15.** Average precision for each query term

7.6 Complexity

Computational complexity is one of the important criteria in examining the perform-ance of algorithms and models. In their original work [2] where LSA is firstly proposed, the authors have mentioned the time complexity of LSA. In particular, the two-mode factor analysis (i.e. SVD) time complexity is $O(N^2 \times k^3)$, where N is the number of terms plus documents (i.e. WSDL files), and k is the number of factors. In this section, we define the LSA complexity problem is composed of two parts – the time complexity and the space complexity. While time complexity examines the time duration needed to conduct SVD for WSDL corpus, space complexity focuses on the memory consumption required for LSA-based WSDL indexing and searching. Both of them are important for the complexity analysis. In what follows, we will elaborate studies on both time and space complexity for general LSA applications.

Theoretically, the time complexity includes design (indexing) time and run (search-ing) time. However, as discussed in Section 3, the LSA searching (query) is based on the classic IR *Cosine Similarity* (i.e. the dot products of two vectors). Therefore, the searching time complexity remains the same as the regular VSM time complexity, i.e. $O(q \times n)$, where q is the number of terms in the user query and n is the number of document. Note that we assume the length of each document vector ($\|v\| = \sqrt{\langle v, v \rangle} = \sqrt{\sum w_i^2}$) has been obtained during the indexing time as it is constant and is independent of any user queries. Since q is relatively much smaller than n, the searching time complexity can thus be reasonably reduced to a linear form $O(n)$ without losing generality. Hence, conceptually, our main concern of time complexity lies in the design time, when the SVD is conducted to construct the rank-reduced matrix for WSDL indexing.

The Single-Vector Lanczos [3] -based LAS2 method is used for calculating the SVD. According to [26], this method generally yields the time complexity

$$O(I \times O(A^T A x) + trp \times O(Ax))$$

where I represents the number of iterations required by a Lanczos (large scale) procedure, which is primarily used for the standard and generalised symmetric eigenvalue problem. The Lanczos (SVD manual) procedure here is used to approximate the eigensystem of $A^T A$, and the trp represents the number of singular triples – singular values in matrix S, its corresponding left singular vectors in matrix T, and its corresponding right singular vectors in matrix D. The SVD algorithms (e.g. JAMA) that apply orthogonal transformations directly to the sparse matrix A often requires tremendous memory consumption. Previous studies show that the Lanczos procedure used in LAS2, however, have less memory consumption that has been reported 'acceptable' in [3].

7.7 Scalability

To particularly examine the time and space complexity of LSA applied in semantic Web service discovery, we have conducted the scalability experiment. Scalability is defined as the capability of the system to maintain or increase the load – large numbers of requests, or interactions among components – without degrading performance (e.g. the server response time) given reasonable resource consumptions (e.g. memory). The key of scalability experiment is thus to simulate the 'changes' and observe the 'behaviour' of the system in response to these changes in terms of time complexity and space complexity. Hence, it is our belief that the scalability measurement reflects both time and space complexity from the empirical perspective. Considering the SVD complexity introduced in Section 7.6, we provide the scalability test scheme as shown in Table 5, where four groups (G1 – G4) of scalability tests are presented. In each group, we specify the Independent Variable (IV) and the Dependent Variable (DV). For example, in G1, we manipulate the number of WSDL files in the corpus as IV, and we expect to measure the time duration for LSA indexing as DV. Likewise, in G4, the number of service consumers who simultaneously submits service requests is manipulated in order to examine the system changes in memory consumption.

Table 5. Scalability test scheme

	Indexing	Searching
Time Complexity DV= Time (milliseconds)	G1 IV = # of WSDL	G2 IV = # of Service Consumers
Space Complexity DV= RAM (megabytes)	G3 IV = # of WSDL	G4 IV = # of Service Consumers

Our prototype is running in one PC configured with the 2G RAM, Pentium 3.59GHz CPU. We use JVM peak committed memory as the measurement for DV RAM consumption. This value is often slightly larger than the actual consumption, but it reflects the true memory reservation for our system. The test results of four groups are illustrated in Figure 16.

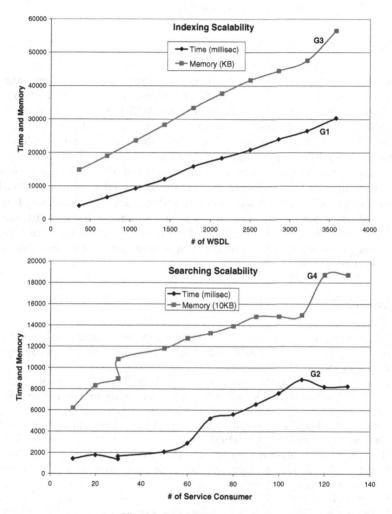

Fig. 16. Scalability test results

From Figure 16, it is clear that, in general, both time complexity and space complexity are linear. In order to quantify the scalability results, we define that

$$scalability = \frac{\Delta dv}{\Delta iv} = \frac{dv_1/dv_0}{iv_1/iv_0},$$

where Δ represents the change part of either dependent variable (dv) or independent variable (iv). Based on this equation, we can obtain the scalability results for four groups of experiments as follows, G1 = *75%*, G2 = *44%*, G3 = *38%*, G4 = *23%*. Since all scalability values are well less than 1, the behaviour of the system is not sensitive to changes in system loads. We can thus reasonably maintain that the LSA prototype for semantic Web service discovery is scalable.

8 Conclusions

In this chapter we provided an introduction to semantic web services, where we briefly discussed three different approaches for adding semantics to Web Services i.e. OWL based Web Services Ontology, WSMO and WSDL-S. We then explained LSA and VSM and their application in information retrieval and studied how it can be applicable to web services discovery. We found that most of current solutions and specifications for SWS rely on ontology building; a task that needs a lot of human (e.g. domain experts) involvement, and hence might not be able to scale very well in face of vast amount of web information and myriad of services providers. To address this issue we proposed a novel way to conduct SWS discovery using LSA. This technology and its existing applications motivate our solution and conceptual model in applying LSA on semantic web services. A prototype system – a service search engine – was developed based on this conceptual model and the experimental results have been evaluated in this chapter.

References

[1] Sajjanhar, A., Hou, J., Zhang, Y.: Algorithm for web services matching. In: Yu, J.X., Lin, X., Lu, H., Zhang, Y. (eds.) APWeb 2004. LNCS, vol. 3007, pp. 665–670. Springer, Heidelberg (2004)

[2] Deerwester, S., Dumais, S., Furnas, G.W., Landauer, T.K., Harshamn, R.: Indexing by Latent Semantic Analysis. Journal of the American Society for Information Science 41, 391–407 (1990)

[3] Berry, M.W.: Large scale singular value computations. International Journal of Supercomputer Applications 6, 13–49 (1992)

[4] Horst, P.: Factor Analysis of Data Matrices: Holt. Rinehart and Winston, Inc. (1965)

[5] Eckart, C., Young, G.: The approximation of one matrix by another of lower rank. Psychometrika 1, 211–218 (1936)

[6] Bartell, B.T., Cottrell, G.W., Belew, R.K.: Latent Semantic Indexing is an Optimal Special Case of Multidimensional Scaling. In: 15th Annual International SIGIR, Denmark (1992)

[7] Caron, J.: Experiments with LSA Scoring: Optimal Rank and Basis, Computer Science Department, University of Colorado at Boulder (2000)

[8] Landauer, T.K., Foltz, P.W., Laham, D.: Introduction to Latent Semantic Analysis. Discourse Processes 25, 259–284 (1998)

[9] Furnas, G.W., Deerwester, S., Dumais, S., Landauer, T.K., Harshamn, R.A., Streeter, L.A., Lochbaum, K.E.: Information Retrieval using a Singular Value Decomposition Model of Latent Semantic Structure (1988)

[10] Skoyles, J.R.: Meaning and context: the implications of LSA (latent semantic analysis) for semantics (2000)

[11] Yu, C., Cuadrado, J., Ceglowski, M., Payne, J.S.: Patterns in Unstructured Data Discovery, Aggregation, and Visualization (2005)

[12] Lin, M.Y., Amor, R., Tempero, E.: A Java reuse repository for Eclipse using LSI. In: Australian Software Engineering Conference (2006)

[13] Ye, Y.: Supporting component-based software development with active component retrieval systems. In: Computer Science, University of Colorado (2001)

[14] Landauer, T., Laham, D., Rehder, R., Schreiner, M.E.: How well can passage meaning be derived without using word order? a comparison of Latent Semantic Analysis and humans. In: 19th Annual Conference of the Cognitive Science Society, Mahwah, NJ. USA (1997)

[15] Kintsch, W.: Predication. Cognitive Science 25, 173–202 (2001)

[16] Wiemer-Hastings, P., Wiemer-Hastings, K., Graesser, A.: How latent is Latent Semantic Analysis? In: Sixteenth International Joint Congress on Artificial Intelligence, San Francisco. US (1999)

[17] Dennis, S.: Introducing word order. In: McNamara, D., Landauer, T., Dennis, S., Kintsch, W. (eds.) LSA: A Road to Meaning. Erlbaum, Mahwah (2005)

[18] Hu, X., Cai, Z., Franceschetti, D., Penumatsa, P., Graesser, A.C., Louwerse, M.M., McNamara, D.S.: LSA: The first dimension and dimensional weighting. In: 25th Annual Conference of the Cognitive Science Society (2003)

[19] Denhière, G., Lemaire, B., Bellisens, C., Jhean, S.: A semantic space for modeling a child semantic memory. In: McNamara, D., Landauer, T., Dennis, S., Kintsch, W. (eds.) A Road to Meaning, Mahwah, NJ (2005)

[20] Kittredge, R., Lehrberger, J.: Sublanguage: Studies of Language in Restricted Semantic Domains. de Gruyter (1982)

[21] Duff, I., Grimes, R., Lewis, J.: Sparse Matrix Test Problems. ACM Transactions on Mathematical Software 15, 1–14 (1989)

[22] Fan, J., Kambhampati, S.: A Snapshot of Public Web Services. ACM SIGMOD Record 34, 24–32 (2005)

[23] Berry, M.W., Drmac, Z., Jessup, E.R.: Matrices, Vector Spaces, and Information Retrieval. SIAM Review 41, 335–362 (1999)

[24] Baeza-Yates, R., Ribeiro-Neto, B.: Modern Information Retrieval. Addison Wesley, Reading (1999)

[25] Dumais, S.T.: Improving the retrieval of information from external sources. Behavior Research Methods, Instruments and Computers 23, 229–236 (1991)

[26] Berry, D.M., Do, T., O'Brien, G.W., Krishna, V., Varadhan, S.: SVDPACKC (Version 1.0) User's Guide, Computer Science Department, Univeristy of Tennessee (1993)

[27] Herrmann, M., Ahtisham Aslam, M., Dalferth, O.: Applying Semantics (WSDL, WSDL-S, OWL) in Service Oriented Architectures (SOA)

[28] Cardoso, J., Sheth, A.P.: Semantic Web Services, Processes and Applications. Springer, Heidelberg (2006)

[29] OWL-S semantic markup of web services – white paper (accessed on June 15, 2007), http://www.daml.org/services/owl-s/1.0/owl-s.html

[30] Marinchev, I., Agre, G.: Semantically Annotating Web Services Using WSMO Technologies. Cybernetics and Information Technologies 5(2) (2005)

Semantic Web Services for Satisfying SOA Requirements

Sami Bhiri[1], Walid Gaaloul[1], Mohsen Rouached[2], and Manfred Hauswirth[1]

[1] Digital Enterprise Research Institute (DERI), National University of Ireland, Galway
IDA Business Park, Galway, Ireland
[2] LORIA-INRIA-UMR 7503
BP 239, F-54506 Vandoeuvre-les-Nancy Cedex, France
{sami.bhiri,walid.gaaloul,manfred.hauswirth}@deri.org,
rouached@loria.fr

Abstract. Service oriented modeling is gaining acceptance among academia and industry as a computing paradigm for business and systems integration. Its strong decoupling between service provision and consumption enables much more flexible and cost-effective integration, within and across organizational boundaries, than existing middleware or workflow systems do. However, it also creates new requirements for handling effective service discovery, dynamic service interoperation and automation support for service composition. Web services have been emerging as the lead implementation of SOA upon the Web. The related technologies define common standards that ensure interoperability between heterogeneous platforms. Nevertheless, they fail in satisfying SOA requirements. Semantic Web services initiatives have then emerged with the objective of providing the foundation to overcome these requirements. The main idea is extending service description with machine interpretable information that software programs can reason over it. This chapter discusses how far Web services and semantic Web services initiatives satisfy SOA requirements.

Keywords: SOA, Web services, Semantic Web services, WSMO, OWL-S, IRS-III, METEOR-S.

1 Introduction

Service oriented modeling has been emerging as a paradigm for business and systems integration. Evolving from oriented object programming and component based modeling; it enables much more flexible and cost-effective integration, within and across organizational boundaries, than existing middleware or workflow systems do.

In recent years, Web services (a.k.a WS) have been emerging as the lead implementation of SOA upon the Web. WS have added a new level of functionality for service description, publication, discovery, composition and coordination extending the role of the Web from a support of information interaction to a middleware for application integration.

Nevertheless, current Web service technologies focus only on a syntactic level which hampers automation support for capability-based discovery, and dynamic

T.S. Dillon et al. (Eds.): Advances in Web Semantics I, LNCS 4891, pp. 374–395, 2008.

service composition and invocation. A common agreement is the need to semantically enrich WS description. Similar to the Semantic Web vision, the idea is making WS description more machine interpretable. A new breed of WS, called Semantic Web Services (a.k.a SWS), has then emerged with a promising potential to satisfy SOA requirements. The machine interpretable description of SWS provides the basis for capability-based service discovery, automatic service composition and dynamic service interoperation.

This chapter consists of three parts. First, we depict the concepts and principles of SOA. We highlight the dynamicity and flexibility it ensures and we discern the challenges it poses. In the second part, we present Web services as the key technology implementing SOA principles. We state in particular the main standards and technologies. And we investigate how far they satisfy SOA requirements. Finally, we exhibit SWS initiatives conceptual models and execution environments and inspect their capability to overcome SOA challenges.

2 Service Oriented Architecture

Service-oriented architecture (SOA) is a hot topic in enterprise computing as many IT professionals see the potential of SOA is dramatically speeding up the application development process. Gartner reports that "By 2008, SOA will be a prevailing software engineering practice, ending the 40-year domination of monolithic software architecture" [1] and that "Through 2008, SOA and web services will be implemented together in more than 75 percent of new SOA or web services projects." [1]. Thus, SOA has received significant attention recently as major computer and software companies such as HP, IBM, Intel, Microsoft, and SAP, have all embraced SOA, as well as government agencies such as DoD (US Department of Defense) and NASA.

In this section, we first remind the SOA genesis. After that, we describe SOA principles, terms and benefits. Thereafter we discuss SOA design and implementation requirements. Indeed, our aim is **to explore how successful existing Web services and semantic Web services implementation are to fulfill these requirements** (i.e. SOA requirements).

2.1 SOA Genesis

SOA can be viewed as an evolutionary computing architecture that closely mirrors the history of the industrial revolution. With SOA, computing architectures are expanding beyond object oriented self-sufficiency and now allowing for highly specialized and interoperable computing consumer/producer relationships [2]. Pre 1980, structured procedural programming was prevalent for assembling well structured software code into a software system. Procedural style APIs focus on the natural ability to solve problems via a functional process. The focus is primarily on how to get from point A to point B. This functional way of solving a problem is often a necessary first step when exploring an unfamiliar problem domain. Between 1980 and 1990, Object Oriented Programming (OOP) evolved and established its dominance in the software industry. OOP focuses on combining elements of the problem domain in the form of objects containing data and methods which help to solve the problem of how to get from point A to point B in a way that will also be good to get to point C (reusability).

However, OOP evolved prior to the common distributed computing environments that we have today. Between 1990 and 2000, enterprise tiered architectures evolved and demonstrated that combining methods with data between tiers worked against scalability and loose coupling of the enterprise system, thus the use of data transfer objects between tiers and the focus on the data model for communication between tiers of the enterprise system. Up to the year 2000, individual computing systems remained relatively self-sufficient.

The pre SOA tiered enterprise architectures and implementations did not provide a good solution for computing specialization and computing interdependence at a business or government level. SOA exploded from the evolution of the tiered enterprise architectures and pressures to provide specialized B2B interoperability. Only under the realm of SOA are the concepts of visibility, service descriptions, interaction, contracts and policies, real world effects, execution contexts, etc. combined to provide the architectures and implementations for the automated computing needs of modern computing consumer/producer relationships. SOA is a computing architecture that allows for complex relationships and specializations of computing services on a global scale.

In other words, service-orientation is a way of sharing functions (typically business functions) in a widespread and flexible way. Thanks to the high level dynamicity and flexibility it promises, SOA has been gaining ground as the key architecture for many kinds of applications like B2B interactions, enterprises application integration and grid computing. Indeed, in B2B applications for instance, enterprises can encapsulate and externalise their business processes as services. They can dynamically look for and interact with other services. They can collaborate on the fly to achieve common goals. And they can even establish dynamically virtual enterprises and create new services from existing ones.

2.2 SOA: Terms and Concepts

Service-orientation, as a means of separating things into independent and logical units, is a very common concept. A service-oriented architecture represents an abstract architectural concept defining an information technology approach or strategy in which applications make use of (perhaps more accurately, rely on) services available in a network such as the World Wide Web. It is an approach to building software systems that is concerned with loose coupling and dynamic binding between components (services) that have been described in a uniform way and that can be discovered and composed. Implementing a service-oriented architecture can involve developing applications that use services, making applications available as services so that other applications can use those services, or both. In fact, one way of looking at an SOA is as an approach to connecting applications (exposed as services) so that they can communicate with (and take advantage of) each other.

The fundamental elements of this computing approach are loosely coupled software components, called services. Services are autonomous platform-independent computational elements that can be described, published, discovered and accessed over network-accessible software module. Loosely coupling means that services interactions are neither hard coded (like in Object Oriented Programming), nor specified at design time (like in Component Based Modelling). On the contrary, services are defined out of any execution context and interact on the fly without prior collaboration agreement.

A service provides a specific function, typically a business function, such as analyzing an individual's credit history or processing a purchase order. It is a mechanism to enable access to one or more capabilities, where the access is provided using a prescribed interface and is exercised consistent with constraints and policies as specified by the service description. A service is provided by an entity – the service provider – for use by others, but the eventual consumers of the service may not be known to the service provider and may demonstrate uses of the service beyond the scope originally conceived by the provider. A service is accessed by means of a service interface, where the interface comprises the specifics of how to access the underlying capabilities. There are no constraints on what constitutes the underlying capability or how access is implemented by the service provider. Thus, the service could carry out its described functionality through one or more automated and/or manual processes that themselves could invoke other available services.

SOA uses the find-bind-execute paradigm as shown in Fig. 1. In this paradigm, service providers register their service in a public registry. This registry is used by consumers to find services that match certain criteria. If the registry has such a service, it provides the consumer with a contract and an endpoint address for that service. In general, entities (people and organizations) offer capabilities and act as service providers. Those with needs who make use of services are referred to as service consumers. In a typical service-based scenario, a provider hosts a network-accessible software module—an implementation of a given service—and defines a service description through which a service is published and made discoverable. A client discovers a service and retrieves the service description directly from the service, possibly from a registry or repository through metadata exchange. The client uses the service description to bind to the provider and invoke the service.

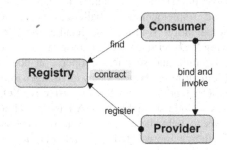

Fig. 1. Find-Bind-Execute paradigm

A service is opaque in that its implementation is typically hidden from the service consumer except for (1) the information and behavior models exposed through the service interface and (2) the information required by service consumers to determine whether a given service is appropriate for their needs. The consequence of invoking a service is a realization of one or more real world effects. These effects may include: (i) information returned in response to a request for that information, (ii) a change to the shared state of defined entities, or (iii) some combination of (i) and (ii). Summarizing up, the following features are intrinsic for services in SOA in the sense that:

1. Services are software components with well-defined interfaces that are implementation-independent. An important aspect of SOA is the separation of the service interface (the what) from its implementation (the how). Such services are consumed by clients that are not concerned with how these services will execute their requests.
2. Services are self-contained and autonomous: The logic governed by a service resides within an explicit boundary. The service has control within this boundary and is not dependent on other services for it to execute its governance. Underlying logic, beyond what is expressed in the descriptions that comprise the contract, is invisible and irrelevant to service requesters.
3. Services are loosely coupled: Services are designed to interact without the need for tight, cross-service dependencies. What distinguishes SOA from other architecture paradigms is loose coupling. Loose coupling means that the client of a service is essentially independent of the service. The way a client (which can be another service) communicates with the service does not depend on the implementation of the service. Significantly, this means that the client does not have to know very much about the service to use it. For instance, the client does not need to know what language the service is coded in or what platform the service runs on.

2.3 SOA: Implementation Challenges

SOA-based integration provides a consistent way to access all applications within a company, and potentially outside the company. However, the value of SOA has perhaps been oversold as a methodology and has often been mistakenly promoted as a technology that will solve almost all IT problems. In this section, we present common design requirements and implementation challenges to ensure an efficient SOA implementation. Implementation solutions to handle these requirements will be discussed in the following Web services and semantic Web services sections.

In order to enable dynamic and seamless cooperation between different systems and organizations, implementing SOA poses new challenges to overcome. The largest barriers to adoption of SOA tend to be establishing effective SOA standards fulfilling the expected promises. Understandably, SOA-based technologies are seeking implementation solutions to help them meet the above promises in the most cost-effective way. The challenge represented by implementing an SOA is the actual implementations may fail to reach some design requirements. More explicitly, these challenges concern mainly (i) the dynamic service interoperation without prior collaboration agreement, (ii) dynamic service discovery and selection based on requester needs, (iii) and automatic service composition to achieve added-value business requirements [1].

Implementation standards developing services in SOA have to consider not simply for immediate benefit, but also for long-term and board benefit. Unlike objects or databases, a service is developed for use by its consumer, which may not be known at the time. To put it in another way, the existence of an individual service is not of much interest unless it is part of a larger collection of services that can be consumed by multiple applications, and out of which new services can be composed and executed. Any collection of services needs common design, discovery, composition, and binding principles since they are typically not all developed at the same time. Thus, any SOA implementation MUST take into account the following design requirements to ensure the described set of SOA benefits and promises:

- *Service Discovery:* Service description should be visible to be discovered and understood by service consumers. Visibility refers to the capacity for those with needs and those with capabilities to be able to see each other [3]. It is the relationship between service consumers and providers that is satisfied when they are able to interact with each other. This is typically done by providing descriptions for such aspects as functions and technical requirements, related constraints and policies, and mechanisms for access or response. This is true for any consumer/provider relationship – including in an application program where one program calls another: without the proper libraries being present the function call cannot complete. In the case of SOA, visibility needs to be emphasized because it is not necessarily obvious how service participants can see each other. Thus, the initiator in a service interaction must be aware of the other parties. Visibility is promoted through the service description which contains the information necessary to interact with the service and describes this in such terms as the service inputs, outputs, and associated semantics. The service description also conveys what is accomplished when the service is invoked and the conditions for using the service. The service description allows prospective consumers to decide if the service is suitable for their current needs and establishes whether a consumer satisfies any requirements of the service provider. The descriptions need to be in a form (or can be transformed to a form) in which their syntax and semantics are widely accessible and understandable. Discovering services need not necessarily be fully automated (one can find many non-technical objections to fully automated discovery), but support for some richer discovery is than necessary. The main challenge of service discovery is using automated means to accurately discover services with minimal user involvement. This requires explicating the semantics of both the service provider and requester. It also involves adding semantic annotations to service definitions. Achieving automated service discovery requires explicitly stating requesters' needs—most likely as goals that correspond to the description of desired services—in some formal request language.

- *Service composition:* SOA provides a new way of application development by composing services. The service-oriented paradigm builds on the notion of composing virtual components into complex behavior. Thus, a consumer can use the functionality offered by multiple providers without worrying about the underlying differences in hardware, operating systems, programming languages, etc. Each service should be designed to satisfy a business task while possibly collaborating with applications or services provided by other entities. For services to interact, they need not share anything but a formal contract that describes each service and defines the terms of information exchange. Therefore a service composition task involves the selection, and interoperation of Web services given a high-level semantic description of an objective using a formal contract.

- ***Interoperability for semantic heterogeneity:*** SOA is applied in an environment where the number of involved actors is more and more heterogeneous and distributed. Consumers and providers communicate and exchange data which may lead to interoperability problems on top of the data mismatch (in structure and meaning). There are three important facts about services that set the basis for their information processing requirements. First, services, especially at the business level, exchange information in the form of messages. Secondly, the information in those messages needs to conform to the enterprise information model and semantics. Third, there is often a transformation between the enterprise semantics and the internal information model of the service. So, the SOA platform must provide: message processing capabilities, integration with existing enterprise information models, definition of messages based on the information model, and information transformation capabilities.

Unlike OOP paradigms, where the focus is on packaging data with operations, the central focus of SOA is the task or business function getting something done. This distinction manifests itself by the fact that an object exposes structure but there is no way to express semantics other than what can be captured as comments in the class definition. SOA emphasizes the need for clear semantics. Especially, in the case of service interaction where the message and information exchanges are across boundaries, a critical issue is the interpretation of the data. This interpretation must be consistent between the participants involved in service interaction. Consistent interpretation is a stronger requirement than merely type (or structural) consistency – the attributes of the data itself must also have a shared basis. For successful exchange of address information, all the participants must have a consistent view of the meaning of the address attributes. The formal descriptions of terms and the relationships between them provide a firm basis for making correct interpretations for elements of information exchanged. Note that, for the most part, it is not expected that service consumers and providers would actually exchange descriptions of terms during their interaction but, rather, would reference existing descriptions – the role of the semantics being a background one – and these references would be included in the service descriptions. Specific domain semantics are beyond the scope of SOA reference model; but there is a requirement that the service interface enable providers and consumers to identify unambiguously those definitions that are relevant to their respective domains [3].

3   Realizing SOA with Web Services

People often think of Web services and Service-Oriented Architecture (SOA) in combination, but they are distinct in an important way. As discussed in the previous section, SOA represents an abstract architectural concept. It's an approach to building software systems that is based on loosely coupled components (services) that have been described in a uniform way and that can be discovered and composed. Web services represent one important approach to realizing SOA. The World Wide Web Consortium (W3C), which has managed the evolution of the SOAP and WSDL specifications, defines Web services as follows:

A software system designed to support interoperable machine-to machine interaction over a network. It has an interface described in a machine-processable format (specifically WSDL). Other systems interact with the Web service in a manner prescribed by its description using SOAP messages, typically conveyed using HTTP with XML serialization in conjunction with other Web-related standards.

Although Web services technology is not the only approach to realizing an SOA, it is one that the IT industry as a whole has enthusiastically embraced. With Web services, the industry is addressing yet again the fundamental challenge that distributed computing has provided for some considerable time: to provide a uniform way of describing components or services within a network, locating them, and accessing them. The difference between the Web services approach and traditional approaches (for example, distributed object technologies such as the Object Management Group – Common Object Request Broker Architecture (OMG CORBA), or Microsoft Distributed Component Object Model (DCOM)) lies in the loose coupling aspects of the architecture. Instead of building applications that result in tightly integrated collections of objects or components, which are well known and understood at development time, the whole approach is much more dynamic and adaptable to change. Another key difference is that through Web services, the IT industry is tackling the problems using technology and specifications that are being developed in an open way, utilizing industry partnerships and broad consortia such as W3C and the Organization for the Advancement of Structured Information Standards (OASIS), and based on standards and technology that are the foundation of the Internet.

3.1 Scope of the Architecture

Web services had its beginnings in mid to late 2000 with the introduction of the first version of XML messaging—SOAP, WSDL 1.1, and an initial version of UDDI [4] as a service registry. This basic set of standards has begun to provide an accepted industry-wide basis for interoperability among software components (Web services) that is independent of network location, in addition to specific implementation details of both the services and their supporting deployment infrastructure. Several key software vendors have provided these implementations, which have already been widely used to address some important business problems.

Developers are looking for enhancements that raise the level and scope of interoperability beyond the basic message exchange, requiring support for interoperation of higher-level infrastructure services. Most commercial applications today are built assuming a specific programming model. They are deployed on platforms (operating systems and middleware) that provide infrastructure services in support of that programming model, hiding complexity, and simplifying the problems that the solution developer has to deal with. For example, middleware typically provides support for transactions, security, or reliable exchange of messages (such as guaranteed, once-only delivery). On the other hand, there is no universally agreed standard middleware, which makes it difficult to construct applications from components that are built using different programming models (such as Microsoft COM, OMG CORBA, or Java 2 Platform, Enterprise Edition (J2EE) Enterprise Java Beans). They bring with them different assumptions about infrastructure services that are required, such as transactions and security. As a consequence, interoperability across distributed heterogeneous platforms (such as .NET and J2EE) presents a difficult problem.

The Web services community has done significant work to address this interoperability issue, and since the introduction of the first Web services, various organizations have introduced other Web services–related specifications. Fig. 2 illustrates a population of the overall SOA stack with current standards and emerging Web services specifications that IBM, Microsoft, and other significant IT companies have developed. The remainder of this part provides a high-level introduction to these Web services specifications that realize more concretely the capabilities that are described in the SOA framework.

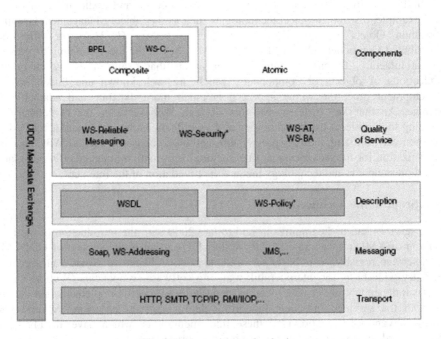

Fig. 2. Web services technologies

3.2 Web Service Transport

Web services are basically an interoperable messaging architecture, and message transport technologies form the foundation of this architecture. Web services are inherently transport neutral. Although you can transport Web services messages by using the ubiquitous Web protocols such as HyperText Transport Protocol (HTTP) or Secure HTTP (HTTPS) to give the widest possible coverage in terms of support for the protocols, you can also transport them over any communications protocol, using proprietary ones if appropriate. Although transport protocols are fundamental to Web services and clearly are a defining factor in the scope of interoperability, the details are generally hidden from the design of Web services.

3.3 Web Service Messaging

The messaging services layer contains the most fundamental Web services specifications and technologies, including eXtensible Markup Language (XML), SOAP, and WS-Addressing [4]. Collectively, these specifications form the basis of interoperable messaging between Web services. XML provides the interoperable format to describe message content between Web services and is the basic language in which the Web services specifications are defined.

SOAP, one of the significant underpinnings of Web services, provides a simple and relatively lightweight mechanism for exchanging structured and typed information between services. SOAP is designed to reduce the cost and complexity of integrating applications that are built on different platforms.

WS-Addressing provides an interoperable, transport-independent way of identifying message senders and receivers that are associated with message exchange. WS-Addressing decouples address information from the specific transport used by providing a mechanism to place the target, source, and other important address information directly within the Web service message. This specification defines XML elements to identify Web services endpoints and to secure end-to-end endpoint identification in messages. This specification enables messaging systems to support message transmission through networks that include processing nodes such as endpoint managers, firewalls, and gateways in a transport neutral manner. WS-Addressing defines two interoperable constructs that convey information that transport protocols and messaging systems typically provide. These constructs normalize this underlying information into a uniform format that can be processed independently of transport or application. These two constructs are endpoint references and message information headers.

3.4 Web Service Description

Service description defines metadata that fully describes the characteristics of services that are deployed on a network. This metadata is important, and it is fundamental to achieving the loose coupling that is associated with SOA. It provides an abstract definition of the information that is necessary to deploy and interact with a service. Web Service Description Language (WSDL) [4] is perhaps the most mature of metadata describing Web services. It allows developers to describe the "functional" characteristics of a Web service—what actions or functions the service performs in terms of the messages it receives and sends. WSDL offers a standard, language-agnostic view of services it offers to clients. It also provides non-invasive future-proofing for existing applications and services and allows interoperability across the various programming paradigms, including CORBA, J2EE, and .NET.

3.5 Web Service Discovery

The Universal Description and Discovery Interface (UDDI) is a widely acknowledged specification of a Web service registry. It defines a metadata aggregation service and specifies protocols for querying and updating a common repository of Web services information. Application developers can query UDDI repositories at well-known locations at design time to ascertain those services that might be compatible with their requirements. After they locate a directory, they can send a series of query requests against the registry to acquire detailed information about Web services (such as who provides them and where they are hosted) and bindings to the implementation. They can then feed this information into an assortment of development time tools to generate the appropriate runtime software and messages required to invoke the required service. Applications can also query UDDI repositories dynamically at runtime. In this scenario, the software that needs to use a service is told at execution time the type of service or interface it requires. Then it searches a UDDI repository for a service that meets its functional requirements, or a well-known partner provides it. The software then uses this information to dynamically access the service. Service discovery (publish/find) plays an important role in an SOA. It is possible to achieve this in other ways, but within a Web services world, UDDI provides a highly functional and flexible standard approach to Web service discovery.

WS-Policy proposes a framework that extends the service description features that WSDL provides. Having more refined service descriptions, qualified by specific WS-policies, supports much more accurate discovery of services that are compatible with the business application that is to be deployed. In a service registry (such as a UDDI registry), queries of WS-Policy-decorated services enable the retrieval of services that support appropriate policies in addition to the required business interface. For example, a query might request all services that support the credit Authorization WSDL interface (port type), use Kerberos for authentication, and have an explicitly stated privacy policy. This allows a service requester to select a service provider based on the quality of the interaction that delivers its business contracts.

3.6 Web Service Composition

Business Process Execution Language for Web services (WS-BPEL) [4] provides a language to specify business processes and how they relate to Web services. This includes specifying how a business process uses Web services to achieve its goal, and it includes specifying Web services that a business process provides. Business processes specified in BPEL are fully executable and are portable between BPEL-conformant tools and environments. A BPEL business process interoperates with the Web services of its partners, whether these Web services are realized based on BPEL or not. Finally, BPEL supports the specification of business protocols between partners and views on complex internal business processes. BPEL supports the specification of a broad spectrum of business processes, from fully executable, complex business processes over more simple business protocols to usage constraints of Web services. It provides a long-running transaction model that allows increasing consistency and reliability of Web service applications. Correlation mechanisms are supported that allow identifying statefull instances of business processes based on business properties. Partners and Web services can be dynamically bound based on service references.

3.7 SOA and Web Services: Need for Semantics

It is clear that Web services standards, both the core and extended specifications, contribute significantly to the ability to create and maintain service-oriented architectures on which to build new enterprise applications. However, despite their success, there still remain important challenges to be addressed in current SOA-based solutions. Service discovery, interoperation and composition are typically part of any development process based on SOA but, nothing is prescribed for effectively supporting these activities.

Discovery: To discover a Web service, the infrastructure should be able to represent the capabilities provided by a Web service and it must be able to recognize the similarity between the capabilities provided and the functionalities requested.

UDDI is the most well-known specification for an XML-based registry of service descriptions on the Web but the descriptions are syntactic only - the meaning is still open to interpretation by the user. Indeed, companies adopting UDDI for internal use have to define their own naming conventions and categorization structure and metadata, which inhibit adoption. Currently keyword-based search is the only means of finding relevant services. UDDI does not allow capability-based discovery of Web services. Support for some richer discovery than keyword-based search is necessary. Semantics bring closer the possibility of switching services dynamically by discovering them at runtime.

WSDL is less suitable for describing the semantics of a Web Service capability. This drawback affects not only the service discovery procedure but also service composition, invocation and interoperation. Indeed, WSDL files contain no information on the semantics of the described operations. It is up to the programmer or engineer to interpret the semantics from available descriptions in natural language. This type of interpretation becomes a challenge because the human factor inhibits automation of service discovery, selection, invocation and composition. Additionally, natural language descriptions are informal and can lead to different interpretations or even to failure to understand. This challenge can be overcome if formal and declarative semantic mark-up is used to complement service descriptions.

Composition: Similarly, service composition is mainly based on the syntactic descriptions provided by WSDL, which are necessary but not sufficient, since the semantics remain implicit and cannot be automatically processed. A number of approaches exist for modeling Web Service composition. Although these Web service composition languages are more suitable than the proprietary languages used in traditional workflow products, they lack the possibility to dynamically bind to Web services at run time. For example, WSCI and WS-BPEL describe how multiple Web services could be composed together to provide a more complex Web service. However, their focus remains on composition at the syntactic level and therefore, does not allow for automatic composition of Web services.

Service requesters have to bind specific services at design time which means they cannot take advantage of the large and constantly changing amount of Web services available. The services have to interoperate with each other seamlessly so that the combined results are a valid solution. Web services must be described and understood in a semantically consistent way in order to resolve terminological ambiguities and misunderstandings, and to avoid the constant revision and redefinition of terms, concepts, and elements of the business. Such inconsistencies make applications not able to talk to each other, and subsequently result in slower response times when changes are needed. Business managers cannot get a clear view of their organization through these multiple un-integrated "languages".

Interoperation: The current Web services infrastructure focuses on syntactic interoperability. Syntactic interoperability allows Web services to identify only the structure of the messages exchanged, but it fails to provide an interpretation of the content of those messages. Indeed, Current standards like XML and XML Schema only solve the mismatch on the syntactical and structural level; solving the mismatch on the semantic level is usually handled on a case-by-case basis (for instance using custom adapters). Mismatches between interaction protocols are not dealt within current standards; semantics of the message exchange sequences are necessary to solve the mismatches on that level.

4 Realizing SOA with Semantic Web Services

4.1 Introduction

Web Services technology based on WSDL, SOAP and UDDI, define common standards that ensure interoperability between heterogeneous platforms. However, although low level interoperability is essential, SOA challenges as discussed in section 1 go beyond data formats and communication protocols interoperability. The purely syntactic focus of WS technologies makes service description non interpretable by the machine which hampers the automation of operations, inherent to SOA, such as service discovery, composition and invocation.

SWS initiatives have emerged with the objective of complementing the interoperability ensured by Web services to deal with data and behavioral heterogeneity along with automation support for capability-based service discovery, and dynamic service composition and invocation. The basic and common principle of these initiatives is extending syntactic service descriptions with a semantic layer the machine can interpret and reason over it. Ontologies play a central role for defining this semantic extension. An ontology is a formal explicit specification of a shared conceptualization [5]. Ontologies define a common vocabulary and formal semantics by providing concepts, and relationships between them. Using a common vocabulary for describing services capability and behaviors ensures interoperability at data level. Formal semantics enables the application of powerful and well proven reasoning based techniques in order to enable capability-based service discovery and automatic service composition.

There are four main SWS initiatives namely WSMO/L/X Framework [6], OWL-S [7], IRS-III framework [8] and METEOR-S system [9]. The first three initiatives separate explicitly between the semantic and syntactic descriptions of a Web service and link them using the concept of grounding that maps abstract concepts and data types of the semantic description to concrete data formats and communication protocols at the

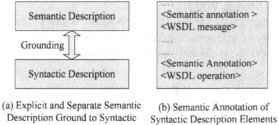

(a) Explicit and Separate Semantic   (b) Semantic Annotation of
Description Ground to Syntactic   Syntactic Description Elements
Description

Fig. 3. Two approaches to semantically extend syntactic Web service descriptions

syntactic level (see Fig. 3 (a)). METOER-S, however, semantically annotate WSDL files by linking their elements to ontology concepts and relations (see Fig. 3 (b)).

While METEOR-S is agnostic as regards to the ontologies used for the semantic annotation, WSMO/L/X, OWL-S and IRS-III can be seen as fully fledged framework with three layers (see Fig. 4): (i) a conceptual model for describing Web services and related information, (ii) a formal language used for defining the conceptual model concepts, relations and axioms, and (iii) an execution environment, as a proof of concepts, showing the use of semantic description for carrying out goal-based service discovery and invocation, and automatic service composition. In the following, we present the conceptual model of each of these initiatives and give an overview of their execution environments.

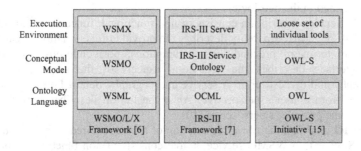

Fig. 4. WSMO/L/X, IRS-III and OWL-S constituent layers

4.2 WSMO/L/X Framework

WSMO [10] is an ontological conceptual model for describing various aspects related to SWS. WSMO refines and extends the Web Service Modeling Framework (WSMF) [11], by developing a set of formal ontology languages. WSMF is based on two complementary principles that WSMO inherits: strong decoupling between the various resources and a strong mediation to ensure the interoperation between these loosely coupled components. While WSMO provides the conceptual model for describing core elements of SWS, WSML [12] provides a formal language for writing, storing and communicating such descriptions.

4.2.1 Conceptual Model: WSMO

Following the main concepts identified in the WSMF, WSMO identifies four top level elements as the main concepts for describing several aspects of SWS, namely ontologies, Web services, goals and mediators (see Fig. 5).

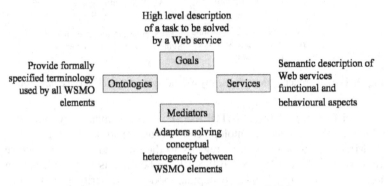

Fig. 5. WSMO top level elements [10]

Ontologies are used as the data model throughout WSMO. All resource descriptions as well as all data interchanged during service usage are based on ontologies. The core elements of an ontology are concepts (the basic entities of the agreed terminology), relations (model interdependencies between several concepts, and instances), instances, and axioms (define complex logical relations between the other elements defined in the ontologies) [6].

WSMO Service description consists of non-functional, functional, and behavioral aspects [10]. A service capability describes the provided functionality. A capability is described in terms of preconditions, assumptions, postconditions and effects. A service interface describes the behavioral aspects of the service in terms of choreography and orchestration. A service choreography details how to interact with the service from a user's perspective. An orchestration describes how the service works from the provider's perspective [10].

A WSMO goal is a high level description of a task required to be solved by Web services. Similar to a WSMO service, a goal consists of non functional properties, a capability describing the user objective and an interface reflecting the user behavior requirements.

Mediation in WSMO aims at resolving mismatches that may arise between different used terminologies (data level), or interaction protocols (process level). WSMO ensures dynamic interoperability by defining mediators during design time that will be used by mediation components during run time to resolve heterogeneity on the fly. A WSMO mediator can be seen as an adapter between WSMO elements defining the necessary mappings and transformations between the linked elements [6]. WSMO defines four types of mediators: OO mediators that resolve terminological mismatches between two ontologies, GG, WG and WW mediators that resolve mismatches respectively between two goals, a service and a goal, and two Web services.

4.2.2 Execution Environment: WSMX

WSMX [13] is an execution environment for dynamic discovery, selection, mediation, invocation and inter-operation of SWS. WSMX is the reference implementation of WSMO and therefore relies on it as conceptual model. A provider can register its service using WSMX in order to make it available to the consumers. A requester can find the Web Services that suit their needs and then invoke them in a transparent way [6].

WSMX exploits semantic service description to support capability-based discovery, not possible to perform having pure syntactic service description. In addition to the classical keyword-based discovery, WSMX supports functional, instance based and Quality of Service (a.k.a QoS) based discovery. Functional discovery reasons over service capabilities by matching them to the user goal capability. WSMX distinguishes different degrees of matching with the required goal [6]. Instance-based discovery considers instance level service descriptions and can dynamically fetch additional information during the discovery process. A Quality of Service based discovery provides a framework which matches specific QoS requirements of the requester with provided SWS.

WSMX implements data and process mediation as distinct components. The Data Mediation component in WSMX deals with heterogeneity problems that can appear at data level. Process level mediation deals with solving interaction protocols mismatches. Both components handle heterogeneity problems by applying the set of mappings rules, between the source and target WSMO element, defined during design-time.

WSMX conceptual architecture defines a distinct component for composition. However, no automatic composition is implemented yet as part of WSMX. Nevertheless, WSMO provides the required foundation for automatic service composition. Indeed, SUPER project [14] has released a composer component enabling automatic WSMO service composition by applying Artificial Intelligence (a.k.a AI) planning techniques.

4.3 OWL-S Initiative

4.3.1 Conceptual Model: OWL-S

OWL-S [7, 15] is an upper ontology for service description based on the Web Ontology Language (OWL) [16]. As shown in Fig. 6, an OWL-S service description consists in three interrelated parts: the service profile, the process model and the grounding. The service profile is used to describe what the service does; the process model is used to describe how the service is used; and the grounding is used to describe how to interact with the service. The service profile and process model are

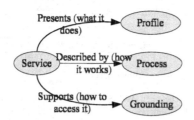

Fig. 6. Top level elements of OWL-S service ontology [7]

abstract descriptions of a service, whereas the grounding specifies how to interact with it by providing the concrete details related to message formats, and communication protocols.

A service profile describes functional, classification and non functional aspects of a service. Similar to WSMO, the capability of a Web service is represented as a transformation from the inputs and the preconditions of the Web service to the set of outputs produced, and the effects that may result from the execution of the service [15]. The classification aspect describes the type of service as specified in a domain-specific taxonomy. Non-functional aspects include service parameters like security, privacy requirements, and Quality of Service properties. OWL-S provides an extensible mechanism that allows the providers and the consumers to define additional service parameters.

The process model provides a more detailed view on how the service is carried out in terms of control and data flow. OWL-S distinguishes between atomic, composite and simple processes. An atomic process corresponds to a single interchange of inputs and outputs between a consumer and a provider. A composite process consists of a set of component processes linked together by control flow and data flow structures. The control flow is described using programming language or workflow constructs such as sequences, conditional branches, parallel branches, and loops. Data flow is the description of how information is acquired and used in subsequent steps in the process [15]. Simple processes can be used to provide abstracted, non invocable views of atomic or composite processes.

The grounding specifies the details of how a service can be accessed. Service grounding allows separating the abstract information described by the process model from the implementation details. Fig. 7 illustrates how the grounding is achieved in OWL-S. It specifies mapping atomic processes into WSDL operations. In addition, it specifies how to translate the messages described as OWL classes and instances to WSDL messages.

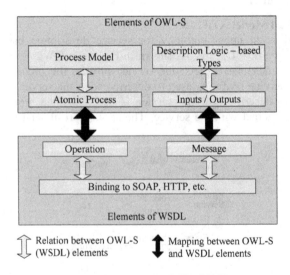

Fig. 7. Grounding in OWL-S [15]

4.3.2 OWL-S Tools

Unlike WSMO, OWL-S does not have a reference implementation like WSMX. Instead, there exists a collection of individual tools like OWL-S Editor [17], OWL-S/UDDI Matchmaker [18], OWL-S Virtual Machine [19], WSDL2OWL-S converter [20] and OWL-S2UDDI converter [18]. In the following we focus on the tools enabling dynamic service discovery and interoperation, and automatic service composition.

The OWL-S/UDDI matchmaker integrates OWL-S capability matching into the UDDI registry. OWL-S2UDDI converter converts OWL-S profile descriptions into corresponding UDDI advertisements, which can then be published in a UDDI registry. The OWL-S/UDDI registry enhances UDDI registry with OWL-S matchmaking functionalities. The matching engine contains five different filters for namespace comparison, word frequency comparison, ontology similarity matching, ontology subsumption matching, and constraint matching [6].

The OWL-S Virtual Machine enables to control the interaction between Web services according to their process models. Unlike WSMO and WSMX, OWL-S conceptual model and implementation do not consider mediators as first class citizens. OWL-S assumes the existence of external mechanisms that can handle heterogeneity at data and process level [6].

Several approaches have been proposed for automatic service composition based on OWL-S description [15]. [21] considers OWL-S process model as abstract workflow which is expanded and refined using automated reasoning machinery. [22, 23] use Hierarchical Task Network (HTN) planning to perform automated Web Service composition. Other planning techniques that have been applied to the composition of OWL-S services, are classical STRIPS-style planning [24], extended estimated-regression planning [25], and Planning as Model Checking [26].

4.4 IRS-III Framework

IRS-III (Internet Reasoning Service) is a framework for creating and executing SWS [8]. It acts as a semantic broker between a client application and deployed Web services by supporting capability-based invocation. A client sends a request encapsulating the desired goal and, by exploiting the semantic description of Web services, IRS-III framework: (a) discovers potentially relevant Web services; (b) selects the set of Web services which best fit the incoming request; (c) mediates any mismatches at the conceptual level; and (d) invokes the selected Web services whilst adhering to any invocation constraints.

4.4.1 Conceptual Model: Service Ontology

IRS-III service ontology defines the conceptual model of IRS-III framework. It extends the core epistemological framework of its previous IRS-II framework [27] by incorporating WSMO conceptual model. Different from WSMO, IRS-III service ontology uses its own ontology language, OCML [28]. While there are some differences between IRS-III and WSMO conceptual models, IRS-III service ontology defines the same concepts for describing SWS namely goals, Web service capability and interface (choreography and orchestration), and mediators.

4.4.2 Execution Environment: IRS-III Server

IRS-III server is the main element of IRS-III framework handling capability-based discovery and dynamic invocation. IRS-III framework includes also IRS-III publishing platform, IRS-III browser and IRS-III API. The publishing platform enables ease publication and deployment of Web services. IRS-III browser provides a goal-centric invocation mechanism to end users. IRS-III API facilitates the integration of IRS-III framework with other SWS platforms.

Similar to WSMX, IRS-III mediation approach consists of defining mediators which provide declarative mappings for solving different types of conceptual mismatches. These mediator models are created at design time and used at runtime. The mediation handler (part of IRS-III server) interprets each type of mediator accordingly during selection, invocation and orchestration of Web services [8].

Like WSMX, IRS-III does not support explicitly automatic service composition; however, it shares with it the same results of SUPER project about automatic service composition by applying AI planning techniques since both of them rely on the same conceptual model WSMO.

4.5 METEOR-S

METEOR-S project addresses the usage of semantics to support the complete lifecycle of Semantic Web processes [9] using four kinds of semantics - data, functional, non-functional and execution semantics. The data semantics describe the data (inputs/outputs) of the Web services. The functional semantics describe the functionality of a Web services (what it does). The non-functional semantics describe the non-functional aspects like Quality of Service and business rules. The execution semantics model the behavior of Web services and processes. Unlike above initiatives, METEOR-S does not define a fully fledged conceptual model for SWS description. It rather follows a light-weight approach by extending WSDL files with semantic annotation. The semantic annotation is achieved by mapping WSDL elements to ontological concepts. WSDL-S [29], METEOR-S specification for WSDL annotation, was one of the main works that influenced SAWSDL [30] the W3C standard for WSDL and XML schema semantic annotation.

METEOR-S framework provides a tool for creating SWS [31], a publication and discovery module [32], a composition module [33] and an execution environment. The GUI based tool provides support for semi-automatic and manual annotation of existing Web services or source code with domain ontologies. The publication and discovery module provides support for semantic publication and discovery of Web services. It provides support for discovery in a federation of registries as well as a semantic publication and discovery layer over UDDI. The composition module consists of two main sub-modules - the constraint analysis and optimization sub-module that deal with correctness and optimization of the process on the basis of QoS constraints. METEOR-S framework doesn't define components dedicated to deal with data and behavioral heterogeneity problems that may arise during services interactions.

4.6 SOA and Semantic Web Services: A Step Forward

The objective of SWS initiatives is providing the means to automate capability-based service discovery, service composition and invocation. The main idea is extending syntactical service description with additional information which can be understood and processed by the machine. Ontologies play a central role in the semantic service description. Using ontologies does not only bring user requirements and service advertisements to common conceptual space, but also helps to apply reasoning mechanisms [9]. Thus software programs are able to understand service descriptions and reason over them.

Indeed, as regards to **service discovery** ontology-enhanced search engine can exploit semantic service descriptions to implement matchmaking techniques, much more powerful than keyword-based ones, based on information retrieval, AI, and software engineering to compute both the syntactical and semantic similarity among service capability descriptions. Regarding **dynamic service interoperation**, software agents can leverage the computer-interpretable service description to understand what input is necessary to the service call, what information will be returned, and how to execute the service automatically. Mediation is a pillar for solving heterogeneity problems on the fly. Data and process mediators have been recognized as first class citizens within WSMO and IRS-III. Concerning **dynamic service composition**, semantic markup of Web services provides the necessary information to select and compose services. Software programs, based on AI planning, software synthesis and model checking, can be written to manipulate these representations, together with a specification of the objectives of the task, to achieve the task automatically [7].

In spite of the undeniable advancement realized by SWS, some issues still remain to be addressed. Indeed the great success of SWS is due to their semantic descriptions that rely on ontologies. However, service providers and requesters may use different terminologies to describe their requirements and services. Therefore, mediation is a key point in all SWS operations (discovery, composition, invocation) in order to resolve the terminological mismatches. Mediators can be seen as adapters at an ontological level enabling to migrate from one conceptualization to another. Consequently SWS initiatives have the same drawbacks as any adapter-based solution. Mediators are often defined at design time manually. In addition mapping between two conceptualizations is not always straightforward. Furthermore, mediators must be maintained each time one or both of the involved ontologies change. Another problems SWS face concerns the computing complexity, especially in terms of response time, of machine reasoning techniques which hamper the application of these techniques in context where soft real-time response is required for user interactions. Furthermore, plans generated by AI planning techniques are relatively simple compared to real composition models, defined by BPEL for instance.

5 Conclusion

In this chapter, we discussed how far Web services technologies and SWS initiatives satisfy SOA requirements. Loose coupling between service provision and consumption has led to new challenges for ensuring seamless and cost effective integration. These challenges concern effective service discovery, dynamic service interoperation, and automation support for service composition. By defining a set of standards, Web services technologies ensure low level interoperability, an essential first step yet not enough. Indeed, human intervention is still heavily required to resolve data and behavioral mismatches, to find the right services, and to select and compose appropriate services. SWS initiatives extend the level of interoperability to deal with data and behavioral heterogeneity using the concept of mediation. They also provide the foundation for (i) capability-based service discovery which is much more efficient than keyword-based one, (ii) dynamic service interoperation and (iii) automatic service composition. In spite of the remarkable results achieved by these initiatives, some open issues still need to be resolved. These issues are mainly related to mediator definition and update, and the computing complexity of machine reasoning techniques.

Acknowledgments. This work was supported by the Lion project supported by Science Foundation Ireland under grant no. SFI/02/CE1/I131, and by SUPER project funded by the EU under grant no. FP6-026850.

References

1. Erl, T.: Service-Oriented Architecture (SOA): Concepts, Technology, and Design. Prentice-Hall, Englewood Cliffs (2005)
2. Service Oriented Architecture Modeling, http://www.soamodeling.org
3. MacKenzie, C.M., Laskey, K., McCabe, F., Brown, P.F., Metz, R., Hamilton, B.A.: OASIS Reference Model for Service Oriented Architecture V 1.0, http://www.oasis-open.org/committees/download.php/19361/soa-rm-cs.pdf
4. Weerawarana, S., Curbera, F., Leymann, F., Storey, T., Ferguson, D.F.: Web services platform architecture: SOAP, WSDL, WS-Policy, WS-Addressing, WS-BPEL, WS-Reliable Messaging, and More. Prentice-Hall, Englewood Cliffs (2005)
5. Gruber, T.R.: A translation approach to portable ontology specifications. Knowledge Acquisition 5, 199–220 (1993)
6. Roman, D., de Bruijn, J., Mocan, A., Lausen, H., Bussler, C., Fensel, D.: WWW: WSMO, WSML, and WSMX in a nutshell. In: 1st Asian Semantic Web Conference, pp. 516–522. Springer, Beijing (2006)
7. Martin, D., Burstein, M., McDermott, D., et al.: OWL-S 1.2 Release, http://www.daml.org/services/owl-s/1.2/
8. Domingue, J., Cabral, L., Galizia, S., Tanasescu, V., Gugliotta, A., Norton, B., Carlos, P.: IRS-III: A broker-based approach to semantic Web services. J. Web Sem. 6(2), 109–132 (2008)
9. Verma, K., Gomadam, K., Sheth, A.P., Miller, J.A., Wu, Z.: The METEOR-S Approach for Configuring and Executing Dynamic Web Processes. LSDSIS technical report, http://lsdis.cs.uga.edu/projects/meteor-s/
10. Roman, D., Lausen, H., Keller, U., et al.: Web Service Modelling Ontology, http://www.wsmo.org/TR/d2/v1.4/
11. Fensel, D., Bussler, C.: The Web Service Modeling Framework (WSMF). Electronic Commerce Research and Applications 1(2), 113–137 (2002)
12. Steinmetz, S., Toma, I.: Web Service Modeling Language, http://www.wsmo.org/TR/d16/d16.1/v1.0/
13. Shafiq, O., Moran, M., Cimpian, E., Mocan, A., Zaremba, M., Fensel, D.: Investigating Semantic Web Service Execution Environments: A comparison between WSMX and OWL-S tools. In: 2nd International Conference on Internet and Web Applications and Services. IEEE Computer Society, Mauritius (2007)

14. Semantic Utilised for Process Management within and between Enterprises, http://www.ip-super.org
15. David, L., Martin, D.L., Burstein, M.H., McDermott, D.V., McIlraith, S.A., Paolucci, M., Sycara, K.P., McGuinness, D.L., Sirin, E., Srinivasan, N.: Bringing Semantics to Web Services with OWL-S. World Wide Web Journal 10(3), 243–277 (2007)
16. McGuinness, D.L., van Harmelen, F.: OWL Web Ontology Language Overview, http://www.w3.org/TR/2004/REC-owl-features-20040210/
17. Elenius, D., Denker, G., Martin, D., et al.: The OWL-S Editor—A development tool for semantic web services. In: 2nd European Semantic Web Conference, pp. 78–92. Springer, Heraklion (2005)
18. Srinivasan, N., Paolucci, M., Sycara, K.: An efficient algorithm for OWL-S based semantic search in UDDI. In: Cardoso, J., Sheth, A.P. (eds.) SWSWPC 2004. LNCS, vol. 3387, pp. 96–110. Springer, Heidelberg (2005)
19. Paolucci, M., Ankolekar, A., Srinivasan, N., et al.: The DAML-S virtual machine. In: 2nd International Semantic Web Conference, pp. 335–350. Springer, Sanibel Island (2003)
20. Paolucci, M., Srinivasan, N., Sycara, K., et al.: Toward a semantic choreography of web services: From WSDL to DAML-S. In: 1st International Conference on Web Services, pp. 22–26. CSREA Press, Las Vegas (2003)
21. McIlraith, S., Son, T.: Adapting golog for composition of semantic web services. In: 8th International Conference on Principles of Knowledge Representation and Reasoning, pp. 482–493. Morgan Kaufmann, Toulouse (2002)
22. Sirin, E., Parsia, B., Wu, D., et al.: HTN Planning for Web Service Composition using SHOP2. Journal of Web Semantics 1(4), 377–396 (2004)
23. Nau, D., Au, T.C., Ilghami, O., et al.: SHOP2: An HTN planning system. J. Artif. Intell. 20, 379–404 (2003)
24. Sheshagiri, M., desJardins, M., Finin, T.: A planner for composing services described in DAML-S. In: Workshop on Web Services and Agent-Based Engineering, Melbourne (2003)
25. McDermott, D.: Estimated-regression planning for interactions with web services. In: 6th International Conference on Artificial Intelligence Planning Systems, Toulouse (2002)
26. Sirin, E., Parsia, B., Hendler, J.: Filtering and selecting semantic web services with interactive composition techniques. IEEE Intell. Syst. 19(4), 42–49 (2004)
27. Motta, J., Domingue, L., Cabral, M.: IRS-II: a framework and infrastructure for semantic Web services. In: Fensel, D., Sycara, K.P., Mylopoulos, J. (eds.) ISWC 2003. LNCS, vol. 2870, pp. 306–318. Springer, Heidelberg (2003)
28. Motta, E.: An overview of the OCML modelling language. In: 8th Workshop on Knowledge Engineering Methods and Languages, Karlsruhe, pp. 21–22 (1998)
29. Miller, J., Verma, K., Rajasekaran, P., Sheth, A., Aggarwal, R., Sivashanmugam, K.: WSDL-S: Adding Semantics to WSDL - White Paper, http://lsdis.cs.uga.edu/library/download/wsdl-s.pdf
30. Kopecky, J., Vitvar, T., Bournez, C., Farrell, F.: SAWSDL: Semantic Annotations for WSDL and XML Schema. IEEE Internet Computing 11, 60–67 (2007)
31. Patil, A., Oundhakar, S., Sheth, A., Verma, K.: METEOR-S Web service Annotation Framework. In: 13th International World Wide Web Conference, pp. 17–22 (2004)
32. Verma, K., Sivashanmugam, K., Sheth, A., Patil, A., Oundhakar, S., Miller, J.: METEOR-S WSDI: A Scalable P2P Infrastructure of Registries for Semantic Publication and Discovery of Web Services. J. Inf. Technol. and Management 6(1), 7–39 (2005)
33. Cardoso, J., Sheth, A.O.: Semantic E-Workflow Composition. J. Intell. Inf. Syst. 21(3), 91–225 (2003)

Author Index

Alves de Medeiros, A.K.   35
Ananthanarayanan, Rema   247

Barker, Adam   81
Besana, Paolo   81
Bhiri, Sami   374
Bodenstaff, Lianne   219
Bundy, Alan   81

Ceravolo, Paolo   219
Chang, Elizabeth   1, 130, 199, 260, 290, 346
Chen-Burger, Yun Heh   81

Damiani, Ernesto   219
de Pinninck, Adrian Perreau   81
Delamer, Ivan M.   276
Dillon, Tharam S.   1, 130, 199, 290
Dupplaw, David   81

Fugazza, Cristiano   219

Gaaloul, Walid   374
Gal, Avigdor   176
Giunchiglia, Fausto   81
Gupta, Ajay   247

Hadzic, Maja   260
Hassan, Fadzil   81
Hauswirth, Manfred   374
Hussain, Farookh Khadeer   199, 290
Hussain, Omar Khadeer   290

Jarrar, Mustafa   7

Kotoulas, Spyros   81

Lambert, David   81
Li, Guo   81

Martinez Lastra, Jose L.   276
McGinnis, Jarred   81
McNeill, Fiona   81
Meersman, Robert   1, 7, 130
Mohania, Mukesh   247

Osman, Nardine   81

Potdar, Vidyasagar   346

Rahayu, Wenny   130
Reed, Karl   219
Robertson, David   81
Rouached, Mohsen   374

Shvaiko, Pavel   176
Siebes, Ronny   81
Sierra, Carles   81
Sycara, Katia   1, 324

Vaculín, Roman   324
van der Aalst, W.M.P.   35
van Harmelen, Frank   81

Walton, Chris   81
Wombacher, Andreas   219
Wongthongtham, Pornpit   199
Wouters, Carlo   130
Wu, Chen   346